W0038107

The Granular State

MATERIALS RESEARCH SOCIETY
SYMPOSIUM PROCEEDINGS VOLUME 627

The Granular State

Symposium held April 24–27, 2000, San Francisco, California, U.S.A.

EDITORS:

Surajit Sen

State University of New York at Buffalo
Buffalo, New York, U.S.A.

Melany L. Hunt

California Institute of Technology
Pasadena, California, U.S.A.

Materials Research Society
Warrendale, Pennsylvania

This material is based upon work supported by the National Science Foundation under Grant No. NSF: DMR-0076869. Any opinions, findings, and conclusions or recommendations expressed in this material are those of the author(s) and do not necessarily reflect the views of the National Science Foundation.

Single article reprints from this publication are available through
University Microfilms Inc., 300 North Zeeb Road, Ann Arbor, Michigan 48106

CODEN: MRSPDH

Published by:

Materials Research Society
506 Keystone Drive
Warrendale, PA 15086
Telephone (724) 779-3003
Fax (724) 779-8313
Web site: http://www.mrs.org/

Library of Congress Cataloging-in-Publication Data

The granular state : symposium held April 24–27, 2000, San Francisco, California, U.S.A. / editors,
    Surajit Sen, Melany L. Hunt
        p.cm.—(Materials Research Society symposium proceedings,
        ISSN 0272-9172 ; v. 627)
        Includes bibliographical references and indexes.
        ISBN 1-55899-535-8
        I. Sen, Surajit II. Hunt, Melany L. III. Materials Research Society symposium proceedings ;
    v. 627
2001

Manufactured in the United States of America

CONTENTS

*Invited Paper

*Invited Paper

GRANULAR FLOWS II

VIBRATED AND ROTATED
GRANULAR MEDIA

*Invited Paper

STRESS DISTRIBUTIONS

FROM AVALANCHES TO SANDCASTLES

*Invited Paper

PREFACE

This volume contains papers presented at Symposium BB, "The Granular State," held April 24–27 at the 2000 MRS Spring Meeting in San Francisco, California.

The study of granular materials is an old but difficult subject that has been studied for many years by the engineering communities and has been a topic of significant attention to many physicists since the late 1980s. This symposium has attempted to bring together a diverse group of researchers from many countries in the Americas, Europe and in Asia who are at the forefront of study on some of the topics that are relevant to the study of the granular state. It is our feeling that each person made their very best effort to actively participate in the symposium by asking questions, by engaging in lively discussions and by contributing to the collection of articles contained in this volume. Many of our colleagues carefully refereed numerous contributions. We are especially indebted to Professor R.P. Behringer, Professor Eric Clement, Professor S.R. Nagel, Professor V. Nesterenko, Dr. Eiichi Fukushima, Dr. Marian Manciu, and Dr. Elie Raphael for their valuable help in preparing this volume.

We thank the participants for spending their valuable time to make this symposium a success and for their authoritative contributions which display the diversity and richness of the granular state. We apologize in advance to our readers for the non-uniformities in styling the front matter of some of the articles and for the typographical and typesetting errors that have surely crept in.

Surajit Sen
Melany L. Hunt

November 2000

ACKNOWLEDGMENTS

We acknowledge the valuable encouragement and support of Dr. Alan J. Hurd of Sandia National Laboratories for playing a pivotal role in making this symposium possible. We are grateful to the staff of both the Materials Research Society and ScholarOne Inc. who worked with great dedication and enthusiasm to make this volume possible. The financial support of the National Science Foundation-Division of Materials Research, Schlumberger Research and Sandia National Laboratories are acknowledged with gratitude.

MATERIALS RESEARCH SOCIETY SYMPOSIUM PROCEEDINGS

MATERIALS RESEARCH SOCIETY SYMPOSIUM PROCEEDINGS

Prior Materials Research Society Symposium Proceedings available by contacting Materials Research Society

Granular Structure

Mat. Res. Soc. Symp. Vol 627 © 2000 Materials Research Society

Jamming in Liquids and Granular Materials

C. S. O'Hern[1,3], S. A. Langer[2], A. J. Liu[1], and S. R. Nagel[3]

[1]Department of Chemistry and Biochemistry,
University of California at Los Angeles, Los Angeles, CA 90095-1569

[2]Information Technology Laboratory, NIST,
Gaithersburg, MD 20899-8910

[3]The James Franck Institute and the Department of Physics,
The University of Chicago, Chicago, Illinois 60637

Many systems can develop a yield stress while in an amorphous state. For example, a supercooled liquid, when cooled sufficiently, forms a glass - an amorphous solid with a yield stress. Another common example is a granular material which will remain solid and not move even under the influence of moderate stresses. This accounts for why piles of grain or sand can exist with a non-zero slope even though gravity is acting to flatten out the upper surface. The solidity in that case is due to the system having become jammed. Similar jamming often inhibits flow out of a hopper or in conduits transporting material across a factory floor. Jamming is a ubiquitous phenomenon occurring in many different systems such as colloidal suspensions, foams and, of course, traffic. We tend to think of the jamming transition as being stress-induced. A "fluid" at constant density (or under a confining pressure) flows if the stress is above the yield stress but becomes stuck in an amorphous configuration if the stress is too low. The idea of temperature, *per se*, does not seem to be crucial to the transition. This makes it seems quite different from the formation of a glass out of a supercooled liquid by lowering the temperature. However, there are similarities between these two types of transitions, aside from the obvious fact that they both have to do with the complete arrest of dynamics and flow. An exploration of these similarities was the subject of a program at the Institute for Theoretical Physics in Santa Barbara held in the Autumn of 1997. A synopsis of this program was published that details some of the interesting ideas now current in that field.[1]

In order to relate the process of jamming by lowering a stress and the process of glass formation by lowering a temperature, Liu and Nagel proposed a generalized "jamming phase diagram" shown in Fig. 1.[2] The vertical axis is the temperature. As the temperature is lowered a liquid becomes a glass. The axis coming out of the plane is the inverse density or inverse pressure axis. As the density or pressure increases, a system becomes so constricted in its phase space that it can no longer move. The horizontal axis is the applied load. (This is the *incompatible* load in the terminology of Cates et al. [3].) For a foam, this could be the shear stress. If this stress were too large the foam would flow. If it is too small, however, the foam will again become stuck or jammed. According to this picture, it is not possible for jamming to occur when the density is too low, the temperature too high, or the external load too great.

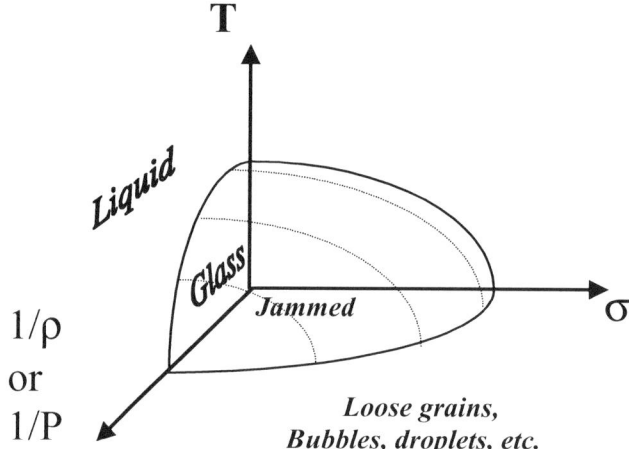

Fig. 1 – A schematic phase diagram that attempts to unify the different ways in which the ability of a system to flow can be lost. After Ref. [2].

The glass transition is normally conceived of as occurring with the load set to zero in the Temperature/Density plane. Likewise jamming is thought of as occurring in the Load/Density plane. By combining the three axes we see that these different phenomena may be related. This diagram is schematic in the sense that the pressure and load axes really just represent two of the different components of the stress tensor. It is also probable that the phase diagram will not be as simple as drawn (e.g., the jamming of a material such as corn starch, which shear thickens, must have reentrant regions). This diagram *is purely schematic* and we do not know the shape of the phase boundaries. Indeed, Weitz and collaborators have proposed that such ideas can give a coherent picture of the behavior of their experimental colloidal systems.[4] They report that the phase boundaries are concave instead of convex as we have drawn in Figure 1.

This diagram suggests that the same types of constraints leading a macroscopic system to jam may also be important for how a liquid becomes frozen into a glass. One consequence of this way of looking at these problems is that it suggests a number of different experiments offering new axes for the study of the glass transition and for jamming: One should look at how thermal motion (due to Brownian motion, vibrations or ordinary temperature) will alter the ability of a system to become jammed; Likewise one might also profitably look at how the application of shear stress changes the glass transition temperature in a supercooled liquid. These are the types of experiments that have not been adequately considered but which clearly have relevance to understanding how these systems are related to one another.

Much work has been done in studying the force transmission in granular matter.[5-20] Although some of this has been focused on identifying and quantifying the force chains themselves, another body of work has studied the distribution of normal forces between particles, denoted by P(F).[7 - 12, 16 - 20] In this work, a histogram is made of the number of contacts which have a given normal force, F, between two particles. This quantity is of central importance for jamming since it deals with how forces, or stresses, propagate within a granular medium. There have been quantitative measures of P(F) from experiments (at boundaries), simulations, and models. The distribution is remarkably robust (occurring for disordered *and* crystalline samples).[12, 20] At high force, there is an exponential tail: $P(F) \propto \exp(-F/F_o)$. At low F, there is a plateau and the distribution is roughly constant below the average force, <F>.

Until now, the question of what P(F) looks like for a liquid or glass has not been asked. However, once we ask that question, we realize that it can be answered exactly in some limits. [21] In liquids, forces depend only on particle separations through $f = -dV/dr$ where V(r) is the pair potential. Thus P(F) dF = G(r) dr, where G(r) dr is the radial distribution function. (G(r) dr is the probability of finding a particle between r and r+dr given a particle at the origin. Aside from phase-space factors, it is simply proportional to the pair distribution function g(r).) In *equilibrium* at temperature T, we have: $g(r) \propto \exp[-V(r)/k_BT]$ in the limit of large potentials (that is, small radii).[22] Thus:

$$P(F) \approx A(\rho,T)\, r^{D-1}\, dr/dF \exp[-V(r)/k_B T\}] \qquad (1)$$

where $A(\rho,T)$ is independent of r and D is the dimension. We can then calculate exactly the force distribution in this $r \to 0$ limit if the pair potential, V(r), is known. For a strongly repulsive core this leads to an essentially exponential tail for P(F): for example using $V(r) = r^{-12}$ we find that $P(F) \propto \exp(-B\, F^{12/13}/T)$, where B is a constant and the power 12/13 derives from the r^{-12} repulsion. This is remarkably similar to the force distribution found in granular materials and gives a new perspective on how to understand the P(F) distributions that appear in those materials and in other *athermal* jammed systems.

The argument just given is good for liquids in equilibrium but does not indicate what would happen as the liquid is driven out of equilibrium as the temperature is lowered into the glass phase. It also does not say how large the forces must be in order to be in the asymptotic regime where these results hold. We have performed molecular dynamics simulations of two-dimensional systems with a variety of inter-particle potentials in order to find the generic non-equilibrium and non-asymptotic behavior of P(F).[21] We have been able to show that we can measure the temperature of a liquid to within 5% by measuring the slope of the high force tail. This indicates that although the above argument is only *exact* in the asymptotic regime it is a reasonable approximation for intermediate forces as well. Previous simulations [23] had shown that the Cartesian components of the force on a atom along the liquid-vapor coexistence line, is also distributed in an exponential fashion. This is related to our results: for high forces, the total force on the particle, which is the vector sum of the normal forces, will be dominated by the highest normal force, which is why the distribution of Cartesian components is also exponential.

We have also studied the P(F) distributions for liquids that have been quenched to temperatures below their glass transition temperatures. Our results taken from ref. [21] show (see Figure 2a) that for liquids at high temperature, the P(F) distribution is essentially exponential over the entire range of forces. However, as the temperature is lowered the low force part of the distribution begins to depart from that behavior and forms a plateau at small forces. As the system is cooled below T_g the plateau turns into a peak which is very similar to what is seen in granular materials. This is shown in Figure 2b where a comparison between these two very different systems (i.e., liquids and granular materials) is made. The agreement is excellent. The fact that in the liquid the P(F) distribution starts to have a plateau at small forces just as the system becomes solidified (i.e., begins to get a yield stress) is suggestive that it may be profitable to think of this transition as a "jamming transition" as was indicated in Figure 1.

Our conclusion is that the formation of a yield stress is accompanied by the formation of a peak in P(F). This peak occurs because of the way forces are balanced in a jammed system. In an unjammed high-temperature state, balance can occur because of particle acceleration as happens in a gas or liquid. In the jammed low-temperature state, since there is little thermal motion, balance must be due to forces acting on opposing sides of the same particle. This means that as a yield stress begins to form, the forces must begin to cluster around some average value in order for force balance to occur.

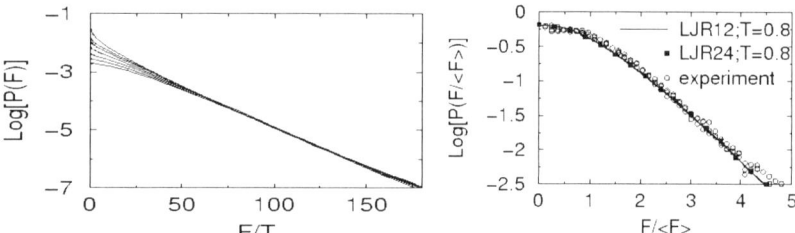

Fig. 2 - (a) P(F) vs. F/T for a mixture of particles with $V(r) \propto r^{-12}$ obtained for seven temperatures above the glass transition in the range 1.2 < T < 5.0 with T decreasing from top to bottom. (b) P(F) vs. F/<F> for two different potentials measured below the glass transition compared with the results from experiments on granular material[12]. The glass has a very similar behavior as does the granular material. Both figures are taken from ref. [21].

We have also studied the behavior of P(F) as the system experiences shear. At high shear rates, where the system is flowing (due to the applied forces) like a liquid, does the peak in P(F) persist or does it disappear as if the system were in a liquid state at an effectively higher temperature? If the peak does disappear, then this is evidence that crossing the phase boundaries of Figure 1 in different directions gives similar physical phenomena and adds credence to the idea of such a jamming phase diagram. So far we have studied [21] the effect of shear in an *athermal* system that was originally proposed by Durian to model foams.[24] We have found that for the jammed system there is a peak in P(F) and that that peak disappears both as the density is decreased (i.e., the system becomes too rarified to support a yield stress) and as the applied stress is increased. This supports the idea of the phase diagram in Figure 1.

One should note that the forces comprising the peak in P(F) are those most relevant to the slow relaxation at the glass transition: Our simulations show that forces in the peak of P(F) come from the particles near the peak in g(r). The peak in g(r) produces the first strong peak in S(q) which, due to de Gennes narrowing, are the slowest modes to decay. Thus the peak in P(F) represents the slowest decaying forces in the system.

Force chains are important for the transmission of forces in a granular media. We are interested in seeing if they also play an important role in the physics of glasses. There has been a lot of work on glasses suggesting that as a liquid is cooled it develops dynamic heterogeneities.[25-30] Yet, despite this work there has not been a direct observation of the heterogeneities. Neutron scattering data did not see any density fluctuations that increase in size as the temperature was lowered into the glass phase.[31] Force chains might be the origin of the dynamic heterogeneities that have been suggested. Since the force chains are one-dimensional objects, they would not necessarily couple to the density. Thus, if they were present in the liquid and grew in size as the temperature was lowered, they could well have been undetected by the neutron scattering experiments.

This work was supported by Support of NSF Grant Nos. DMR-9722646 (CSO,SRN), CHE-9624090 (CSO,AJL), and PHY-9407194 (SAL,AJL,SRN).

References:
[1]   S. R. Nagel, "Jam Session, Santa Barbara, 1997," Europhysics News **29** #2, 58-59 (March/April 1998).
[2]   A. J. Liu and S. R. Nagel, "Jamming is not just cool any more," Nature **396**, 21-22 (1998).
[3]   M. E. Cates, J. P. Wittmer, J.-P. Bouchaud and P. Claudin, "Jamming, force chains, and fragile matter," Phys. Rev. Lett. **81**, 1841-1844 (1998).
[4]   V. Trappe, V. Prasad, P. N. Segre, L. Cipelletti, and D. A. Weitz, "Jamming transition in suspensions of weakly attractive colloidal particles," Bull. Am. Phys. Soc., (2000).
[5]   P. Dantu, "Étude expérimentale d'un milieu pulvérent," Annales des Ponts et Chaussées **IV**, 193-202 (1967).
[6]   T. Travers, M. Ammi, D. Bideau, A. Gervois, J. C. Messager, and J. P. Troadec, "Mechanical size effects in 2d granular media," Journal de Physique, **49**, 939-948 (1988).
[7]   C.-h. Liu, S. R. Nagel, D. A. Schecter, S. N. Coppersmith, S. Majumdar, O. Narayan, T. A. Witten, "Force Fluctuations in Bead Packs," Science **269**, 513-515 (1995).
[8]   S. N. Coppersmith, C. H. Liu, S. Majumdar, O. Narayan, and T. A. Witten, "Model for force fluctuations in bead packs," Physical Review E, **53**, 4673-4685 (1996).
[9]   F. Radjai, M. Jean, J.-J. Moreau and S. Roux, "Force Distributions in Dense Two-Dimensional Granular Systems," Phys. Rev. Lett. **77**, 274-277 (1996).
[10]   C. Thornton, "Force Transmission in Granular Media," KONA **15**, 81-90 (1997).
[11]   S. Luding, "Stress distribution in static two-dimensional granular model media in the absence of friction," Phys. Rev. E **55**, 4720-4729 (1997).

[12]   D. M. Mueth, H. M. Jaeger, and S. R. Nagel, "Force Distribution in a Granular Medium," Phys. Rev. E **57**, 3164-3169 (1998).

[13]   C. T. Veje, D. W. Howell, and R. P. Behringer, "Kinematics of a two-dimensional granular Couette experiment at the transition to shearing," Phys. Rev. E **59**, 739-745 (1999);

          D. Howell, R. P. Behringer and C. Veje, "Stress Fluctuations in a 2D Granular Couette Experiment: A Continuous Transition," Phys. Rev. Lett. **82**, 5241-5244 (1999).

[14]   R. S. Farr, J. R. Melrose and R. C. Ball, "Kinetic Theory of Jamming in Hard-Sphere Startup Flows," Phys. Rev. E **55**, 7203-7211 (1997);

          J. R. Melrose and R. C. Ball, "The pathological behavior of sheared hard spheres with hydrodynamic interaction," Europhys. Lett. **32**, 535-540 (1995).

[15]   M. E. Cates, J. P. Wittmer, J. -P. Bouchaud and P. Claudin, "Jamming and Static Stress Transmission in Granular Materials," Chaos **9**, 511-522 (1999).

[16]   G. Lovoll, K. J. Maaloy, and E. G. Fekkoy, "Force measurements on static granular materials," Phys. Rev. E **60**, 5872 (1999).

[17]   M. G. Sexton, J. E. S. Socolar, and D. G. Schaeffer, "Force distribution in a scaler model for noncohesive granular material," Phys. Rev. E **60**, 1999 (1999).

[18]   H. A. Makse, D. L. Johnson and L. M. Schwartz, "Packing of Compressible Granular Materials," Phys. Rev. Lett. **84**, 4160-4163 (2000).

[19]   E. C. Longhi, N. Easwar, and N. Menon, "Force fluctuations in a flowing granular material," Bull. Am. Phys. Soc., (2000).

[20]   D. L. Blair, N. W. Mueggenburg, A. H. Marshall, H. M. Jaeger, and S. R. Nagel, (unpublished).

[21]   C. S. O'Hern, S. A. Langer, A. J. Liu and S. R. Nagel (preprint).

[22]   B. Widom, "Potential-Distribution Theory and the Statistical Mechanics of Fluids," J. Phys. Chem. **86**, 869-872 (1982).

[23]   J. G. Powles and R. F. Fowler, "A simple property of a simple liquid," Molecular Physics **62**, 1079-1084 (1987).

[24]   D. J. Durian, "Foam mechanics at the bubble scale," Phys. Rev. Lett., **75**, 4780-4783 (1995);

          D. J. Durian, "Bubble-scale model of foam mechanics: Melting, nonlinear behavior, and avalanches," Phys. Rev. E, **55**, 1739-1751 (1997).

[25]   K. Schmidtrohr and H. W. Spiess, "Nature of Nonexponential Loss of Correlation Above the Glass Transition Investigated By Multidimensional Nmr," Phys. Rev. Lett. **66**, 3020-3023 (1991).

[26]   F. Fujara, B. Geil, H. Sillescu, and G. Fleischer, "Translational and Rotational Diffusion in Supercooled Orthoterphenyl Close to the Glass Transition," Zeitschrift Fur Physik B-Cond. Matt. **88**, 195-204 (1992);

          A. Kasper, E. Bartsch, and H. Sillescu, "Self-diffusion in concentrated colloid suspensions studied by digital video microscopy of core-shell tracer particles," Langmuir, **14**, 5004-5010 (1998).

[27]   M. T. Cicerone and M. D. Ediger, "Relaxation of Spatially Heterogeneous Dynamic Domains in Supercooled Ortho-Terphenyl," J. Chem. Phys. **103**, 5684-5692 (1995);

M. T. Cicerone and M. D. Ediger, "Enhanced Translation of Probe Molecules in Supercooled O-Terphenyl - Signature of Spatially Heterogeneous Dynamics," J. Chem. Phys. **104**, 7210-7218 (1996);

C. Y. Wang and M. D. Ediger, "How long do regions of different dynamics persist in supercooled o-terphenyl?," J. Phys. Chem. B, **103**, 4177-4184 (1999).

[28] U. Tracht, M. Wilhelm, A. Heuer, H. Feng, K. SchmidtRohr, and H. W. Spiess, "Length scale of dynamic heterogeneities at the glass transition determined by multidimensional nuclear magnetic resonance," Phys. Rev. Lett., **81**, 2727-2730 (1998).

[29] B. Schiener, R. Bohmer, A. Loidl, and R. V. Chamberlin, "Nonresonant Spectral Hole Burning in the Slow Dielectric Response of Supercooled Liquids," Science, **274**, 752-754 (1996).

[30] W. K. Kegel and A. van Blaaderen, "Direct observation of dynamical heterogeneities in colloidal hard-sphere suspensions," Science, **287**, 290-293 (2000).

[31] R. L. Leheny, N. Menon, S. R. Nagel, D. L. Price, K. Suzuya, and P. Thiyagarajan, "Structural Studies of an Organic Liquid through the Glass Transition," J. Chem. Phys. **105**, 7783-7794 (1996).

Mat. Res. Soc. Symp. Proc. Vol. 627 © 2000 Materials Research Society

A Low-Frequency Forced Torsion Pendulum for the Measurement of the Mechanical Properties of Granular Media

G. D'Anna
IGA, Ecole Polytechnique Fédérale de Lausanne,
CH-1015 Lausanne, Switzerland

ABSTRACT

We present a forced torsion pendulum designed to measure the dynamic moduli of granular media. In the method, the oscillating probe of the pendulum is immersed into various materials, such as sand, glass beads, snow. The apparatus operates at low-frequency and provides information about the quasi-static mechanical properties of the medium. In particular, a peak in the losses is ascribed to friction and cohesion between the grains, and a measure of the macroscopic failure limit can be obtained. As an example the effect of moisture-induced ageing in small glass beads, and the effect of sintering of ice grains in snow, are shown.

INTRODUCTION

In the experiments, the oscillating cylindrical probe of a forced torsion pendulum is immersed at a given depth into a large bucket containing the granular material (figure 1 left). The pendulum is forced into torsion oscillation by a time dependent torque $T(t)=T_o\exp(i\omega t)$, of frequency (1 Hz) well below the natural frequency of the pendulum, and the angular displacement, $\alpha(t)$, is detected optically. A lock-in measures the complex frequency response of the pendulum, $G=T/\alpha$. In a typical experiment we record the

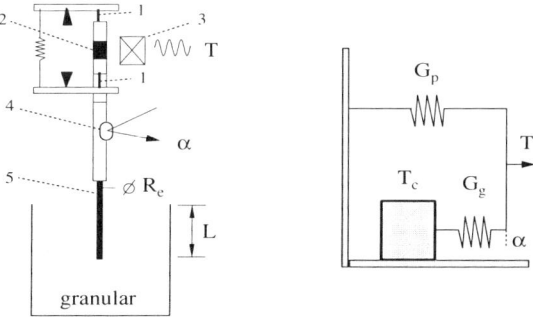

Figure 1. Left: Sketch of the forced torsion pendulum immersed into a granular medium. 1=suspension wires; 2= permanent magnet; 3= external coils; 4=mirror; 5= probe; Right: Rheological model.

argument, $\arg(G_n)$, and the absolute modulus, $|G_n|$, of the nth harmonic of the response (usually $n=1,3,5$), as a function of the amplitude of the applied torque T_o. For the pendulum not immersed the response reduces to $T=G_p\alpha$, where $G_p=18\ 10^{-3}$ Nm/rad is the torsion constant of the suspension wires. When immersed, various tribological processes [1] occurring within the granular medium, as well as at the grain-probe interface, cause energy dissipation. The damping properties of an oscillating system are well represented by the loss factor η, which is the dimensionless ratio of dissipated to stored energy per cycle, defined by $\eta=Q/2\pi W$. For a linear system $G\equiv G_1$, and $\tan[\arg(G_1)]=\eta$. For a strongly non-linear system as is the case here, as an approximation of the loss factor η we display either the quantity $\tan[\arg(G_1)]$ or $\tan[\arg(G)]$ with $G = G_1+G_3+G_5$.

The expected mechanical behaviour of the pendulum's response can be estimated using a simple rheological model [2] (figure 1 right). The spring G_p represents the suspension wires of the pendulum, while the other branch represents the granular medium, characterised by a Coulomb-Amontons slide unit with critical torque T_c and a spring of torsion constant G_g. The torsion constant G_g represents the elastic response of the granular medium to very low applied torque. On the other hand, the critical torque can be seen as the torque for which the maximum possible shear stress is reached somewhere within the granular medium, or at the grain-probe interface. Beyond this limit, friction and cohesion forces are overcome, and energy dissipation sets on. As an approximation of the real process, the pendulum can be seen as sliding in the granular medium, in a similar way as a macroscopic body slides over a large surface. The precise nature of the "mesoscopic" processes occurring beyond T_c will depend on the specific system under question.

For the rheological model the damping energy Q is the surface of the hysteresis loop, and we define the strain energy W as the energy stored in the two springs during loading (other definitions are possible). Adopting the notation $\Delta=G_g/G_p$ we obtain [3] a loss factor which displays a peak as a function of T_o, with maximum occurring at about $T_o^*\approx T_c(2+\Delta)/\Delta$, and of magnitude $\eta^*=(2\Delta/\pi)/[(1+\Delta)^{1/2}+1]$ which has the limits $\eta^*=2\sqrt{\Delta}/\pi$ for $\Delta\gg 1$ and $\eta^*=\Delta/\pi$ for $\Delta\ll 1$. (Defining W as the area under the secant modulus gives a peak magnitude $\eta^*=(4/\Delta\pi)[(1+\Delta)^{1/2}-1]^2$ which saturates to $\eta^*=4/\pi$ for $\Delta\gg 1$.) The modulus $|G|$ displays a step between G_p for $T_o\gg T_o^*$ and G_p+G_g for $T_o\ll T_o^*$, which corresponds to a modulus defect of Δ.

RESULTS

Glass beads

Figure 2 (left) shows data obtained in a granular system composed of glass beads of diameter 1.1 ± 0.05 mm with polished smooth surfaces. The cylindrical probe is covered by a layer of beads glued by an epoxy, giving an effective radius of $R_e=1.5$ mm, and is immersed at different depths L. The probe was pushed into the granular medium fluidised by strong external vibrations, and measurements were taken after a fixed time from the fluidisation. We systematically observed a mechanical behaviour with a pronounced loss peak, noted T_o^* in figure 2, and a step in the modulus between the two levels, noted G_{low} and G_p, close to the rheological behaviour. These measurements were conducted in ambient humidity of 40%, but no difference was observed in the dry system. There was also no difference by decreasing or increasing amplitude. In visu observations at the sample surface revealed that at high torque amplitude, multiple grain slip events occurred randomly in a large area around the probe, and a localised failure surface was not observed. In general, when a layer of grains is solidly bound to the probe surface, the mechanism controlling the slide unit in the rheological model seems to be a delocalised failure process, controlled by the inhomogeneous force distribution between the grains opposing the rotation of the pendulum.

For glass beads with sizes less than about 200 µm, moisture-induced ageing effects were observed at ambient humidity. Figure 2 (right) shows the experiments in glass beads with diameter 70µm±10µm, at ambient humidity of 40%. The probe was covered by glued beads, giving R_e=1.75 mm, and L=19 mm. According to Bocquet et al. [4], water capillary bridges nucleate at nanometric surface asperities between glass beads, causing a logarithmic time-dependent friction force. To control the ageing effect experimentally, we exploited the fact that capillary bridges are destroyed during the fluidisation of the granular. In figure 2 (right) the loss peak and the modulus are measured at different ageing times, t_a, where t_a is the time elapsed from the fluidisation procedure. For this particular experiment, the fluidisation-induced desegregation procedure was repeated before each point on a curve, so that a single curve was obtained strictly after the same ageing time. During the imposed ageing time, the pendulum continuously oscillates at the given torque amplitude. The inset of figure 2 (right) shows T_o^* as a function of t_a. The data align on a straight line, showing a logarithmic time-dependence of the form T_o^*=c+pln t_a/t_o, typical of the moisture-induced ageing effect. Surprisingly, the pendulum method is sensitive to the quasi-static ageing even if it oscillates at 1 Hz.

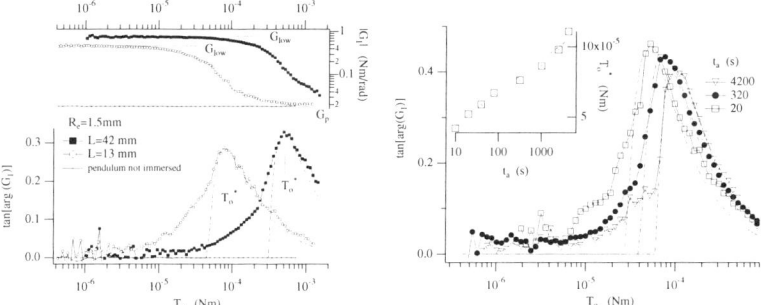

Figure 2. *Left: Data obtained with the pendulum immersed into glass beads of 1.1 mm diameter. A layer of grains is glued onto the probe surface. The point-dashed lines on the loss data are fitting-lines obtained from the rheological model; Right: Data obtained in glass beads of 70 µm diameter at ambient humidity 40%, for different ageing times t_a. Inset: The position of the loss peak T_o^* versus the ageing time, with the typical logarithmic time-dependence.*

We also studied the dependence of the loss peak with the geometrical parameters of the experiment [3]. The conclusion is that data can be summarised by an empirical relation $T_o^* \propto L^2 R_e^2$. The validity of this empirical relation is restricted to systems where macroscopic load-independent cohesion forces are absent or negligible, which is the case for the large glass beads used here. In the snow system (see below), solid bond cohesion forces between ice grains dominate and a different empirical relation is observed. In presence of moisture-induced ageing effects, as observed in glass beads of small size, the empirical relation remains valid provided these effects are properly taken into account, e.g., if the position of the peak is measured at the same ageing time.

Sand

Figure 3 shows data obtained in sand composed of multifaceted quartz grains with homogenous size of about 160μm±32μm, the flat faces of the grains appearing polished and very clean. A cylindrical copper probe of radius R=2 mm was immersed a depths of about 60 mm. In this case the surface of the probe was kept clean. The experiment was conducted in ambient humidity of about 45%, and small ageing effects were observed. In visu inspection revealed that a localised fracture arose, apparently, at the grain-probe interface for applied torque above the peak. Considering high harmonics of the response (n=3 and 5), corrections to the magnitude of arg(G) are relevant in the region of the peak. The position of the peak and the modulus are not affected.

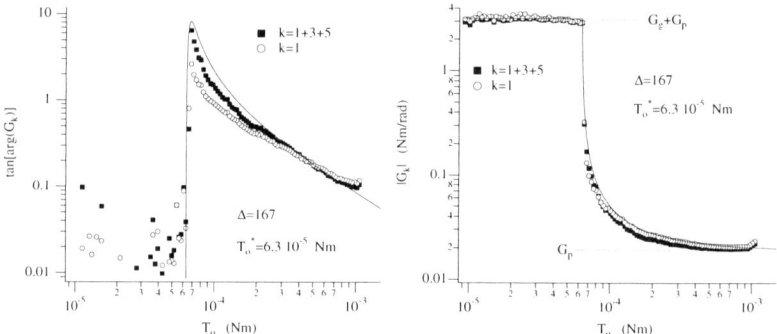

Figure 3. *Data for a metallic cylindrical probe with clean surface, immersed in natural sand. The effect of high harmonics on the response at the fundamental frequency is shown. The continuos lines are fitting lines obtained using the rheological model. Δ and T_o* are free parameters for the fitting to the modulus only.*

Snow

Figure 4 shows data measured in granular snow, taken at different sintering times. For this system the oscillating probe was a copper cylinder of radius 1.5 mm and length 20 mm, terminating in a cone of height 2 mm. Experiments were performed in a cold room, at –13°±0.1°. Data were obtained in snow samples composed of two-week-old grains. The snow was desegregated and gently poured into a large cylindrical aluminium container. The density was 270 kg/m³. The pendulum was lowered and the probe was immersed into the sample in such a way that the conical part was at about 30 mm from the surface. The speed of the immersion was of the order of 5 mm/sec, i.e., at sufficiently high speed for the bonds between the ice grains colliding with the probe to be broken. These grains were pushed aside by the conical point. As discussed for example by Fukue [5], one expects the formation of a zone of high density snow in front of the conical section, and a similar layer along the cylindrical section. The high density snow layer is strongly bound to the cold, metallic probe surface, and the pendulum method measures properties intrinsic to the snow system, instead of grain-probe interface properties.

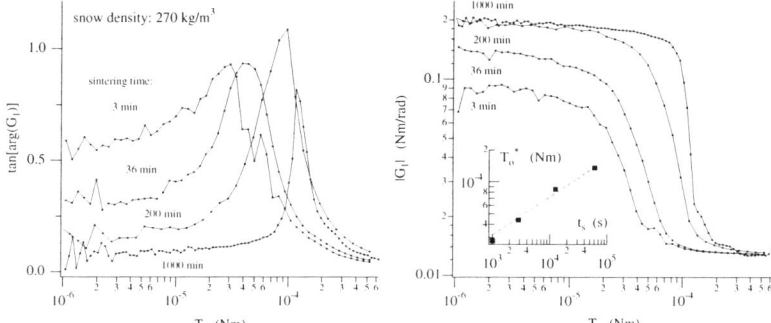

Figure 4. *Effect of the snow sintering on the losses (left) and modulus (right) of initially desegregated, two-weak-old grains. Inset: The position of the loss peak versus the sintering time at which the peak is observed. The observed time evolution is $t_s^{0.36}$. The position of the peak is a measure of the snow shear strength, and the pendulum method can be used in field experiments in very low-density snow.*

In the experiment, immediately after the preparation (about 3 min) the probe was immersed into the sample and a first measurement as a function of T_o (by decreasing the torque) was obtained. A single measurement takes about 30 min. Successive measurements were then obtained after 36 min, 200 min and 1000 min, respectively. We observed a large loss peak and a step in the modulus, in analogy to the rheological model. In this system the main tribological mechanism is the cohesion provided by solid bonds between ice grains, which grow very fast at the contact points. Therefore, the pendulum provides a measure of the snow shear strength. (We underline, however, that the validity of the picture is restricted to the conditions of our experiment. In other conditions, the visco-plastic deformation of the ice bonds, as well as the thin layer of water formed by frictional heating must be considered. In particular, the losses measured by the pendulum close to 0 C° can be much larger than for pure friction processes, due to stress-induced melting of the ice bonds.)

The inset of figure 4 shows the position of the loss peak, T_o^*, as a function of the sintering time t_s, taken at the time when the peak is observed. (E.g., even if the first measurement starts only 3 min after the sample preparation, the peak is observed about 16 min later.) The dashed line in the inset is the best fit to the data of the form $T_o^* = At_s^d$, with d =0.36. Accordingly, the time evolution of the snow shear strength can be approximated by a $t_s^{0.36}$ time-dependence. This is indeed close to the predicted $t_s^{0.25}$ time-dependence for the growth of ice bonds by Colbeck [6].

DISCUSSION

The pendulum response displays a characteristic mechanical behaviour, with a peak in the loss factor and a step in the absolute modulus. The position of the peak can be seen as a measure of the critical torque of a simple rheological model, thus a measure of a macroscopic threshold above which dissipation initiates in the granular medium.

However, the physical signification of the measured critical torque depends on details of the system considered. When a layer of grains is solidly bound to the probe surface (either glued or naturally bounded), the dissipation is due to **slip events** between grains. In glass beads, slip events are controlled mostly by friction mechanisms (notice that capillary effects have to be included in the friction component [4]), while in snow, slip events result from the fracture of solid bonds between ice grains. Because of the inhomogeneous force distribution in the system, slip events arise randomly in a large volume of the medium, and each event has repercussions at large distance via force-chains. As a consequence, the critical torque, or the position of the loss peak, is a parameter sensitive to static properties of the system, e.g., ageing or sintering. When the probe surface is clean and the grain-surface friction is low enough, a localised failure arises at this interface.

The immersed pendulum is a simple method which has the potential for further developments in the understanding of granular physics at the mesosocopic level [7]. By its operation principles, the pendulum can measure the mechanical properties of many systems, including fresh snow, and can be used in the field.

ACKNOWLEDGMENTS

The experiments in snow were conducted at the Eidgenössisches Institut für Schnee- und Lawinenforschung, Davos, and we thank M. Schneebeli and J. Schweizer for the assistance provided and for many helpful discussions.

REFERENCES

1. F. P. Bowden and D. Tabor, *The Friction and Lubrication of Solids* (Clarendon Press, Oxford, ed. 4, 1986).
2. B. J. Lazan, *Damping of Materials and Members in Structural Mechanics* (Pergamon, Oxford, 1968).
3. G. D'Anna, *to be published.*
4. L. Bocquet, E. Charlaix, S. Ciliberto, and J. Crassous, *Nature* **396**, 735 (1998); J. Crassous, L. Bocquet, S. Ciliberto, and C. Laroche, *Europhys. Lett.* **47**, 562 (1999).
5. M. Fukue, *Mechanical Performance of Snow Under Loading* (Tokai University Press, Tokyo, 1979).
6. S. C. Colbeck, *J. Appl. Physics* **84**, 4585 (1998).
7. For a review see: H. M. Jaeger, S. R. Nagel, and R. P. Behringer, *Rev. Mod. Phys.* **68**, 1259 (1996).

Mat. Res. Soc. Symp. Proc. Vol. 627 © 2000 Materials Research Society

Investigation of particulate systems using optical pathlength spectroscopy

Gabriel Popescu and Aristide Dogariu
School of Optics/CREOL, University of Central Florida
Orlando, FL 32816, U.S.A

ABSTRACT

In many industrial applications involving granular media, knowledge about the structural transformations suffered during the industrial process is desirable. Optical techniques are noninvasive, fast, and versatile tools for monitoring such transformations. We have recently introduced optical path-length spectroscopy as a new technique for random media investigation. The principle of the method is to use a partially coherent source in a Michelson interferometer, where the fields from a reference mirror and the sample are combined to obtain an interference signal. When the system under investigation is a multiple-scattering medium, by tuning the optical length of the reference arm, the optical path-length probability density of light backscattered from the sample is obtained. This distribution carries information about the structural details of the medium. In the present paper, we apply the technique of optical path-length spectroscopy to investigate inhomogeneous distributions of particulate dielectrics such as ceramics and powders. The experiments are performed on suspensions of systems with different solid loads, as well as on powders and suspensions of particles with different sizes. We show that the methodology is highly sensitive to changes in volume concentration and particle size and, therefore, it can be successfully used for real-time monitoring. In addition, the technique is fiber optic-based and has all the advantages associated with the inherent versatility.

INTRODUCTION

Particulate systems such as suspensions of small particles and powders are intensively used in various industrial processes. Therefore, the problem of characterizing such media in terms of their concentration and particle sizes becomes an important task. In this paper, we apply a recently implemented optical technique for particulate system investigation [1]. The method, referred to as optical pathlength spectroscopy, relies on the principle of the interference with low-coherence light and makes use of optical fibers to transmit and receive the light. The new technique has a great potential for on-line monitoring, since is completely noninvasive in the sense that no preparation (such as dilution) of the samples are required prior to measurements. In the following, the principle of the optical pathlength spectroscopy as well as a number of experimental results are presented.

OPTICAL PATHLENGTH SPECTROSCOPY FOR PARTICULATE SYSTEMS INVESTIGATION

Multiple light scattering techniques are intensively used to obtain optical characteristics of different diffusive systems and a broad range of applications is being developed. A multiple scattering regime is usually associated with waves propagation through optically dense random systems and is commonly described in terms of a diffusion equation. This is an approximation for energy transport where isotropic elastic scattering and wave propagation at a constant group velocity are considered, while the polarization and interference effects are neglected [2]. The diffusive wave propagation depends on characteristics of the specific scattering geometry and is characterized by the probability density $P(s)$ of optical pathlengths through the medium.

In general, $P(s)$ can be theoretically estimated for different experimental configurations but, so far, direct experimental studies regarding optical pathlengths distribution have been limited to investigations of temporal broadening of short light pulses that propagate diffusively [3, 4]. The time t necessary for the optical wave to propagate along a path of length s is simply given by $t=s/v$, where v is the average velocity of energy transport. The steady state transport mean free path l_t relates to the dynamic diffusion coefficient D by considering a constant energy transport velocity $v=3D/l_t$. In steady state conditions, l_t depends on both number density of scatterers and size and shape of each individual scatterer: $l_t=[n\sigma_s(1-g)]^{-1}$, where n is the number density of scatterers, σ_s is the cross section of a single scattering event and g is the average cosine of the scattering angle.

In this paper, we present a novel approach to investigate the properties of granulate systems. Based on the principle of low-coherence interferometry, the new technique called optical pathlengths spectroscopy (OPS) infers directly the pathlengths distribution $P(s)$ of waves backscattered in a specific geometry. From the shape of this distribution, the transport mean free path of the medium can be determined. A typical OPS signal consists of backscattered intensity contributions corresponding to waves scattered along closed loops that have the same optical pathlengths and, in addition, have the total momentum transfer of these waves equal to $4\pi/\lambda$ (backscattering). This is somehow similar to the information obtained in time-resolved reflectance measurements where the diffusion approximation makes a reasonable description of the experimental data.

In low-coherence interferometry, light from a broad bandwidth source is first split into probe and reference beams which are both retroreflected from a targeted scattering medium and from a reference mirror, respectively, and are subsequently recombined to generate an interference signal, as illustrated in Fig. 1.

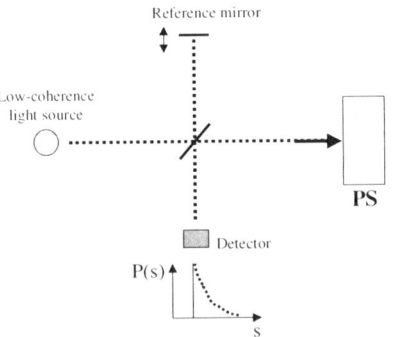

Fig 1 *Experimental setup for optical pathlength spectroscopy; PS- particulate system.*

Assuming quasi-monochromatic optical fields $(\Delta\lambda/\lambda <<1)$, the detected intensity has the simple form of $I_d=I_s+I_{ref}+2(Is \cdot Iref)^{1/2} cos(2\pi\Delta s/\lambda)$, I_d, I_s, and I_{ref} are the detected, scattered, and reference intensity, respectively. The optical path difference between the scattered and reference field is denoted by Δs and is the central wavelength. Two conditions are needed in order to obtain fringes of interference: i) Δs to be a multiple of wavelength and ii) $|\Delta s| < l_{coh}$ where l_{coh} is the coherence length of the source. In our OPS configuration, I_s corresponds to the reflectance of a multiple scattering medium.

It should be noted that, as long as the waves propagate with a constant average velocity, even for stationary sources, the waves characterized by the same optical pathlengths, can be considered as emitted at the same moment t_0. Thus the pathlengths probability density does not change if a steady-state source is replaced by a short pulse emitted at t_0. In our interferometer, only the class of waves that have traveled an optical distance which corresponds to the length of the reference arm is able to produce fringes and is, therefore, detected. In the optical pathlength domain, the interferometer acts as a bandpass filter with a bandwidth given by the coherence length of the source. Accordingly, the shorter the coherence length, the narrower the optical pathlengths interval of backscattered light that will produce detectable fringes. Now, if we let the reference mirror sweep the reference arm, waves with different optical pathlengths through the medium are detected and an optical pathlength distribution is reconstructed. Experimentally, the OPS approach is limited by the fact that the signal corresponding to long paths within the medium is weak and a large dynamic range is needed for accurate measurements of pathlengths distribution tails. However, as opposed to dynamic techniques, the measurements can be extended over longer periods of time and there is no need for sophisticated time-of-flight configurations. The low-coherence interferometer has a cw light source with a coherence length of *10 μm* and a broad wavelength band centered at *1300 nm*.

In the following we will use the optical pathlength distribution to obtain information about granular media. In our experiments, both liquid and solid systems have been investigated.

EXPERIMENTAL RESULTS

In order to investigate the potential of the method described earlier for particulate systems characterization, we performed OPS measurements on suspensions of particles of different concentrations. Fig. 2 shows the OPS data obtained from polystyrene particles suspended in water at different concentrations, as indicated. It can be seen that the experimental signals vary drastically with concentration. Due to the high dynamic range of the measurement, a broad range of suspension densities can be experimentally investigated, which should be of interest for online monitoring. In addition, one can note that for the system of lowest concentration, the experimental data approach an exponential decay, which is a mark of the single scattering regime [5]. As the concentration increases, the pathlength-resolved curves exhibit a more complex behavior, which tells that the light multiply scattered in the medium becomes important in the overall balance of the backscattered signal. For highly concentrated media, the photon transport is well described by the diffusion equation, as described earlier, which allows evaluating the scattering properties of the medium in terms of its transport mean free path. The results prove that OPS is able to monitor precisely any transformations that have as effect a particle concentration variation in the medium.

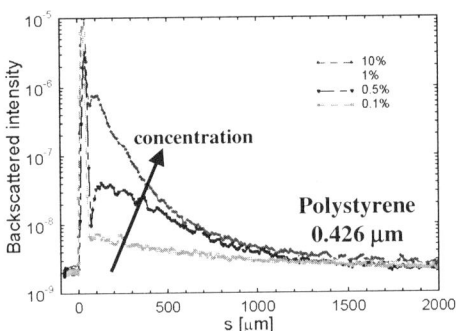

Fig. 2 *OPS data for polystyrene suspensions of different concentrations, as indicated.*

We apply further the OPS technique to powders, which is another class of particulate systems, of large interest for various applications. The experiments were performed on alumina powders of different particle sizes. Fig. 3 shows the OPS data obtained on alumina powders made of particles with two different sizes. The test fiber was immersed deep into the powder for each measurement. This geometry corresponds to the infinite configuration used for the light diffusion in random media. It can be seen that the pathlength distributions associated with the two powders exhibit dissimilar features, especially in the early time range (close to the origin of pathlengths). However, the tails of the curves are parallel, which tells that, for long times of traveling, the light undergoes a diffusive regime for both particle sizes. In addition, the fact that the tails are not overlapping proves that the diffusion constants for the powders are different [6]. The inset of Fig. 3 shows an experimental curve fitted with the diffusion model discussed earlier.

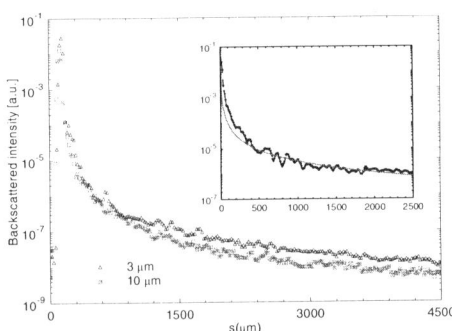

Fig. 3 *OPS data from powders with particle sizes of 3 and 10 μm. The inset shows a fit with the diffusion model*

It can be seen the OPS is able to sense difference in the particle sizes for solid samples. This should be useful for many sensing applications in industrial environment and it can be used to further investigate the fundamental phenomenon of wave transport in powders.

CONCLUSIONS

We presented the principle of optical pathlength spectroscopy as a new technique for particulate systems charcaterization. The results presented here suggest that the method is sensitive to changes in both density and particle size of the particulate systems such as colloidal suspensions or dry powders. Applications that require monitoring such properties could benefit from the versatility of this measurement.

REFERENCES

1. G. Popescu and A. Dogariu, *Opt. Lett.*, **24**, 7, 442-444 (1999).
2. A. Ishimaru, Wave Propagation and Scattering in Random Media (Academic, New York, 1971), Vol. 1, Chap. 9.
3. F. Liu, K.M. Yoo, R.R. Alfano, *Opt. Lett.*, **18**, 6 (1993).
4. K.M. Yoo, F. Liu, and R.R. Alfano, *Phys. Rev. Lett.*, **64**, 22 (1990).
5. J. M. Schmitt, A. Knuttel, and R. F. Bonner, *Appl. Optics*, **32**, 30, 1993.
6. G. Popescu, C. Mujat, and A. Dogariu, Phys. Rev. E, **61**, 4 (2000).

Mat. Res. Soc. Symp. Proc. Vol. 627 © 2000 Materials Research Society

Material Analysis by Ultrasonic Atomic Force Microscopy

Chiaki Miyasaka, Lily Jia, and Bernhard R. Tittmann
Department of Engineering Science and Mechanics,
The Pennsylvania State University,
227 Hammond Building, University Park, PA 16802, USA

ABSTRACT

Spray-dried ceramic powders (*e.g.*, Al_2O_3) are composed of a plurality of granules, each of which, includes ceramic particles and organic binders. It is assumed that the binders become concentrated in the surface layer of the granule in accordance with its type or its volume mixed into a ceramic portion of the granule. However, evidence to prove the assumption was limited because conventional microscopes were not able to clearly visualize the segregation. This paper presents a technique for imaging detailed structure of the spray-dried ceramic powders with the ultrasonic-atomic force microscope (U-AFM). The distribution of binder vis-a-vis Al_2O_3 particles is highly resolved with good contrast. The distribution was confirmed by nano-indentation. Thus, the U-AFM is shown to be a useful diagnostic tool for the development of approaches to spray-dried process evaluation.

INTRODUCTION

Spray drying is a process used in the manufacture of ceramic powder [1]. In the process, some organic binders must be mixed into a fluid feed material, which is so-called slurry. The slurry is pumped to an atomizer located in a drying chamber for making spherical shaped powders. Therefore, the ceramic powder comprises granules, each having ceramic portions and binder portions. It is assumed that some types of binders generally move to the surface of the granule and become a concentrated binder layer of the granule. This segregation is considered to cause processing defects (*e.g.*, voids and formation of irregularly shaped granules) [2]. However, since the distribution behavior of the binders of the slurry is difficult to visualize by conventional microscopes, the assumption has not been proved with strongly supported scientific evidence.

Recently, ultrasonic techniques have been integrated into an atomic force microscope (AFM) for forming amplitude and phase images besides topographic images to discriminate materials [3]. When emitting ultrasonic waves from a transducer into a material, the material vibrates. The vibration of the material is unique to its elastic properties. When the AFM cantilever scans a surface of a specimen, the tip of the cantilever senses a series of displacements caused by the vibration. Then, the series of displacements is converted to electric signals representing pixels of an image. The more amplitude or phase caused by the vibration is significantly different, the more the difference of elastic properties of materials are large. Furthermore, the difference of amplitude gives a good contrast of an image formed by the U-AFM [4]. The quality of the U-AFM image is usually better than that of the AFM image for the slurry droplet for the spray-dried ceramics [5]. By means of this approach, the binder portions and the ceramic portions within the slurry droplet are easily discriminated. Moreover, the distribution of the binders can be confirmed when using a nano-indenter. Furthermore, utilizing values obtained by the nano-indenter data can be compared to U-AFM data, which is by contrast is acquired non-destructively.

EXPERIMENTAL INVESTIGATION

Specimen Preparation
Three types of specimens were prepared for the present experiment. One of them was a pure alumina droplet, and the others were slurry droplets. One of the slurry droplets included 4% of poly-vinyl alcohol (hereinafter called simply "PVA"). Another included 3% of acylic emulsion binder (hereinafter called simply "EB"). Each specimen was made by embedding the granules into epoxy. After the curing of the epoxy, the embedded specimens were mechanically polished. The shapes of specimens were rectangular, and their X, Y and Z dimensions were about 4 mm x 4 mm x 2 mm.

U-AFM imaging
Fig. 1 is a schematic diagram of the U-AFM. One specimen was mounted on the front surface of a transducer generating ultrasonic waves up to 0.5 MHz. The cantilever gently contacts the specimen. The drive frequency selected to visualize images in this experiment was in the frequency range from 140 MHz to 150 MHz.

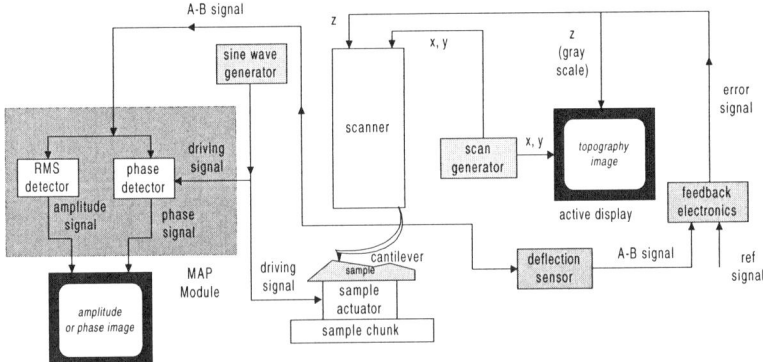

Fig.1 Schematic diagram of the UAFM

Figs. 2(a), 2(b), and 2(c) are amplitude images formed by the U-AFM, and showing pure alumina granules, an alumina ceramic granule including 4% of the PVA, and an alumina ceramic granule including 3% of the EB. In Fig. 2(a), each granule of alumina ceramics was highly resolved. The contrast difference within the granule was caused by its shape. That is, the topographic difference was enhanced by the force modulation mode. In Fig. 2(b), the dark contrast concentrated to the edge is assumed to be the distribution of the PVA. In Fig. 2(c), portions observed in dark contrast are also assumed to be the distribution of the EB. However, it seems that the EB is uniformly distributed.

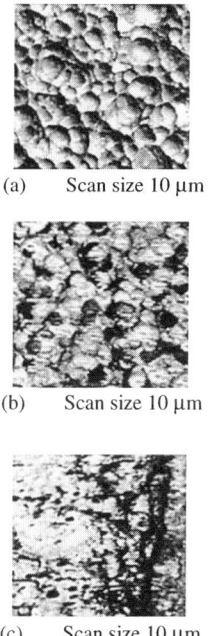

(a)   Scan size 10 μm

(b)   Scan size 10 μm

(c)   Scan size 10 μm

Fig. 2 U-AFM image (a) Pure almina ceramics, (b) Almina ceramics including 4%, of PVA, (c) Almina ceramics including EB 3%

Nano-indenter

As can be appreciated, the more the material under examination contains the binders, the softer and more compliant it is. Therefore, the nano-indenter (*i.e.*, apparatus to obtain information of elastic properties for a small area by giving a nano-meter size of indentation depth to a material) can be applied to evaluate whether the dark contrast is the distribution of the binders or not. In this experiment, the ultra micro-indentation system (model: UMIS-2000: CSIRO) was used.

The Berkovich indenter was loaded at three points of each granule, wherein the points are respectively located from 10, 20, and 40 µm from the edge of the specimen. These points were determined by the UFM images. The radius of the indentation area is approximately 10 µm. Five granules in accordance with ceramic types were chosen. Then, averaged hardness values were obtained. Hardness (H) and toughness index (E*) of each point is shown in Tables I, II, wherein H and E* are expressed in Eqs (1) and (2).

$$H = F/A = F/kh^2 \qquad\qquad (1)$$

Where F is the indentation force, A is the projected area of indentation, k is a constant determined by the shape of the indentation tip, and h is the penetration depth.

$$E^* = E/(1 - \nu^2) \qquad\qquad (2)$$

Where E is elastic modulus, and ν is Poisson's ratio.

Table I Hardness of granules including binders

Type of Granule	Hardness		
	Edge	Inside	Center
Alumina + EB (3%)	0.29GPa	0.26GPa	0.30GPa
Alumina + PVA (4%)	0.19GPa	0.27GPa	0.32GPa

Table II Toughness index of granules including binders

Type of Granule	Toughness		
	Edge	Inside	Center
Alumina + EB (3%)	10.8GPa	14.0GPa	14.6GPa
Alumina + PVA (4%)	6.3GPa	9.3GPa	13.2GPa

For the granule having the EB, there are no significant changes of hardness among the portions given the indentations. For the granule having the PVA, hardness increases from the edge toward the center. Therefore, in the case of the granules, it was confirmed that the contrast of the images obtained by the U-AFM was representing the distribution of the binders referring from values of Table I.

M. C. Bhardwaj reported values of Young's module of green alumina ceramics by the non-contact and air-coupled transducer [6]. The toughness index can be calculated from these values when assuming Poisson's ratio is 0.27. The toughness index ranges from 5.0 GPa to 15.0 GPa. Therefore, the values in Table II were considered to be reasonable.

CONCLUSIONS

The binder distributions within the granules were visualized by the U-AFM, and the contrast distributions of the images were consistent with the values obtained by the nano-indenter. Thus, the U-AFM is shown to be a useful diagnostic tool for the development of approaches to spray-dried process evaluation.

ACKNOWLEDGMENT

The authors are indebted to Mr. M. Mandanas and Prof. G. L. Messing of Material Research Lab in the Pennsylvania State University for their assistance in preparing specimens to carry out this study.

REFERENCES

1.   S. J. Lukasiewicsz, "Spray-Drying Ceramic Powders", Journal of the American Ceramics Society, Vol. 72, No.4, 1989, pp. 617-624

2.   Y. Zhang, M. Kawasaki, K. Ando, Z. Kato, N.Uchida and K.Uematsu, "Surface Segregation of PVA during Drying of PVA-Water-Al$_2$O$_3$ Slurry", Journal of the Ceramic Society of Japan, Vol. 100, No.8, 1992, pp. 1070-1073

3.   U. Rabe, and W. Arnold, "Acoustic Microscopy by Atomic Force Microscopy", Applied Physics Letters, Vol.64, No.12, 1994, pp. 1493-1495

4.   W. Gao, B. R. Tittmann, and C. Miyasaka, "Contrast Mechanism of Ultrasonic-Atomic Force Microscope", presented at *IEEE 1999 Ultrasonic Symposium* (1999)

5.   L. Jia, M. Mandanas, C, Miyasaka, B. R. Tittmann, and G. L.Messing, "Analysis of Organic Binder Distribution in Spray Dried Granules by Ultrasonic-Atomic Force Miceroscopy," Proceedings of SPIE's International Symposium on Nondestructive Evaluation Techniques for Aging Infrastructure & Manufacturing, 1999, pp. 270-281

6.   M. C. Bhardwaj, "Nondestructive testing of green ceramics by contact-free ultrasound", ULTRAN's sales materials.

Granular Flows I

Mat. Res. Soc. Symp. Proc. Vol. 627 © 2000 Materials Research Society

Nuclear Magnetic Resonance Studies of Granular Flows – Current Status

Stephen A. Altobelli, Arvind Caprihan, Eiichi Fukushima, Joseph D. Seymour
New Mexico Resonance, Albuquerque, NM 87108, U.S.A.

ABSTRACT

Nuclear magnetic resonance (NMR) is a non-intrusive method that can characterize not only the particulate density but also velocity and velocity fluctuation parameters. A survey of all the known NMR measurements of granular flow will be followed by a brief description of NMR as it applies to granular flow. Two new experiments, both involving flows in partially filled rotating horizontal cylinders, will be described. First, the effect of a stationary blade to suppress the azimuthal velocity of particles being brought up and deposited into the flowing layer on flow-velocity profiles will be studied. Suppressing the azimuthal velocity reduces the deviation of the velocity profile from a quadratic dependence on the height above the rigid layer. Second, a new NMR scheme will be presented that yields spatial distributions of collisional correlation times for macroscopic particles undergoing granular flow. It is based on Pulsed-Gradient Spin-Echo strategy that is commonly used to measure molecular diffusion in liquids. The scheme will be demonstrated with an example from shear flow in a partially-filled horizontal cylinder. Spatially resolved collisional correlation times and velocity fluctuation intensities are derived from the measurements and have values of ~1 ms and ~10^{-3} m^2/s^2, respectively, at the center of the free surface for 2 mm particles in a 70 mm diameter cylinder rotating at 2.36 rad/s.

INTRODUCTION

Granular flow occurs in natural phenomena such as avalanches and formation of sand dunes while the property of flowing sand has also been used in construction of large structures in ancient times. Past studies of such flows have mostly been of their bulk properties such as overall behavior, flow velocities that can be measured optically, acoustic (noise) measurements, pressure determination, and so on. Detailed studies of such flows on a microscopic scale have been difficult because the opacity of nearly all granular materials prevents direct observation inside the flow. Traditional methods of obtaining information from inside flows include radioactive tracer and visual examination, both incurring some obvious problems. X-ray CT is a modern method that can yield information about static structure while diffusing wave (optical) spectroscopy yields dynamic parameters without spatial localization of information.

Nuclear magnetic resonance (NMR) imaging is a non-invasive technique that is a useful and unique tool to measure flow parameters in liquid and liquid-solid multiphase flows. [Fukushima, 1999] Even though granular flows seems to be ill-suited for NMRI because the technique is very difficult to apply to solid samples, there have been a few applications of NMRI to flows of particles that contain sufficient liquid components to give rise to adequate NMR signals.

A virtue of NMR, besides that of being able to overcome the opacity of the granular matter, is its versatility; many parameters that are useful for granular flow studies can be measured by NMR. These parameters include components along any direction of velocity, particulate density, velocity fluctuations and particulate diffusion coefficient. For granular flows, parameters measured by NMR represent temporal and spatial averages. By changing the scale of the averaging process, we can statistically characterize distributions that result in these averages. The addition of imaging to NMR (to do NMRI) allows us to map out spatial heterogeneities of such averages such as diffusion coefficient and velocity on a scale much coarser than individual particles but fine compared to the overall system.

BASIC CONCEPTS OF NMR THAT ARE USEFUL FOR GRANULAR FLOW STUDIES

We defer the topic of NMR physics to the reviews [for example, Fukushima, 1999] except for a cursory description that will allow us to describe the results in this article. NMR stands for nuclear magnetic resonance. Certain atomic nuclei precess in a magnetic field at a frequency that is proportional to both the strength of the magnetic field B and the gyromagnetic ratio γ that is a unique property of the particular atomic nuclei such as a proton or ^{19}F. The fact that it is possible to detect the frequency distribution of precessing nuclei makes NMR possible. Thus, if the precession frequency that can be measured is ω,

$$\omega = \gamma B. \tag{1}$$

In order to understand NMR imaging, commonly called MRI, we note that the experimenter has control of the magnetic field and, therefore, B can be made to depend on location in space. If a unique value of B can be assigned to every point in space, there would be a one-to-one relation between ω and position. This can be done, at least, in one dimension with, for example, a magnetic field intensity B that varies linearly with distance x, i.e.,

$$B(x) = Gx, \tag{2}$$

where G is the gradient of the magnetic field, $G \equiv dB/dx$. Then, the measured frequency ω becomes a function of x so

$$\omega(x) = \gamma B(x) = \gamma Gx \tag{3}$$

and when a certain signal is measured as having a frequency ω, we know that it arises from spins located at position x. In this way, we can generate a one-dimensional image. Two and three dimensional images can be obtained by extension even though the actual implementation of the NMR experiment becomes quite complex.

NMR is also capable of yielding velocities. Suppose in the above 1-D example, the coordinate depends on time, i.e., the nucleus moves. Suppose, further, that we will examine the motion for a short time so the coordinate can be expanded in a Taylor series. Then, (3) can be rewritten

$$\omega(x(t)) = \gamma B(x(t)) = \gamma Gx(t) = \gamma G[x_o + vt + \tfrac{1}{2}at^2 + \dots], \qquad (4)$$

where x_o is the initial coordinate, v is the velocity which will be taken to be constant in the short interval, and a is acceleration.

When we do an experiment and observe a quantity oscillating at frequency ω, we can integrate it to get the change in phase angle ϕ. Thus,

$$\phi = \int \omega(t)dt = \int \gamma G[x_o + vt + \tfrac{1}{2}at^2 + \dots]dt = \gamma v \int Gt \ dt \equiv \gamma v m_1. \qquad (5)$$

Here, we chose the time dependence of G to be bipolar so the integral of the first term of the expansion is zero and ignored the third term by assuming we are going to consider a very short time interval. Now, the phase of the NMR signal that we can measure is proportional to the velocity v. Therefore, we can measure the velocity knowing γ and m_1.

The last NMR parameter we will discuss is diffusion coefficient. The NMR signal arises from a large number (that approaches Avogadro's number) of nuclei precessing as per Eq. (1) in a magnetic field B. Suppose that $B(x) = Gx$. G is no longer bipolar so the integral involving x_o is not nulled. In the absence of motion, the phase after time t depends on the location x_o of each particle, i.e., now,

$$\phi = \int \omega(t)dt = \int \gamma Gx_o dt \qquad (6)$$

so the nuclei at different locations x_o accumulate phase at different rates and the total signal from all nuclei dephases in time, decreasing to zero. However, if we can reverse the gradient polarity G to -G for the same duration, the spins that dephased should rephase by exactly the same amount and we should recover the entire signal that dephased while the gradient was G. In other words, a nucleus at location x_o will dephase by

$$\phi(\tau) = \gamma Gx_o \tau \qquad (7)$$

over time τ and then rephase by $-\gamma Gx_o \tau$ over the next interval τ so there will be no net phase shift for *all* nuclei, resulting in no signal loss. This is called a spin-echo and the signal simply reappears from baseline.

Suppose that some of the particles undergo random motion along x during this experiment. Then they will not have the same x_o during the time the gradient is first G and then -G so the net phase will not become zero over time 2τ. Those nuclei would not have the same phase (=0) as the nuclei that were stationary, leading to loss of signal amplitude caused by the random motion. In this way, it is possible to measure diffusive motion by NMR.

The particles that can be studied by NMR are of two kinds – natural particles containing liquids and artificial particles. In the former category are seeds such as mustard, poppy, and sesame, while the latter include vitamin and other pharmaceutical pills as well as sugar beads with imbibed liquid [Porion, et al., 1999].

The flow chamber will have to be made of electrical insulators so that the radiofrequency magnetic field can penetrate it. The most practical material to use in constructing granular flow systems that can be studied by NMR is acrylic, polycarbonate, some other plastic, or glass.

GRANULAR FLOW SYSTEMS STUDIED BY NMR

The granular flow system that has been studied most by NMR, to date, is the partially-filled horizontal rotating drum. NMR permits the study of a transverse slices near the center of a long cylinder, so end effects can be avoided. Dynamic measurements, i.e., those made while the flow is taking place, of velocity and concentration as well as static monitoring of dynamic processes, e.g., axial and radial segregation, were made early on [Nakagawa, et al., 1993; Nakagawa, 1994; Metcalf & Shattuck, 1996]. Vibrating granular systems are more difficult to handle, experimentally, but initial forays have been made in the study of convection and diffusion of particles in containers during a single vertical shake [Ehrichs, et al., 1995; Kuperman, et al., 1995; Knight, et al., 1996; Kuperman, 1996]. A study has also been reported of a continuously shaking vertical granular system [Caprihan, et al., 1997].

Most recently, granular particles being sheared in a vertical Couette were studied by a combination of NMRI, x-ray tomography, and visual observation [Mueth, et al., 2000]. The velocity profile across the gap was found to be determined by a slip contribution arising from the layering of the relatively large (in terms of the Couette gap) identical spherical particles in addition to the contribution from randomly packed particles. The latter contribution gave rise to a linear normalized velocity gradient which implies a Gaussian velocity profile across the gap.

To date, other granular flow geometries, such as chute flow and hopper flow, have not been investigated by NMR, to our knowledge.

RECENT EXPERIMENTS ON FLOW-VELOCITY PROFILES IN HORIZONTAL ROTATING CYLINDER

The velocity depth-profile of granular flows in a slowly rotating horizontal rotating cylinder is a smoothly varying function of the height above the bottom of the flowing layer. It can be fit to a quadratic function of the height except near the free surface where it falls below the quadratic, especially when the rotation frequency of the cylinder is increased [Nakagawa, et al, 1997a]. It was conjectured in that work that the departure from quadratic near the free surface might be caused, at least in part, by inertial effects, i.e., the azimuthal component of velocity for the particles that are brought up into the sliding layer by the rigid-body rotation around the bottom of the rotating cylinder.

A recent experiment performed at New Mexico Resonance by Christian Heine of RWTH-Aachen, Germany, confirms this conjecture by the use of a flat stationary blade that is placed at the top of the flowing layer to suppress the azimuthal component of the velocity for those particles being brought up into the flowing layer by the rigid-body rotation along the lower part of the rotating cylinder. The blade is fixed in the laboratory frame and its support enters the rotating cylinder at the center of one of the cylinder ends through a bushing. For each rotation speed of the cylinder, the blade is repositioned to flatten out the initial slope of the free surface at the uphill end.

The sketches below show cross-sectional views of the cylinder that is 50% full of particles undergoing counter-clockwise rotation. The left sketch is the cylinder without the blade and shows (exaggerated) the S-shaped free surface that is formed at the higher rotation rates. The right sketch shows the effect of the stationary blade which flattens the free surface.

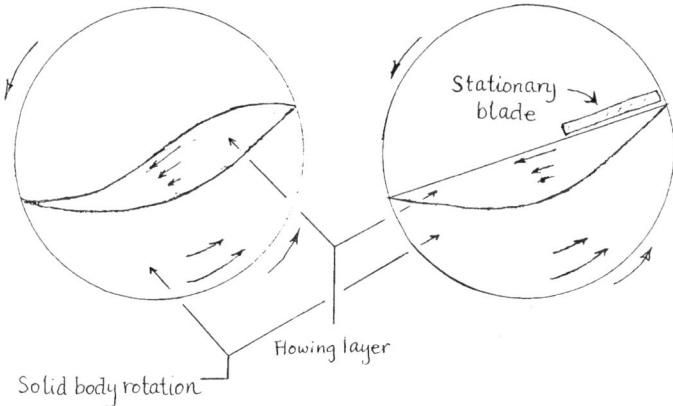

The system studied was an acrylic cylinder whose inside dimensions are 88 mm diameter and 560 mm long. The particles studied were spherical mustard seeds of diameter approximately 1.5 mm.

The flow velocity profile as a function of the depth in a direction orthogonal to the free surface was measured for a variety of conditions such as presence or absence of the blade, the blade position, and rotation speed. In order to measure the effect of the presence of the blade at the same effective rotation rate, the actual rotation rates of the cylinder with the blade had to be adjusted to be higher than that without the blade because the pressure of the blade on the particles caused additional slippage of the beads against the cylinder wall.

It was found that the deviation from the quadratic dependence mentioned above was reduced by the blade at the same effective rotation rate in every case. This finding supports the conjecture that the deviation from the quadratic dependence of the velocity profile near the free surface is caused by inertial effects. This further implies that the velocity profile of this granular flow system is most likely to be determined by a single mechanism rather than a combination of mechanisms.

MEASUREMENT OF COLLISIONAL PARAMETERS BY NMR

It has already been mentioned that an advantage of NMR over many other techniques is the versatility that allows measurements of many different aspects of any given system. One important such capability, especially useful in granular flow or any other motions involving fluctuations, is to measure particulate diffusion parameters and collisional correlation times of macroscopic particles undergoing granular flow.

In a recent study [Seymour, et al., 2000], pulsed gradient spin-echoes were used to probe random motion of particles in the shear flow regime of a 3D flow of particles in a partially filled horizontal rotating cylinder. This is accomplished by using two different radio-frequency and gradient pulse sequences so that one detects the effects of all relative motions between particles being examined while the other detects only those relative motions that are random in a time frame set by the experiment. The difference between these two observations represents the coherent or correlated component of the motion *for that time period* that is under the control of the experimenter.

Thus, by changing the time frame of the observation, we are able to assign a decay rate to the coherence, i.e., the longer the observation window, the less likely that the motion between any particles will remain coherent. If an exponential form of velocity fluctuation correlation is assumed, i.e.,

$$\langle u(t)u(0)\rangle = \langle u^2\rangle \exp(-t/\tau_c), \tag{8}$$

where $u(t)$ is the velocity at time t, then τ_c, the autocorrelation time constant, can be easily measured.

The system studied was a 70 mm inside diameter, 245 mm long acrylic cylinder half full of 2 mm diameter oil-filled plastic beads. The rotation rate considered here is 2.36 rad/s which corresponds to 135 degrees of rotation per second. The flow takes place in a convex lens-shaped layer with the flat part of the lens corresponding to the free surface. The particles enter this flowing layer after being brought around from the lower end of the flowing layer by the rigid-body rotation near the bottom of the cylinder as already mentioned in connection with the velocity profile measurement. The average flow is virtually parallel to the free surface.

In this work, only the axial component of motion will be described. The axial motion is the simplest component to consider because of the absence of average axial velocity. Thus, all axial motion becomes random, if the observation window is made long enough.

The primary parameter extracted from the experiment is the axial component of the apparent diffusion coefficient as a function of depth for any given rotation rate. Depending on the pulse sequence used, the apparent diffusion coefficient is due to either all motion or only to motion that is random over the measurement time window which, in this experiment, varied between 2.6 and 8.7 ms. For all measurement times in this range, it was found that the two apparent diffusion coefficients are identical in the lowest region of the flowing layer, implying that all axial motion is random in the time scale used. [They are trivially identical in the rigid-body rotation part because there is no relative motion between the particles.]

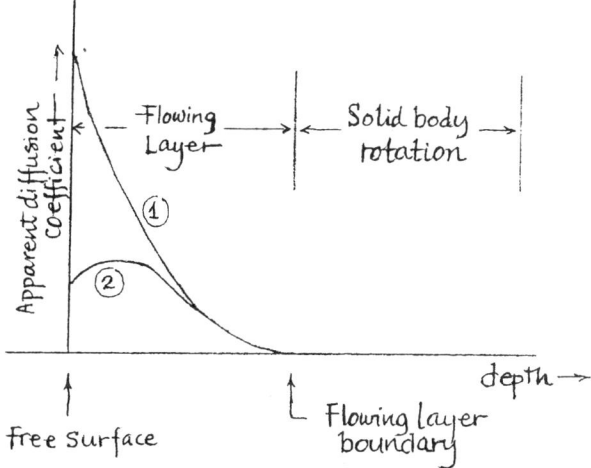

The sketch above shows how the two apparent diffusion coefficient varies as a function of depth along a line perpendicular to the free surface and going through the center of the cylinder. Curve #1 is the apparent diffusion coefficient that is derived from measured signal attenuation caused by all fluctuations while curve #2 represents attenuation caused only by motions that are random on the time frame of the NMR experiment.

At some height above the bottom of the flowing layer, the curves diverge, revealing the presence of motion that is not random on the time scale of the observation window. This means that the collision times become longer at shallower depths in the flowing layer, the increasing mean-free-path winning out over the increase in granular temperature as one gets closer to the free surface, a finding supported by computer simulations [Nakagawa, 1997a].

Even with the limited range of the observation time, we conclude that τ_c is of the order of 1 ms and the velocity fluctuation intensity $<u^2>$ is of the order to 10^{-3} m^2/s^2 in the center of the free surface at this rotation rate and becomes shorter at increasing depth in the flowing layer.

This technique is applicable, of course, to other granular flows, regardless of the kind of interparticle interactions, provided measurable NMR signals exist and the correlation times are within range of available observation window. At present, the upper limit for the observation window for these particles is about 200 ms because of the spin-spin relaxation time whereas the lower limit is around a few milliseconds because of hardware limitations.

CONCLUSIONS

We conclude that NMR is an excellent non-invasive technology to spatially resolve granular flow parameters from within the flowing region. In this way, we have concluded that the flow profile in a partially filled horizontal rotating cylinder is quadratic as a function of depth and is probably determined by a single mechanism. We have also measured collisional correlation times as a function of location in the same granular medium in a half filled horizontal cylinder. The axial component of velocity has a correlation time of about 1 ms and velocity fluctuation of about 10^{-3} m^2/s^2 at the center of the free surface for 2 mm particles flowing in a 70 mm inside diameter acrylic tube, rotating at 2.36 red/s.

ACKNOWLEDGMENTS

Some of this work was performed by C. Heine of RWTH-Aachen; his stay at New Mexico Resonance was made possible by the support of Professor B. Bluemich. Discussions with our colleagues D. O. Kuethe of New Mexico Resonance and M. Nakagawa of Colorado School of Mines are gratefully acknowledged. This work was sponsored, in part, by the Engineering Research Program of the Office of Basic Energy Sciences, United States Department of Energy, under Grant No. DE-FG03-98ER14912 but this sponsorship does not constitute an endorsement by the USDOE of the views expressed in this work.

REFERENCES AND BIBLIOGRAPHY

All of the literature on NMR studies of granular flow, known to the authors at present, are listed below in chronological order. This list also provides the references in the text, cited by the first author and the year of publication. If the citations by first author and the year are still not unique, lower case letters are appended to the year to correspond to the appearance of the references in this list.

M. Nakagawa, S. A. Altobelli, A. Caprihan, E. Fukushima, and E.-K. Jeong, "Non-invasive Measurements of Granular Flows by Magnetic Resonance Imaging," *Experiments in Fluids*, 1993, 16:54-60.

M. Nakagawa, "Axial segregation of granular flows in a horizontal rotating cylinder," *Chem. Engng. Sci.*, 1994; 49:2540-2544.

E. E. Ehrichs, H. M. Jaeger, G. S. Karczmar, J. B. Knight, V. Yu. Kuperman, and S. R. Nagel, "Granular Convection Observed by Magnetic Resonance Imaging," *Science*, 1995; 267:1632-34.

V. Yu. Kuperman, E. E. Ehrichs, H. M. Jaeger, and G. S. Karczmar, "A new technique for differentiating between diffusion and flow in granular media using magnetic resonance imaging," *Rev. Sci. Instrum.* 1995; 66:4350-55.

H. M. Jaeger, S. R. Nagel, and R. P. Behringer, "The Physics of Granular Materials," *Phys. Today,* 1996 (4); 32-38.

J. B. Knight, E. E. Ehrichs, V. Yu. Kuperman, J. K. Flint, H. M. Jaeger, and S. R. Nagel, "Experimental study of granular convection," *Phys. Rev. E*, 1996; 54:5726-38 .

V. Yu. Kuperman, "Nuclear Magnetic Resonance Measurements of Diffusion in Granular Media," *Phys. Rev. Letters*, 1996; 77: 1178-81.

G. Metcalf, M. Shattuck, "Pattern formation during mixing and segregation of flowing granular materials," *Physica A.*, 1996; 233:709-17.

K. M. Hill, A. Caprihan, and J. Kakalios, "Bulk Segregation in Rotated Granular Materials Measured by Magnetic Resonance Imaging," *Phys. Rev. Letters*, 1997; 78: 50-53.

J. B. Knight, "External boundaries and internal shear bands in granular convection," *Phys. Rev. E,* 1997; 55: 6016-23.

M. Nakagawa, S. A. Altobelli, A. Caprihan, and E. Fukushima, "NMR measurement and approximate derivation of the velocity depth-profile of granular flow in a rotating, partially filled, horizontal cylinder," in Powders and Grains 97, Proceedings of the Third International Conference on Powders & Grains, edited by R. P. Behringer and J. T. Jenkins (A. A. Balkema, Rotterdam, 1997); pp. 447-450.

H. A. Cheng, S. A. Altobelli, A. Caprihan, and E. Fukushima, "NMR and mechanical measurements of the collisional dissipation of granular flow in a rotating, partially filled, horizontal cylinder," in Powders and Grains 97, Proceedings of the Third International Conference on Powders & Grains, edited by R. P. Behringer and J. T. Jenkins (A. A. Balkema, Rotterdam, 1997); pp. 463-465.

K. M. Hill, J. Kakalios, K. Yamane, Y. Tsuji, and A. Caprihan, "Dynamic angle of repose as a function of mixture concentration: Results from MRI experiments and DEM simulations," in Powders and Grains 97, Proceedings of the Third International Conference on Powders & Grains, edited by R. P. Behringer and J. T. Jenkins (A. A. Balkema, Rotterdam, 1997); pp. 483-486.

A. Caprihan, E. Fukushima, A. D. Rosato, and M. Kos, "Magnetic Resonance Imaging of Vibrating Granular Beds by Spatial Imaging," *Rev. Sci. Instrum.*, 1997; 68:4217-4220.

M. Nakagawa, S. A. Altobelli, A. Caprihan, and E. Fukushima, "An MRI study: Axial migration of radially segregated core of granular mixture in a horizontal rotating cylinder," *Chemical Engineering Science*, 1997; 52:4423-4428.

K. M. Hill, A. Caprihan, and J. Kakalios, "Axial segregation of granular media rotated in a drum mixer: Pattern evolution," *Phys. Rev. E*, 1997; 56:4386-4393.

K. Yamane, M. Nakagawa, S. A. Altobelli, T. Tanaka, and Y. Tsuji, "Steady Particulate Flows in a Horizontal Rotating Cylinder," *Phys. Fluids*, 1998; 10:1419-1427.

M. Nakagawa, J. L. Moss, and S. A. Altobelli, "Segregation of granular particles in a nearly packed rotating cylinder: A new insightfor axial segregation," in Physics of Dry Granular Media, edited by H. J. Herrmann, H. J. Hovi, and S. Luding, Kluwer Academic, 1998; pp. 703-710.

E. Fukushima, "Nuclear Magnetic Resonance as a Tool to Study Flow," *Annu. Rev. Fluid Mech.*, 1999; 31: 95-123.

P. Porion, N. Sommier, and P. Evesque, "Mixing and Segregation of Solids in a Turbulence Mixer Studied by M.R.I.," presented at International Union of Theoretical and Applied Mechanics Conference on Segregation in Granular Flows, Cape May, New Jersey, June 6-9, 1999.

J. D. Seymour, A. Caprihan, S. A. Altobelli, and E. Fukushima, "Pulsed Gradient Spin Echo Nuclear Magnetic Resonance Imaging of Diffusion in Granular Flow," *Phys. Rev. Lett.*, 2000; 84:266-269.

A. Caprihan and J. D. Seymour, "Correlation Time and Diffusion Coefficient Imaging: Application to a Granular Flow System," *J. Magn. Reson.*, 2000; accepted for publication.

D. M. Mueth, G. F. Debregeas, G. S. Karczmar, P. J. Eng, S. R. Nagel, and H. M. Jaeger, "Signatures of granular microstructure in dense shear flows," submitted to *Nature*, 2000.

Mat. Res. Soc. Symp. Proc. Vol 627 © 2000 Materials Research Society

Avalanche Dynamics: Influence of the Granular Packing Size

M. A. Aguirre[*], A. Calvo[*], I. Ippolito[**].
[*]Grupo de Medios Porosos, Facultad de Ingeniería, Universidad de Buenos Aires. Paseo Colón 850, (1063) Buenos Aires, Argentina.
maaguir@fi.uba.ar.
[**]Groupe Matière Condensée et Matériaux, U.M.R. 6626, Université de Rennes 1, Campus de Beaulieu, 35042 Rennes Cedex, France.

ABSTRACT

We present an experimental study about the stability and dynamics of a granular packing and the influence of the packing size on the avalanche mass and on its characteristic angles: maximum angle of stability and angle of repose.

For a system with a large enough number of granular layers only N_c superficial layers contribute to the avalanche mass. It can be observed that N_c strongly depends on the packing size. Experiments are done on monosize glass beads with packings of different sizes.

INTRODUCTION

Surface flow of granular material and avalanche dynamics present different regimes depending on the packing size. Three regimes are observed when a box filled with grains is tilted very slowly [1].

First, there is a build up period where many small rearrangements occur on the free surface of the packing, then as the angle increases a second regime is found where inertia effects are not negligible and some big rearrangements appear. Finally, when the maximum angle of stability θ_M is reached a big avalanche is produced (figure 1).

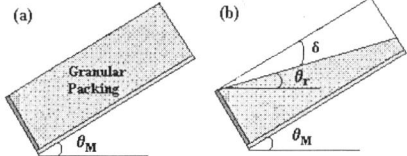

Figure 1. *(a) Scheme of the packing just before the avalanche begins at θ_M, the maximum angle of stability. (b) Scheme of final configuration for a packing with a large enough number of layers. θ_r is the angle of repose where the avalanche stops. Angle δ is the difference between θ_M and θ_r and the wedge it forms (white zone) corresponds to the mass displaced during the avalanche.*

The avalanche takes place at θ_M, the maximum angle of stability. This large sliding decreases the slope of the free surface of the packing until a second critical angle is reached: the angle of repose, θ_r. It can be observed, that the avalanche size is proportional to a wedge of angle $\delta = \theta_M - \theta_r$.

As was said, all regimes strongly depend on the granular packing size and in this work we present its influence on the critical angles and on the avalanche mass M, characteristic parameters of the avalanche regime.

EXPERIMENTAL PROCEDURE

We present an experiment on a granular packing in a contained geometry: a box filled with a certain number of layers of glass beads is tilted very slowly up to the threshold of instability where a big avalanche is produced.

The packing is obtained by filling up the box with glass beads of mean diameter $d = (2.2 \pm 0.2)$mm. The same type of beads is used to prepare the rough bottom. Experiments are performed under a controlled relative humidity of 50 %.

The box is inclined at a rate of 0.3° per minute. When the critical angle θ_M is reached a large avalanche is detected. The avalanche mass, M, the maximum angle of stability, θ_M and the angle δ between the plane and the final free surface are measured. Finally, the angle of repose is computed as: $\theta_r = \theta_M - \delta$. In order to make a good determination of these quantities, 10 experiments were performed in identical conditions.

It should be noticed, that for systems with a few numbers of layers, the final free surface of the packing, after the avalanche, does not take a clear profile and cannot be measured. Nevertheless, for system with more than 10 layers, the final free surface is almost flat and it can be assumed that it forms a wedge of angle δ that is directly measured with a goniometer.

In a previous work [2], we observed that for a system with more than 13 layers, the mass displaced by the avalanche does not depend on the height of the packing and therefore always involves a same quantity, N_c, of superficial layers. Later we have verified the dependence of N_c on the packing length [3]. Those results were obtained using granular packings contained in two boxes of different sizes. One of the system had length $L = 320$mm and width $W = 250$mm and another had length $L = 640$mm and width $W = 125$mm, both with the same area A. New experiments were done in a box length $L = 484$mm and width $W = 125$mm presenting a different covered area.

EXPERIMENTAL RESULTS

The avalanche regime was thoroughly studied in previous works [2] and it was observed that θ_M values fluctuate within a certain range, typically 3°, with corresponding fluctuations in the relative avalanche size δ. These fluctuations are due to different initial packing fraction [4, 5, 6]. However, within these fluctuations, a strong linear correlation is found between the relative size of a given avalanche and the corresponding values of θ_M and δ. Results had showed [2] that once the avalanche starts at some angle θ_M the process evolves displacing out of the packing a number of layers such that the free surface always reaches the same angle θ_r, independent of θ_M and M values. In other words, θ_r appears to be an intrinsic parameter of the granular medium.

The maximum angle of stability presents the same behavior as a function of the packing height independent of the system. As in previous works [2,5], the same three regimes due to dilatancy and packing fraction effects were also observed in this case. But the system with less covered area was less stable presenting a smaller value of $<\theta_M>$ (see Table I). In all cases, to get results that are independent of the constraints due to the rough bottom of the system, we find that the total number of layers $N > 10$. Also, $<\theta_M>$ is independent of N.

Table I. Dimensions and mean values of characteristic angles and N_c for the different used packings.

Packing	L (cm)	A (cm^2)	$<\theta_M>$	$<\theta_r>$	$<\delta>$	N_c
I	32	800	26.0°±0.2°	21.2°±0.4°	4.7°±0.5°	13±1
II	48.4	605	23.7°±0.3°	19.9°±0.4°	3.8°±0.7°	13±1
III	64	800	25°±1°	20.4°±0.8°	4.7°±0.6°	26±1

In all systems, $<M>$ presented the same behavior as a function of N as was reported elsewhere [2]: it increases with N until $<M>$ gets independent of N reaching an asymptotic value $\langle M_{asymptotic} \rangle$ involving the same N_c superficial layers. Unlike other systems where the rough bottom is made of more irregular grains [6], in these granular packings $\langle M_{asymptotic} \rangle$ is reached for $N > N_c$. Taking into account the final profile of the free surface in systems with more than N_c layer, it can be stated that $\langle M_{asymptotic} \rangle$ occupies half of the parallelepiped of height corresponding to N_c layer: $\langle M_{asymptotic} \rangle = \dfrac{mN_c}{2}$, where m is the mass of one layer having a packing fraction $C_f = 0.7$.

For an easier comparison of the results obtained for the three different boxes we have plotted (figure 2) the ratio, F, between the avalanche mass M and the total packing mass, M_T, as a function of N/N_c.

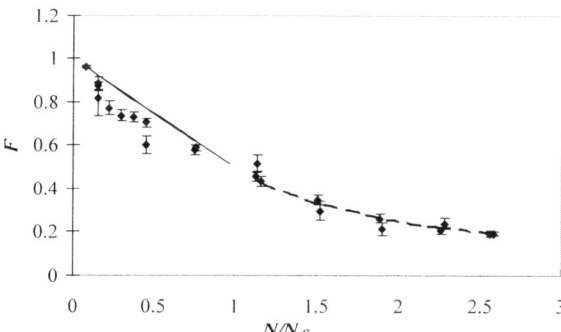

Figure 2. *Fraction of avalanche mass as a function of the number of layers scaled with N_c for all systems.*

We observe that both sets of data collapse in the same curve, displaying a decreasing behavior. Considering that $M_T = mN$ and $\langle M_{asymptotic} \rangle = \dfrac{mN_c}{2}$ [2] we obtain:
$F = \dfrac{\langle M_{asymptotic} \rangle}{M_T} = \dfrac{1}{2}\dfrac{N_c}{N}$ for $N/N_c > 1$. This variation is drawn with a dash line showing an excellent agreement with experimental data.

For $N/N_c < 1$ and assuming a constant value of $<\delta>$ for all N and not only for $N > 10$ as actually occurs, the fraction F will vary as $F = \dfrac{<M>}{M_T} = 1 - \dfrac{1}{2}\dfrac{N}{N_c}$. This variation is shown in figure 2 with solid line. Experimental values are well fitted for $N > 10$ ($N/N_c^{I} > 0.77$; $N/N_c^{II} > 0.77$, $N/N_c^{III} > 0.38$ [see Table I for specifics on Packings I, II and III]). As was expected, for $N < 10$ the experimental points are not fitted by the equation. In fact, the values are below the curve, which indicates that the rough bottom affects the dynamics of the system, producing a larger retention of mass.

As in a previous work [2], δ and thereby $\theta_r = \theta_M - \delta$ have reached a constant value for $N > 10$. It can be observed in Table I that $<\delta>$ reaches the same constant value for both boxes with equal area [3]. This means, that after the packing destabilizes it evolves toward the same $<\delta>$ which explains the existence of different N_c values depending on the system length. Figure 3 shows that if δ takes the same value independently of the box length then the avalanche has affected a larger number of layers in the longer box. When the wedge is reached ($N > 10$): $N_c = \dfrac{L\tan(\delta)}{d}$ and as $<\delta>$ is constant in for boxes I and III, then $N_c \propto L$, which explains why $N_c^{III} = 26$, $N_c^{I} = 13$.

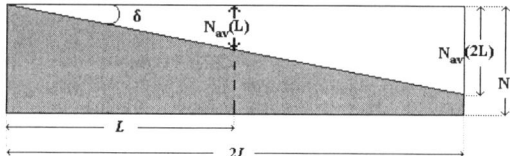

Figure 3. *Visualization of the number of layers (N_{av}) affected by the avalanche. For a fix N layer packing of equal covered area. δ is almost the same for both granular packings but not the number superficial layers involved in M.*

The different $<\delta>$ value for the box with different area can be explained taking into account the avalanche mass: $\tan\langle\delta\rangle = \dfrac{2d < M_{asymtotic} >}{mL}$, where m depends on the area: $m = C_1 dA\rho$ where ρ is the glass density. The latter result indicates that the avalanche dynamics is not only determined by the number of layers of the system but strongly depends on its length and width. Values of δ were obtained using the above expression and a good agreement with direct measurements is found (see Table II), which corroborates that effectively, for system with many layers, it is possible to describe the final profile of the free surface as a wedge of angle δ.

Table II. Characteristics of the different packings used and mean values of $M_{asymtotic}$ and δ obtained by direct measurement and calculated using the value of $<M_{asymptotic}>$.

Packing	L (cm)	m (g)	$<M_{asymtotic}>$	$<\delta>_{measured}$	$<\delta>_{calculated}$
I	32	230	1524±70	4.7°±0.5°	5.2°±0.2°
II	48.4	174	1143±168	3.8°±0.7°	3.4°±0.5°
III	64	230	2593±470	4.7°±0.6°	4.4°±0.8°

An important observation is that for $N > 10$ all critical angles $<\theta_M>$, $<\delta>$ and $<\theta_r>$ are independent of N.

CONCLUSIONS

The same qualitative variation of θ_M with N due to dilatancy and packing fraction effects, was observed for all the studied systems. Variations of the angle of repose θ_r with the system sizes are negligible.

In packings with a rough bottom made of regular grains (spheres) and more than N_c layers, the avalanches always involve the same number N_c of superficial layers and therefore $<M>$ takes a constant value $\langle M_{asymptotic} \rangle$. The mass displaced by the avalanche can be contained in a wedge of angle δ, width W and length L. For smaller systems, the avalanche regime depends critically on the thickness of the packing: all layers are involved in the avalanche and its mass is proportional to the number of layers of the packing.

ACKNOWLEDGMENTS

This work was supported by Ecos-Sud A97-E03, PICS CNRS-CONICET 561 and SECyT TI-07 UBA.

REFERENCES

1. N. Nerone, M. A. Aguirre, A. Calvo, I. Ippolito, D. Bideau, Physica A, **283**, 218-222 (2000).
2. M. A. Aguirre, N. Nerone, A. Calvo, I. Ippolito and D. Bideau, Phys. Rev. E **62**, p. 738 (2000).
3. M A. Aguirre, N. Nerone, A. Calvo, D. Bideau, I. Ippolito, in Traffic and Granular Flow'99: Social, Traffic and Granular Dynamics, edited by D. Helbing, H. J. Herrmann, M. Schreckenberg, and D. E. Wolf. (Springer, Berlin, 2000) p. 489.
4. O. Pouliquen and N. Renault, J. Phys. II France **6**, p. 923-935 (1996).
5. P. Evesque, D. Fargeix, P. Habib., M.P. Luong and P. Porion, Phys. Rev. E **47** p. 2326-2332 (1993).
6. M. A. Aguirre, N. Nerone, A. Calvo, I. Ippolito, D. Bideau. Granular Matter. Proceedings of the Workshop of the Consortium of the Americas on Interdisciplinary Science on Sparsely Connected Systems: Porous and Granular Materials, 2000 (submitted).

Mat. Res. Soc. Symp. Proc. Vol. 627 © 2000 Materials Research Society

The Dynamics of Avalanching Bead Chains

Michael Bretz* and Alberto G. Rojo
Department of Physics, University of Michigan, Ann Arbor, MI 48109

ABSTRACT

Avalanching studies were performed on 2D beds of a unique granular material - uniform length, 2.06 mm diameter flexible metal bead chain. The dynamical behavior of the slowly rotating vertical chain beds was separately characterized using chain lengths of 1, 2, 5, 9 and 13 beads, respectively. Digital images were acquired at 30-second intervals and analyzed to provide comparisons of mean profiles, slope angle variations, avalanche mass distributions and surface step discontinuities. This study is designed to probe how elongation, flexing and rolling affect avalanching through interlocking and polymer-like behavior.

INTRODUCTION

Over the last decade the physics community has substantially advanced the understanding of granular state dynamics. The avalanching process, vibrational flows, mixing properties, jamming, friction and stick-slip motion in spherical grain beds have by now been extensively studied [1]. These efforts have been complemented by experimental and modeling research work using non-spherical shaped grains such as sand [2], elongated rice [3], superconducting vortices [4], water droplets [5], etc. Here, we use a new granular material, short lengths of flexible metal bead chain, to explore variations in 2D avalanching behavior when single beads of length $L = 1$ take on unique new characteristics as L increases. Our bed tray is large but fixed in size, for chain interlocking overrides experimental size effects in importance. The data was acquired via digital imaging, similar to that for a slowly ramping tray of spherical grains [6].

Although single beads cut from a chain possess the standard grain packing and flow behavior of dry, low friction spherical grains, $L = 2$ chains are appreciably different in behavior. They have a single roll axis and can expand to interlock with neighboring chains on either side. The $L = 3$ chains, additionally, possess a flexibility in shape that allows for better interpenetration and interlocking with its neighbors. By $L = 5$ the chains can form semicircles leading to bed voids or they can form curved microcrystals within the bead bed (Figure 1). The 9-chains can actually snake, caterpillar or hoop down an avalanche slope. After runout many loops are retained, further enlarging the voids and thereby lowering the overall bed density. Finally, 13-chains are flexible enough to start taking on a polymer-like character, overhang heavily, and avalanche in clumps. These changing properties with length call for an extensive investigation of bead chain dynamics, which we commence with the present study on chain bed surface profiles.

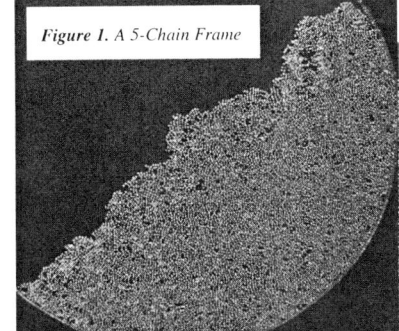

Figure 1. A 5-Chain Frame

EXPERIMENTAL DESCRIPTION

Chain lengths were hand clipped from a 1 kilometer spool of the smallest chain readily available. The 2.06 ± 0.1mm diameter, 25.1 milligram #2 nickel coated brass beads are attached with nickel wire segments. When fully compressed the beads touch and when extended they lengthen by a factor of 1.39. A circle can form with 8 beads minimum. Clipped wire ends recede into their respective bead interiors where they are trapped. The holes on the chain segment ends do not contact the experimental plates, except for 1-chains that can rotate freely. The spherical beads have seams that may occasionally catch on surface irregularities.

A ½ meter diameter, ¼" thick aluminum disk with a ¼" thick glass faceplate and rear axial mount was used in the rotating avalanche experiment. The faceplate was carefully spaced with a 37 cm diameter O-ring and held under compression with four brackets through holes in the disk. A synchronous motor friction-coupled to the disk's circumference turned it at 9.0 deg/min about its horizontal axis. Beads were washed with alcohol and acetone and then manipulated with tweezers. Enough bead chains were used to give filling factors 35% - 45% when vertical, and the disk was rotated for an hour before data acquisition. A Sony DVD video camera positioned on a tripod some 10' away recorded single frames every 30 seconds during the (typically) 24-hour runs in a dimmed room. The Mini-DV cassette images were transferred to Quicktime movies and recorded on compact disks for analysis with a Windows-NT PC. We used a Scion Image analysis program [7] to import movie segments and highlight the individual bed surface profiles, with Matlab used for the data analysis.

Care was required in preparing the disk and faceplate surfaces and in positioning the chamber gap. On a finely sanded disk surface beads would hang up on irregularities. When hand polished, avalanching was further disrupted, as the bead seams caught on micro scratches. Waxing likewise proved fruitless. Eventually, we learned that rubbing both surfaces with powdered graphite eliminated wall frictional problems and minimized interbead friction as well. Although darkened noticeably, the presence of graphite actually improved the experimental visual contrast.

Granular 2D experiments are subject to disruption by strong buckling effects if the gap width is set improperly [8]. Too narrow, and grains jam between the walls, too wide and beads move to opposite surfaces compacting the bed, and more critically, leading to added steric hindrance. Both effects will increase locking and change bead dynamics appreciably. The gap was set to 2.10-2.15 mm and monitored carefully to assure the absence of readily apparent bead jamming or lattice buckling patterns in the bed.

ANALYSIS AND RESULTS

A Scion Image line crawler program was written to convert most of each surface profile into one line of the 700x950 pixel images generated from 950 frame movie segments (Figure 2a). The uphill-moving crawler converted any backtracking (overhangs) to vertical risers. From repose to critical angle only 3 or 4 rotated frames were recorded so there is little correlation between frames. A few profiles from the Chain-13 image are shown in Figure 2b. Risers are quite well developed here and strongly affect avalanche slope angles. We observe angles that are *smaller* and profiles that are *smoother* uphill from large risers than they are downhill (to the left of them).

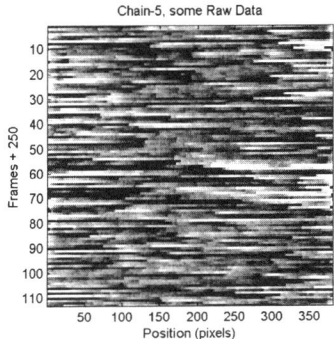

Figure 2a. *Time Development (time goes up, Whiter is higher x 2)*

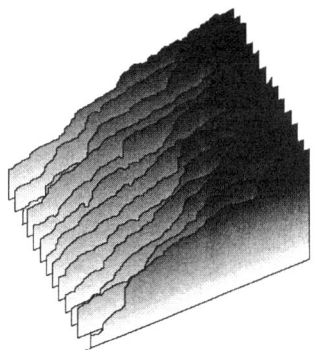

Figure 2b. *Assorted Chain-13 Profiles (aspect ratio ≈ 2)*

Mean Profiles and Slope Angles

An overall measure of bead chain features is captured in the mean profile as shown in Figure 3a for 9-chains. The mean surface profile is not straight along the bed (except for 1-chains and 2-chains), but bulges near the center of rotation. This variation reflects the mentioned effect of the overhangs on slope angle. The dashed lines in the figure represent rms deviations in the 5 - chain mean profile, σ, while the insert shows how σ grows with L. The growth is not uniform, but plateaus at twice σ for single beads before increasing sharply for 13-chains.

The analysis of our slope angles is complicated by chain interlocking and timing effects. A plot of sorted bed angles that were best fit over the entire bed length (solid line) is given in Figure 3b. Also included are respective plots for best fitting slopes made over shorter (150

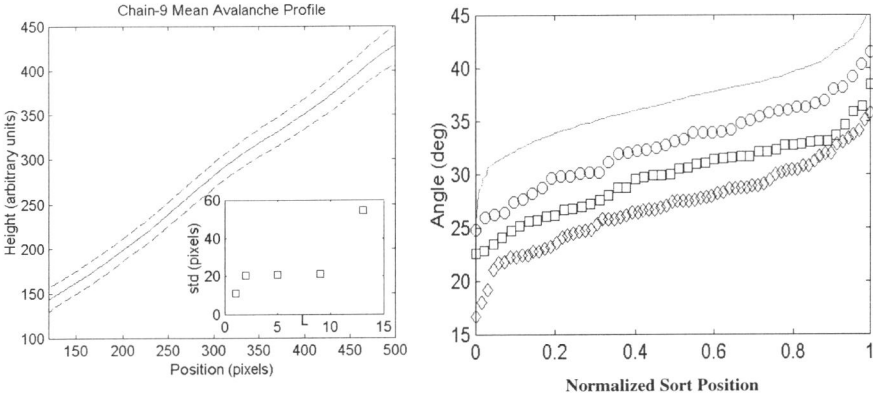

Figure 3a. *Mean Profile Inset: σ vs. chain length, L*

Figure 3b. *Sorted 5-chain slope angles*

pixel) bed portions within the lower third of the bed (O), middle third ( ) and upper third (diamond). Point counts of these various runs were normalized for comparison purposes, with just enough of the very best fits used to be representative of existing straight regions on the slope profiles. Generally, there appear more good fits and low angle profiles as one moves uphill.

As expected, the overall fit results are distributed over a wide angular range as the experimental disk slowly turned while acquiring data. The results from regions of the lowest and highest angles, however, are underrepresented because avalanches occur suddenly and the camera exposure interval (1/2 min.) makes it less likely to capture critical profiles. Likewise, the smallest angles are underrepresented since the turning disk steepens all angles between camera frames. Said conversely, one either catches a frame of the system close to critical *or* just after avalanching, so statistically they share a frame. Away for these extremes (the lowest and highest 10% of angles measured) we expect and observe a linear trend of angle with sort row. Angles of repose, θ_r, and critical angles, θ_c, are thus determined for the various curves from straight line extrapolations to values 10% *past* the vertical axes (with errors of ~ ±1°).

For interpretation, a composite diagram of how critical and repose angles vary with chain length is given in Figure 4. The dot-dashed lines represent θ_c (Δ) and θ_r (inverse Δ) obtained from extrapolating the distribution of overall linear fit angles as in Figure 3b. The (blue) solid lines and (red) dashed lines represent θ_c and θ_r obtained from the best fits of the low and high regions of the surface profile, respectively (same symbols as in Fig 3b). For 1-chains all critical and repose angles are in fair agreement since the slope profiles are almost straight with little surface roughness. However, the θ_c and θ_r for longer chains become dependent on where the short best fits were taken along the bead bed profile. The dotted curve was obtained from fits high on the avalanche slope and might best represent critical angles, while repose angles are better characterized by the solid line of fits from data

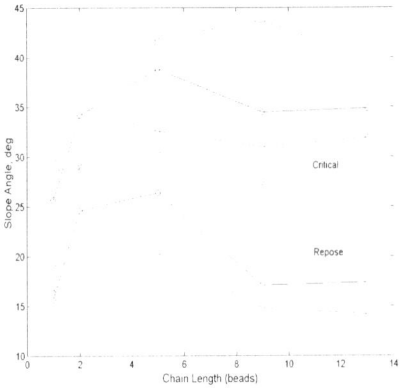

Figure 4. The Critical and Repose Angles

obtained low down on the slope. However, overall best fits only represent average trends since the surface profiles can be remarkably rough in places.

Both critical and repose angles are fairly small for the single metal beads (27° and 16°, respectively), but both increase dramatically for 2-chains. Evidently, the appearance in 2-chains of interlocking, elongation, and its breaking of roll symmetry are of overriding importance for avalanching behavior. As chains grow longer the critical angle asymptotes toward ~32°, demonstrating that the added presence of chain curvature softens asymmetry effects. Repose angles, on the other hand, after growing to 26°, relax all the way back to 17°, about that for single beads. This trend is consistent with our observation of chain looping during avalanching that seems to mimic single bead-like rolling. For the longer chains especially, there exist other features in addition to these analyzed regions of relatively uniform slope.

Vertical Riser Statistics and Avalanche Mass Distributions

Risers (step discontinuities) of all sizes are present along the bed surface profiles, most due to overhangs. A riser is defined as a vertical jump h > 6 pixels (>1.25% of frame height), with size increasing as long as the nearby pixels jump as well. These show up as discontinuities in the data (Figure 2a). Figure 5 presents a log-log rank ordering plot that has been generated by sorting 5-chain risers for size and then plotting size, h, versus row number, n. (Rank ordering is equivalent to a cumulative size distribution and is useful for emphasizing the infrequent largest events [9].) The largest steps tend to appear either well uphill or downhill, as one might expect from inertial considerations alone. Smaller jumps are distributed randomly along the avalanche slopes (with several

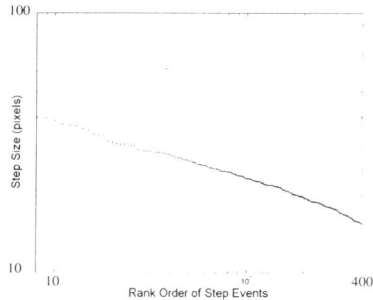

Figure 5. 5-chain Step Riser Rank vs Sizes (x2)

appearing per frame). For the 5-chain data shown, we find a power law dependence of the form $h(L,n) \sim n^{-\eta(L)}$ with $\eta = 0.35$ extending over more than 1 decade in rank order. Other L-chains are consistent with this η, although those fits are not as extensive. The maximum riser height increases monotonically with L (not shown) as the system appears to be approaching criticality at some undetermined L, beyond which a very different polymer-like behavior should set in.

A final question is whether or not avalanche size still follows power law statistics as L increases. For that, we looked at the avalanche mass distributions. Before and after displacements were determined by carefully computer rotating a frame forward in time and then subtracting it from the next frame in the sequence (assuming a constant bed density). Excellent overlap was found when there was no mass displacement, subtraction errors being at the ±1 pixel level. Mass gain is proportional to *excess area,* mass loss was associated with *area deficit.*

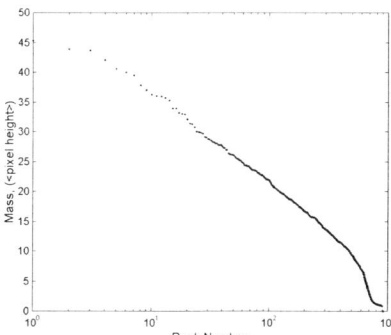

Figure 6. Rank Ordering Distributions of 9-chain avalanche Masses

Corrections were made when chains disappeared from view in their runout, or tumbling into the picture image at topside. Also, frequent slumping occurred as higher parts as the bed repacked during rotation. Comparisons having no appreciable activity over the lower 2/3 of the bed, but movement higher up were discarded as mere slumping events.

The mass rank order distributions for the 9-chains data is presented on the semilog plot of Figure 6 (mass units are in average pixels of height per pixel of width). The largest ranks contain frames with no movement and can be treated as a baseline value. The curve is linear over 2 decades in rank number, giving a mass number-size relation that is logarithmic, not power law as might be expected [2]. However, a low power law exponent, α, cannot be excluded since $\log x \sim \lim((1-x^{-\alpha})/\alpha)$ as $\alpha \to 0$. Also, any extra missing mass from the very largest risers will add upward curvature to the data line. Thus, our logarithmic result is still tentative.

CONCLUSIONS

We have analyzed mean profiles, slope angles, discontinuity statistics and avalanche mass for various length metal bead chains. Taken together, the picture is that interlocking and rolling constraints immediately increase bed slope angles, but the emergence of curvature with L relaxes these angles again, while also dramatically contouring the surface. These surface features and the avalanche mass sizes appear to both be scale invariant. Further analysis of the present data and new studies with additional bead lengths and gap widths are ongoing.

REFERENCES

*Corresponding author
1. *See e.g. Powders and Grains*, edited by R.T. Behringer and J.T. Jenkins (Balkema, Rotterdam, 1997).
2. J. Feder, Fractals **3** (3), 431(1995).
3. A. Malthe-Sørenssen, J. Feder, K. Christensen, V. Frette and T. Jøssang, Phys. Rev. Lett. **83**, 764(1999).
4. S. Field, J. Witt, F. Nori and X. Ling, Phys. Rev. Lett. **74**, 1206(1995).
5. B. Plourde, F. Nori and M. Bretz, Phys. Rev. Lett. **71**, 2749(1993).
6. M. Bretz, J. B. Cunningham, P.L. Kurcynski and F. Nori, Phys. Rev. Lett. **69**, 2431(1992).
7. Scion Image is a free Windows image analysis program distributed by Scion Corp. Fredrick, MD, 21703.
8. S. Neser, C. Bechinger, P. Leiderer and T. Palberg, Phys. Rev. Lett., **79**, 2348(1997).
9. For an application of Rank Ordering statistics to earthquakes see D. Sornette, L. Knopoff, Y.Y. Kagan, and C. Vanneste, J. Geo. Res., **101**, 13883(1996).

Mat. Res. Soc. Symp. Proc. Vol 627 © 2000 Materials Research Society

AVALANCHE STRATIFICATION - EXPERIMENTAL TESTS OF THE "METASTABLE WEDGE" AND "CONTINUOUS FLOW" MODELS

M. E. SWANSON, M. LANDREMAN, J. MICHEL and J. KAKALIOS
The University of Minnesota, School of Physics and Astronomy, Minneapolis, MN 55455

ABSTRACT

When an initially homogeneous binary mixture of granular media such as fine and coarse sand is poured near the closed edge of a "quasi-two-dimensional" Hele-Shaw cell consisting of two vertical transparent plates held a narrow distance apart, the mixture spontaneously forms alternating segregated layers. Experimental measurements of this stratification effect are reported in order to determine which model, one which suggests that segregation only occurs when the granular material contained within a metastable heap between the critical and maximum angle of repose avalanches down the free surface, or one for which the segregation results from smaller particles becoming trapped in the top surface and being removed from the moving layer during continuous flow. The result reported here indicate that the Metastable Wedge model provides a natural explanation for the initial mixed zone which precedes the formation of the layers, while the Continuous Flow model explains the observed upward moving kink of segregated material for higher granular flux rates, and that both mechansims are necessary in order to understand the observed pairing of segregated layersfor intermediate flow rates and cell separations.

INTRODUCTION

When a homogeneous binary mixture of two different granular materials, such as sand and sugar, are poured between two vertical plates held a narrow distance apart, stratification can result, with the sand and sugar separating into alternating layers along the top surface of the quasi-two-dimensional sandpile [1-3], as illustrated in fig. 1. It has been suggested that this stratification effect may be responsible for striation patterns in aeolian sandstone, which serves as reservoirs into which hydrocarbon deposits migrate over geological time scales [4-6]. An elucidation of the mechanisms by which layering occurs in granular systems may therefore improve our understanding of the formation and permeability properties of these geological structures, as well as industrial processes which require that mixtures poured from a hopper remain homogeneous [7]. The stratification pattern is sensitive to both the separation of the vertical walls of the Hele-Shaw cell and the rate at which the granular mixture is poured into the cell, which results in a non-trivial phase diagram for the stratification effect [3].

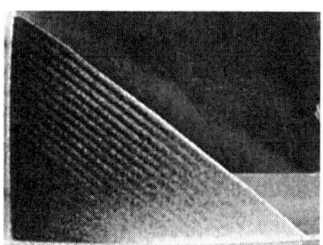

Fig. 1: Digital images of the avalanche stratification banding pattern for a 50/50 mixture of sand and sugar poured along the closed edge of a vertical Hele-Shaw cell.

MODELS FOR AVALANCHE STRATIFICATION

Two distinct models have been proposed to account for the avalanche stratification effect. The "Metastable Wedge" model involves periodic avalanching of granular material between the maximum and critical angles of repose [3, 8-11]. As granular material is poured into the Hele-Shaw cell, a quasi-two-dimensional sandpile forms at the angle of repose of the mixture. Inelastic collisions cause some of the grains rolling down the top surface to come to rest before reaching the bottom plate, forming a metastable heap, as indicated schematically in fig. 2a. When the maximum angle of stability is exceeded, the material in the heap avalanches down the free surface, returning the top of the pile to the critical angle of repose. The strong velocity gradient normal to the surface of the flowing pile leads to shear dilation [12] which in turn enables separation or sieving of the smaller or denser granular material to fall into small gaps or voids in the flowing surface beneath the larger or less dense material in the flowing layer [9-11]. An alternative explanation requires the larger and more faceted granular material to have a larger angle of repose compared to the smaller and smoother material. This difference between angles of repose of the two materials leads to separation of the two species as they roll down the free surface, with the smaller material becoming fixed in irregularities of the granular surface [2, 13-18]. As shown in fig. 2b, the stratification of the granular mixture is an ongoing process during the movement of the material along the free surface, and we consequently term this the "Continuous Flow" model. Numerical simulations of avalanche stratification using either model are able to faithfully recreate an observed avalanche straification pattern for a given flow rate [3,13,16]. In this paper we report experimental measurements of the avalanche stratification effect which indicate that the Metastable Wedge model provides a good explanation for the initiation of the stratification, while the Continuous Flow model can account for the subsequent development of the striation pattern.

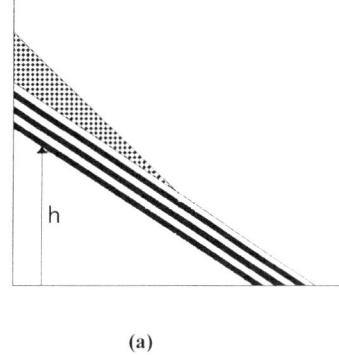

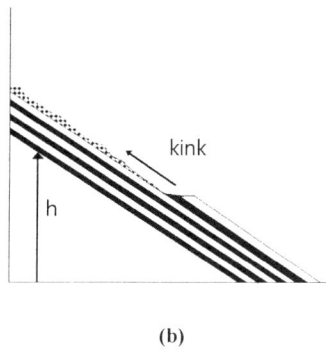

(a)   (b)

Fig. 2a: Sketch of the avalanche stratification in the Metastable Wedge model. The incoming granular material is poured from the upper left hand corner. Granular media brought to rest due to inelastic collisions on the free surface is indicated on the upper left hand portion of the pile.

Fig. 2b: Sketch of the avalanche stratification in the Continuous Flow model. The incoming granular material is poured from the upper left hand corner. The downward flow of incoming granular material and the uphill motion of the kink are indicated.

EXPERIMENTAL TECHNIQUES

Our experimental setup, similar to that employed by Makse and co-workers [2], consists of two vertical Plexiglas sheets of area 10.5 inches by 8 inches are mounted parallel to each other on a horizontal base plate. A 50/50 mixture by mass of black sand (roughly spherical with an average diameter 0.4 mm, density 2.6 mg/mm^3) and white sugar (roughly cylindrical, with an average long axis of 0.8 mm, density 1.5 mg/mm^3) is poured against the closed edge of this Hele-Shaw cell using a flux box, which consists of a cylindrical reservoir with a rectangular aperture at the bottom. The location of two of the orthogonal sides of the aperture may be varied using micrometer adjusters. The width of the aperture is always maintained to equal the plate spacing of the Hele-Shaw cell. By varying the length of the aperture for a fixed plate spacing the flow rate of granular material poured into the cell can be changed. For studies of the influence of the plate separation on the stratification pattern the length of the aperture is also varied for different plate spacings in order to maintain the same flux of granular media pouring into the cell. This flux box replaces the titrating bulb with a rotating stopcock and fixed aperature, used in previous studies of avalanche stratification [3]. The granular mixture is premixed by stirring prior to pouring. Samples of the granular mixture exiting the flow bulb onto a flat surface confirm that the granular material does not undergo segregation either exiting the bulb or while falling into the cell. The static angle of repose of the sand is 37.3° and of the sugar is 40°, while the angle of repose of sand on sugar is 40.3° and of sugar on sand is 39.8°. Digital images of the stratification pattern are taken with a monochrome charge coupled device (CCD) camera (Cohu 4910) in conjunction with a Scion LG-3 frame grabber and a Power Macintosh 7100/80. Data image analysis is performed using the public domain program Image from the NIH.

EXPERIMENTAL RESULTS

An important clue regarding the mechanism underlying the stratification effect is the fact that the striated layers do not occur immeadiately upon pouring the granular mixture into the Hele-shaw cell. As indicated in fig. 1, there is a "dead-zone" of mixed sand and sugar in the bottom corner which extends vertically up $h_{dz} \sim 10$ cm before the segregated bands appear. The volume of this dead-zone depends on the plate separation and flow rate. In the Continuous Flow model, there is no reason why stratification should not begin as soon as the granular material is poured into the cell. On the other hand, an initial mixed region is required by the Metastable Wedge model. As indicated in fig. 2a, the formation of a heap above the critical angle of repose is a necessary precondition for layering to occur. This heap will only form if the length of the top surface is long enough so that the incoming granular material suffers enough inelastic collisions to come to rest before reaching the bottom of the pile. If the length is insufficient to halt the rolling grains, then the granular material remains homogenously mixed as the pile grows. To test this proposal we have inserted a plastic spacer in the lower left hand corner of the Hele-Shaw cell, of the dimensions of the dead zone as in fig. 1, where a thin layer of granular material is glued to the top of the wedge. As the image in fig. 3 indicates, stratified layers form immediately as the mixture is poured into the cell. When a spacer whose height is only half that of the dead zone is employed, a partial dead zone is observed, with the layering beginning at the same height h as before. Moreover, as shown in fig. 4, when the kinetic energy of the incoming granular flux is increased, by varying the height of the flux box above the Hele-Shaw cell (without a plastic spacer), then the height of the dead zone h_{dz} similarly increases, indicating that a longer free surface is needed to bring the more energetic granular mixture to rest before a metastable wedge, and hence segregated layers can form. The slope of the data in fig. 4 gives the energy loss per distance for granular material flowing along the free

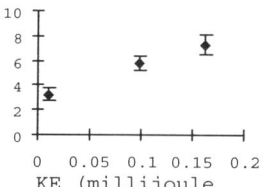

KE (millijoule

Fig.3: Digital image of avalanche stratification with plastic spacer wedge, the height of the dead zone in fig. 1, placed in lower left hand corner of Hele-Shaw cell prior to pouring granular mixture into cell.

Fig. 4: Plot of the height of the observed dead zone against the initial kinetic energy of the granular mixture poured in to the Hele-Shaw cell (varied by increasing the height above the top surface of the sandpile that the granular material is poured).

surface of the sandpile, which is $\sim 3.33 \times 10^{-5}$ Joule/cm for the sand/sugar mixture studied here.

Once the stratification has begun, the Continuous Flow model accounts in a natural manner for the observation of a kink of segregated material which forms at the bottom of the pile (fig. 2b), and evolves upward (in the opposite direction to the flow of granular media) until reaching the top of the pile [13]. As explained in this model, the larger grains of the incoming granular media avalanching down the top surface will reach the bottom plate first, due to: (i) that for the same initial velocity (which is determined to first order by the height from which the mixture is poured into the Hele-Shaw cell) the more massive particles will have a greater kinetic energy and hence travel further before stopping, and (ii) it being more difficult for the larger grains to become trapped in surface irregularities in the surface profile than the smaller particles [13,16,17]. During the avalanche, the smaller grains come to rest, providing a static layer and enabling more small particles to continue rolling down to the bottom of the pile. At the base of the pile, a well-defined kink develops where the flow is stopped. As more granular material is introduced, the kink continues to develop, moving up the sandpile surface until it reaches the top, at which point a pair of layers has been deposited. Experimental observations and numerical simulations have confirmed this kink mechanism for layer growth in the limit of high granular flux rates [16,17]

For very low flow rates, or very wide plate spacings, the incoming granular material has more interactions with the top surface of the sandpile than with other avalanching grains. In these situations the observations of the stratification effect are consistent with the Metastable Wedge model. For higher flux rates, interactions in the flowing layer dominate and the Continuous Flow model provides a better explanation for the layer growth, and in particular the upward moving kink of segregated material. It is interesting to note that there is a regime of plate separations and flow rates for which both models are necessary in order to account for the surprising pairing of stratified layers [3]. As shown in fig. 5, for plate separations of ~ 0.55 cm and flow rates of ~ 1 g/sec, a clear pairing of the the striated layers, most notably of the darker sand particles is observed. The image in fig. 5 has been rotated by the angle of repose of the sandpile, so that the top free surface will be horizontal. This pairing is confirmed by Fourier analysis of the digital images, which display two peaks in the image intensity structure function. Digital images of the top flowing surface of the sandpile in the Hele-Shaw cell while the paired segregated layers are forming are shown in fig. 6. The five successive images were obtained sequentially in time, with the top image being the earliest. The granular mixture is poured into the cell, far to the right of the images, and fig. 6 shows a region in the center of the top surface. The development of a small metastable wedge is evident in the top image in fig. 6, while in the next image an upward moving kink of stratified material is seen developing from the left. The arrow the center inage indicates the highest point uphill that the kink reaches (the furthest to the right) before stopping. That is, for this plate spoacing and flow rate, the upward kink does not extend all the way to the top of the sandpile. Further deposition of granular material then begins the development of another metastable wedge, which then avalanches, leading to a second thin layer of sand closer to the lower layer than if either the metastable wedge or continuous flow mechanisms alone were responsible for the striation pattern.

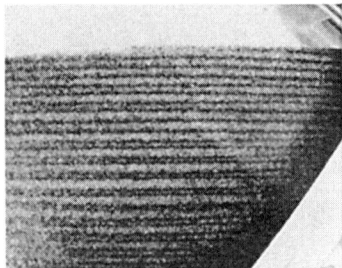

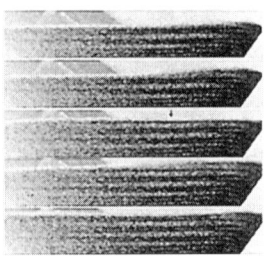

Fig. 5: Digital image of the avalanche stratification pattern as in fig. 1, for a plate separation of 0.55 cm and a flow rate of 1 g/sec, showing clear pairing of adjacent layers of dark sand.

Fig. 6: Successive time images of the top surface of the sandpile in the Hele-Shaw cell as granular media is poured, during the formation of a pair of stratified layers of sand (dark material). The arrow in the middle trace indicates the height to which the upward moving kink reaches.

In summary, experimental measurements of the avalanche stratification effect observed when a mixture of sand and sugar is poured against the closed edge of a vertical Hele-Shaw cell have been performed in order to determine whether the Metastable Wedge or Continuous Flow models provide a better explanation for this phenomenon. The Metastable Wedge model provides a natural explanation for the observed initial mixed region which precedes the formation of striation layers, while the Continuous Flow model can account for the upward moving kink of segregated material present during layer deposition. The Metastable Wedge model applies in the case of granular mixtures poured into the cell at very low flux rates while for higher flow artes and narrower plate separations, the Continuous Flow model applies. Both mechanisms must be invoked to explain the observed the pairing of stratified layers present for intermediate plate separations and flow rates.

We gratefully acknowledge experimental assistance and helpful comments from J. P. Koeppe and M. Enz. This work was supported by the University of Minnesota through the Undergraduate Research Opportunity Program and the NSF-REU program.

REFERENCES

1.   J. C. Williams, Powder Tech. **2**, 13 (1968).
2.   Hernan A. Makse, Shlomo Havlin, Peter B. King and H. Eugene Stanley, Nature **386**, 379 (1997).
3.   J. P. Koeppe, M. Enz and J. Kakalios, Phys. Rev. E **58**, R4104 (1998).
4.   *Aeolian Geomorphology: An Introduction*, by Ian Livingstone and Andrew Warren (Addison Wesley Longman Lmt., Singapore) 1996; *Sedimentary Structures*, 2nd Ed., by J. D. Collinson and D. B. Thompson (Unwin Hyman Ltd., London) 1989; J. R. L. Allen, *Sedimentary Structures: Their Character and Physical Basis* , (Elsevier, Amsterdam, 1982).
5.   C. Paola, S. M. Wiele and M. A. Reinhart, Sedimentology, **36**, 47 (1989); R. C. Arnott and Bryce M. Hand, J. of Sedimentary Petrology, **59**, 1062 (1989).
6.   G. V. Middleton, *Mechanics of Sediment Movement* (S.E.P.M. Providence,RI) (1984); Geo. Assoc. of Canada, special paper no. **7**, 253 (1970).
7.   R. M. Nedderman, U. Tuzun, S. B. Savage and G. T. Houlsby, Chem. Eng. Science **37**, 1597 (1982).
8.   S. B. Savage, *Developments in Engineering Mechanics*, ed. by A. P. S. Selvadurai (Elsevier Science Publishers, Amsterdam) p. 347 (1987); S. B. Savage and C. K. K. Lun, J. Fluid Mech. **189**, 311 (1988).
9.   K. Ridgway and R. Rupp, Powder Tech. **4**, 195 (1970).
10.   J. A. Drahun and J. Bridgwater, Powder Tech. **36**, 39 (1983).
11.   P. Y. Julien, Y. Q. Lan and Y. Raslan, Proc. of the Third Intl. Conf. on Powders and Grains, ed. by Robert P. Behringer and James T. Jenkins (A.A. Balkema, Rotterdam) p. 487 (1997).
12.   R. A. Bagnold, *The Physics of Blown Sand and Desert Dunes* (Chapman and Hall, London) (1941).
13.   Hernan A. Makse, Pierre Cizeau and H. Eugene Stanley, Phys. Rev. Lett. **78**, 3298 (1997); Hernan A. Makse, Phys. Rev. E **56**, 7008 (1997).

14. T. Boutreux and P.-G. deGennes, J. Phys. (France) **6**, 1295 (1996); T. Boutreux, Eur. Phys. J. B **6**, 419 (1998); T. Boutreux, H. A. Makse and P.-G. deGennes, Eur. Phys. J. B **9**, 105 (1999).

15. Y. Grasselli and H. J. Herrmann, Granular Media **1**, 43 (1998); Hernan A. Makse and Hans J. Herrmann, Europhys. Lett. **43**, 1 (1998).

16. Hernan A. Makse, Robin C. Ball, H. Eugene Stanley and Steven Warr, Phys. Rev. E **58**, 3357 (1998).

17. P. Cizeau, H. A. Makse and H. Eugene Stanley, Phys. Rev. E **59**, 4408 (1999).

18. S. N. Dorogovtsev and J. F. F. Mendes, Phys. Rev. E **61**, 2909 (2000).

Mat. Res. Soc. Symp. Proc. Vol 627 © 2000 Materials Research Society

A Simple Description of Thick Avalanches at the Surface of a Granular Material

Achod Aradian, Elie Raphaël and Pierre-Gilles de Gennes

Laboratoire de Physique de la Matière Condensée, URA n°792 du C.N.R.S., Collège de France,
11 place Marcelin Berthelot, 75231 Paris Cedex 05, France
e-mail: Achod.Aradian@college-de-france.fr, Elie.Raphael@college-de-france.fr, Pierre-
Gilles.deGennes@espci.fr

ABSTRACT

Some years ago, Bouchaud *et al.* introduced a phenomenological model to describe surface
flows of granular materials [J. Phys. Fr. I **4**, 1383 (1994)]. According to this model, one can
distinguish between a static phase and a rolling phase that are able to exchange grains through an
erosion/accretion mechanism. Later, Boutreux *et al.* [Phys. Rev. E **58**, 4692 (1998)] proposed a
modification of the exchange term in order to describe thicker flows where saturation effects are
present. However, these two approaches assumed that the downhill convection velocity of the
grains is constant inside the rolling phase, a hypothesis that is not verified experimentally. We
have recently modified the above models by introducing a velocity profile in the flow, and
analyzed the physical consequences of this modification in the simple situation of an avalanche
in an open cell. We here emphasize the physical predictions of our model, and show, in
particular, that the thickness of the avalanche depends strongly on the velocity profile.

A SIMPLE MODEL FOR THICK AVALANCHES

Onset of avalanches

It is a well-known that the top surface of a sand heap need not be horizontal, unlike that of a
stagnant liquid. However, there exists an upper limit to the slope of the top surface, and the angle
between this maximum slope and the horizontal is known, for non-cohesive material, as the
Coulomb critical angle θ_{max}. Above this angle, the material becomes unstable, and an avalanche

at the surface might occur. The Coulomb angle is related to the friction properties of the material through $\tan \theta_{max} = \mu_i$, where μ_i is an internal friction coefficient [1].

As of today, the physical picture associated with the onset of the avalanche is still obscure. One could imagine a local scenario in which the dislodgement of some unstable grains leads by amplification to a global avalanche (see for instance [2]). Alternately, one can think of a delocalized mechanism [3], in which a thin slice of material is destabilized and starts to slide as a whole. In the present paper, we will focus on the latter point of view.

It has been recently suggested [3] that the thickness of the initial gliding layer should be of the order of _, the mesh size of the contact force network [4]. For simple grain shapes (spheroidal), one expects $\xi \sim 5\text{-}10$ grain diameter d. The angle at which the avalanche process actually starts would then be of the order of $\theta_m + \xi / L$, where L is the size of the free surface. At the moment of onset, our picture is that this initial layer starts to slip, and is rapidly fluidized by the collisions with the underlying heap, therefore generating a layer of rolling grains on the whole surface.

Now that we have proposed a description of the initial situation, we may turn to the model scheme accounting for the further evolution of the avalanche.

Saturation effects for thick avalanches

Some years ago, Bouchaud, Cates, Ravi Prakash and Edwards introduced a model to describe surface flows of granular materials [5]. The model assumes a rather sharp distinction between *immobile* particles and *rolling* particles and, accordingly, introduces the following two important physical quantities (see Figure 1): the local height of immobile particles h(x,t) (where x denotes the horizontal coordinate[1] and t the time), and the local amount of rolling particles R(x,t). The time evolution of h(x,t) is written in the form

$$\frac{\partial h}{\partial t} = \gamma R (\theta_n - \theta) \tag{1}$$

where $\theta \cup \tan \theta = \partial h / \partial x$ is the local slope, γ a characteristic frequency and θ_n the neutral angle of grains at which erosion of the immobile grains balances accretion of the rolling grains. For the

---

[1] In this article, we restrict ourselves to two-dimensional sandpiles.

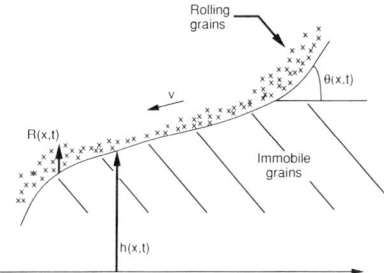

Figure 1. *The basic assumption of the BCRE picture is that there is a sharp distinction between immobile grains with a profile* $h(x,t)$, *and rolling particles with a local amount* $R(x,t)$. *The immobile grains consitute the "static phase" and the rolling ones the "rolling phase". The local slope of the static profile is called* $\theta(x,t)$.

rolling particles, Bouchaud and co-workers wrote a convection-diffusion equation [5] that was later simplified by de Gennes as [6]

$$\frac{fR}{ft} = v\frac{fR}{fx} - \frac{fh}{ft}$$

(2)

where v is the downhill typical velocity of the flow, assumed to be constant.

According to the Bouchaud-Cates-Ravi Prakash-Edwards (BCRE) model, fh/ft is linear in R (see Eq. 1). This is natural at small R, when all the rolling grains interact with the immobile particles. But as explained in Refs. [7,8], this cannot hold when R becomes larger than a given *saturation length* ξ', since the grains in the upper part of the rolling phase are no longer in contact with the immobile grains. The length ξ' is expected[2] to be of the order of a few grain diameters d. This led Boutreux, Raphaël, and de Gennes to propose [8] a modified *saturated* version of the BCRE Eq. 1, valid for thick surface flows and of the form

$$\frac{fh}{ft} = v_{uh}(\theta_n - \theta) \qquad (R \gg \xi'),$$

(3)

where v_{uh} is defined by $v_{uh} \ldots \gamma\xi'$. The constant v_{uh} has the dimensions of a velocity (where the subscript "uh" stands for uphill).

The description of thick avalanches modelized by Eq. 3 was discussed in Ref. [8]. However, one might encounter situations where the local amount R of rolling particles is rather large except in some regions of space where it takes values smaller than ξ'. For such cases, various

---

[2] One expects ξ' to be somewhat smaller than ξ, see [17].

"generalized" forms of the BCRE equations valid both in the large and small R limit, and able to handle intermediate values have been proposed [7,9,10]. As we will be mainly concerned with thick flows, we will here use the saturated form (Eq. 3).

Velocity profiles in thick flows

We now consider the hypothesis made in Eq. 2 that the downhill typical convection velocity of the rolling grains v is constant. As a matter of fact, v might vary for two reasons.

First, v depends on the local slope $\partial h / \partial x$ of the static bed, reflecting that the mean convection velocity should increase as the sandpile is further tilted. However, in the situation we are going to consider, the slope should never depart from θ_n by more than a few degrees, so that the variations of v originating in this may reasonably be taken to be negligible.

Second, v might as well depend on the local amount of rolling particles R. This dependence is quite natural, since as soon as the thickness of the flow exceeds a few grain diameters, one would expect a velocity gradient perpendicular to the flow to establish. Such a possibility was already considered by Bouchaud *et al.* [10], but, to our best knowledge, not further studied. We think that taking this velocity gradient into account does lead to an improvement of the model description of avalanches. In the forthcoming sections we will analyse the physical consequences of this modification.

If analyticity is assumed, we can expand v(R) in powers of R, and considering only the first two contributions to be significant, we write:

$$v(R) = v_0 + \Gamma R \tag{4}$$

with Γ a constant, homogeneous to a shear rate, and v_0 a constant velocity. When R becomes small, Eq. 4 tells us that v(R) becomes constant [$v(R) \blacklozenge v_0$]. Physically, this velocity should correspond to the typical rolling velocity of a single grain on a bed of immobile grains. For simple grain shapes (spheroidal) and average levels of inelastic collisions, one expects this velocity v_0 to scale as $\sqrt{gd}$ (where g is the gravity) [6,11,12]. Similarly, the shear rate Γ is expected to scale as $\sqrt{gd} / d \cup \sqrt{g/d}$. We can therefore rewrite Eq. 4 as:

$$v(R) = \Gamma(R + d) \tag{5}$$

We note that v_0 becomes negligible compared to ΓR as soon as R exceeds a few grain diameters.

In our approach, the typical velocity v(R) depends *linearly* on the local rolling height R. Such a form is in part motivated by the recent work of Douady *et al.* [13] (see also our concluding remarks). It is also supported by the experimental results of Rajchenbach *et al.* who carried measurements in a rotating drum [14,2]. These authors have found linear velocity profiles in the surface flow, with a shear rate Γ independent of the thickness of the flow. However, in other experiments of chute flows carried out on rough inclined planes, Azanza *et al.* [15] and Pouliquen [16] have observed that the mean velocity (averaged on cross-sections) scales as a power-law of the thickness with an exponent about 3/2. In the following we will mainly focus on the linear form of Eq. 5, since it allows us to give explicit analytical solutions, and shall discuss the changes that occur in the case of a power-law velocity later on.

In the next section, we will derive the governing equations from the saturated BCRE equations and the above considerations on the velocity profile inside the flow.

Governing equations

We may define a reduced profile $\tilde{h}$, deduced from h by substracting the "neutral" profile $\theta_n x$: $\tilde{h}(x,t) = h(x,t) - \theta_n x$. Using Eqs. 2, 3, and 5, we easily obtain the following system:

$$\frac{f\tilde{h}}{ft} = -v_{uh}\frac{f\tilde{h}}{fx}, \tag{6}$$

$$\frac{fR}{ft} = \Gamma(R+d)\frac{fR}{fx} + v_{uh}\frac{f\tilde{h}}{fx}. \tag{7}$$

In our approach, these equations are the governing equations for surface avalanches displaying linear velocity profiles.

An important point is that we must have R > 0 for Eqs. 6 and 7 to be valid. If we reach R=0 in a certain spatial domain, then Eq. 6 must be replaced in that domain by $f\tilde{h}/ft = 0$.

CASE OF AN OPEN SYSTEM

Physical situation

Let us consider a cell, of dimension L, partially filled with monodisperse grains of diameter d, as shown on Figure 2. The heap has an initial uniform slope θ_{max}, the Coulomb angle of the

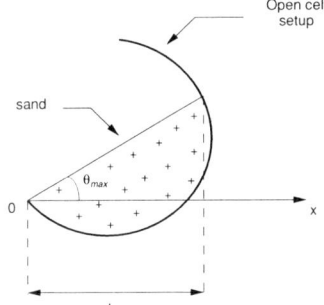

Open cell setup

sand

Figure 2 *Example of an open cell, so as to let the rolling material flow out at the bottom. We suppose that the avalanche starts precisely at* $\theta = \theta_{max}$.

material. The origin of the x-axis is taken at the bottom of the cell, and the orientation of the axis is such that the slope of the heap is positive.

Suppose that an avalanche has just started in the cell (as described above), so that we have at time t=0 a layer of rolling grains in the whole cell, of thickness $\sim \xi$ greater than the saturation length ξ'. We may thus use the saturated equations 6 and 7 from the beginning of the avalanche. As the rolling population will rapidly grow and become independent of the initial thickness ξ (for ξ small), we can as well consider the initial condition on R to be $R(x, t = 0) = 0$. We also know the initial value of $\widetilde{h}$:

$$\widetilde{h}(x, t = 0) = (\theta_{max} - \theta_n)x \; .. \, \eta x \qquad (8)$$

where η is defined as the (positive) difference between the Coulomb angle and the neutral angle. We have additional conditions in our system, due to the boundaries. At the top of the cell, there is no input of rolling species, so that we impose $R(x = L, t) = 0$ at any time $t \, ? \, 0$. Another condition arises from the fact that grains fall off the cell at the bottom and cannot accumulate there, giving $\widetilde{h}(x = 0, t) = 0$ at any time $t \, ? \, 0$.

Uphill wave in the static phase

Equation. 6 can be readily solved along with the initial and boundary conditions on $\widetilde{h}$ to give:

$$\widetilde{h}(x, t) = \eta(\, x - v_{uh}t) \, H(x - v_{uh}t) \qquad \text{for } 0 \le x \le L, \qquad (9)$$

where H denotes the Heavyside unit step function [H(u)=1 if u>0, H(u)=0 otherwise] [3]. This result corresponds to the uphill propagation (at constant speed v_{uh}) of a surface wave on the static phase. Let us call $x_{uh}(t)$ the time-dependent position of the wavefront, given by $x_{uh}(t) = v_{uh} t$.

The wave starts from the bottom of the cell at time t=0 and reaches the upper end at a time t_2 defined by

$$t_2 .. L / v_{uh}. \tag{10}$$

The profile of the static phase at a given time t, smaller than t_2, is pictured on Figure 3, and can be described as follows: ahead of the wavefront [$x_{uh}(t) \le x \le L$], the profile is linear with a slope given by the initial angle θ_{max} (since $\tilde{h} = \eta v_{uh}$). Behind the wavefront [$0 \le x \le x_{uh}(t)$], the slope has decreased and reached the neutral angle θ_n ($\tilde{h} = 0$). For times t ? t_2, the slope of the static phase inside the cell is uniformly equal to the final value θ_n, which is thus the angle of repose of our specific open cell system.[4]

Downhill convection of rolling grains

Substituting Eq. 9 into the evolution equation for R (Eq. 7) gives:

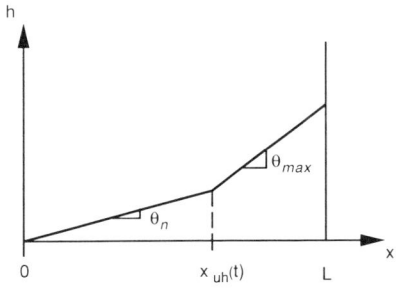

h

Figure 3. *The profile of the static phase for $0 < t < t_2$. At the left of $x_{uh}(t)$, the slope has relaxed to its final value θ_n. On the right of it, it still retains the initial angle θ_{max}.*

[3] Please note that Eq. 9 corrects Eq. 13 of Ref. [17], which contains a misprint.

[4] Boutreux *et al.* have shown in Ref. [8] that the notion of "repose angle" is not an intrinsic property of the material, but depends explicitly on the cell geometry.

$$\frac{fR}{ft} - \Gamma(R + d)\frac{fR}{fx} = \eta v_{uh} H(x - v_{uh} t). \tag{11}$$

Equation 11 is a non-linear convection equation. The rolling species are thus convected downhill, with a convection velocity that depends on the local rolling thickness R. In the spatial region $x > v_{uh} t$, the right-hand side (which links the evolution of R to that of $\tilde{h}$) plays the role of a source term, leading to an amplification of the avalanche. On the contrary, for $x \leq v_{uh} t$, the r.h.s. of Eq. 11 goes to zero, so that the material flowing through the surface $x = v_{uh} t$ from uphill is convected without amplification or damping.

Temporal evolution of the avalanche

To obtain a full analytical description of the avalanche evolution, one must solve Eq. 11 along with the two boundary conditions $R(x = L, t) = 0$ and $\tilde{h}(x = 0, t) = 0$. This has been done in Ref. [17], using the method of characteristics. In the present paper, however, we will focus on the essentiel physical features predicted by the model.

The influence of the bottom boundary condition $\tilde{h}(x = 0, t) = 0$ progressively invades the whole cell with a velocity v_{uh} (see Eq. 9 above). In a similar way, the influence of the top boundary condition $R(x = L, t) = 0$ propagates progressively from the top to the bottom of the cell (one can show that this propagation is uniformly accelerated, a direct consequence of the non-linearity in Eq. 11). The propagation of these boundary condition controls the different stages of the avalanche evolution. Figure 4, for example, illustrates what happens in the first stage of the avalanche. In the top region corresponding to $x_{dh}(t) \leq x$, the flow is under the influence of the boundary condition at $x = L$. On the contrary, the bottom region [$x \leq x_{uh}(t)$] is controlled by the boundary condition at $x = 0$, and the evolution equation region for R displays no more amplification (since the local slope has relaxed to its repose value). In the central region [$x_{uh}(t) \leq x \leq x_{dh}(t)$], the boundary conditions have no influence, and the rolling phase grows linearly with time.The overall shape of the rolling phase at the different stages of the avalanche is of course very dependent on the boundary condition for R at the top of the cell, but also on the condition for $\tilde{h}$ at the bottom, since the evolution of $\tilde{h}$ and R are coupled.

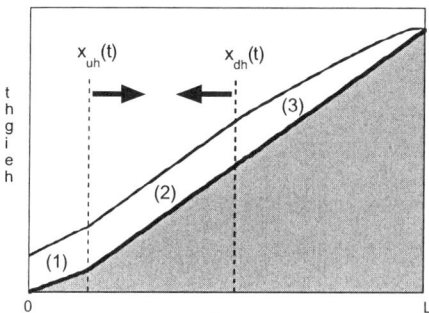

Figure 4. . *"Snapshot" of the avalanche during its first stage, from the full analytical solution: rolling phase (light) and static phase (dark). $x_{uh}(t)$ and $x_{dh}(t)$ represent the extension of boundary effects in the cell (respectively originating from the bottom and the top of the cell). The arrows symbolize the respective motion of the fronts. For clarity purposes, the thickness of the rolling part was augmented relatively to the static one.*

DISCUSSION AND SIMPLE CHECKS

From the full analytical solution of Eqs. 6 and 7 (presented in [17]), we can extract the following interesting physical quantities characterizing the avalanche evolution.

Avalanche duration

From the onset of the avalanche at $t = 0$ till time $t = t_2 = L / v_{uh}$, there exists a spatial region where the local slope of the pile has not yet relaxed to its repose value, thus leading to a net creation of rolling grains. In contrast, for times greater than t_2, the slope of the static part is θ_n everywhere in the cell, and no amplification of the moving grains can take place; the rolling phase existing at time t_2 is then simply convected downwards. Hence, the last grains to fall off the cell are those that had left the top end of the cell at time t_2. This event, which brings the avalanche to an end, occurs at time

$$t_{end} = t_2 + \frac{L}{\Gamma d}. \tag{12}$$

Equation 12 gives the overall avalanche duration.

Predictions for the maximum thickness of the avalanche

Another interesting quantity is the maximum thickness R_{max} reached by the avalanche in the course of its evolution. If we consider flows displaying a linear velocity profile, the analytical expression for R_{max} is [17]

$$R_{max} = -d - \frac{v_{uh}}{\Gamma} + \sqrt{\left[d + \frac{v_{uh}}{\Gamma} \right]^2 + 2\eta \frac{v_{uh}}{\Gamma} L}. \qquad (13)$$

For large values of L, R_{max} scales as:

$$R_{max} \cup \sqrt{2\eta \frac{v_{uh}}{\Gamma} L}, \qquad (14)$$

that is, as the square root of the system size L.

Let us give some numerical applications of this last expression. For the case of a standard laboratory experiment, with L=1 m, d=1 mm, v_{uh} / Γ = 3d and η = 0.1 rad, we find R_{max} =2.45 cm. In the case of a system at the scale of a desert dune, made of fine sand, we take L= 10 m, d=0.2 mm and, with others parameters unchanged, we get R_{max} =3.46 cm. One has to notice that R_{max} is quite small, even for large systems as a sand dune.

It is interesting to contrast this result with the work of Boutreux et al. [8], who carried out the same calculation in an open cell configuration, but with a constant downhill convection velocity v(R)~ v_0 (instead of Eq. 4). They found $R_{max} \cup \eta L$. For the two above examples, this formula leads to maximum amplitudes of respectively 10 cm and 1 m. The effect of the velocity gradient is thus to considerably limit the amplitude of avalanches, especially for large systems.

In the beginning of this article, we quoted the work of Azanza et al. [15] and of Pouliquen [16] who find that the average speed for a granular material flowing on a rough plane is related to its thickness through a power-law relation $v(R) \cup \Gamma R^{\alpha}$ with an exponent α close to 3/2.[5] If

---

[5] As pointed out by Pouliquen [16], the influence of the rough underlying plane on the behaviour of the flow is rather complex and not yet fully understood. Our model, which assumes

we use such a power-law relation in our model, we find that R_{max} is expected to be of the order of

$$R_{max} \cup \left[(\alpha+1)\eta \frac{v_{uh}}{\Gamma} L \right]^{\frac{1}{\alpha+1}}. \tag{15}$$

Note that R_{max} diminishes as α increases. In particular, for $\alpha = 3/2$, Eq. 15 can be rewritten as $R_{max} \cup (5\eta L v_{uh} / 2\Gamma)^{2/5}$.

Loss of material at the bottom of the cell

The loss of material at the bottom edge of the cell might be measured experimentally and compared with the following theoretical prediction. This loss corresponds to the flow rate at the bottom of the cell, and is given by:

$$Q(x=0,t) = \int_{0}^{R(x=0,t)} v(z)\,dz = \frac{\Gamma}{2} R(x=0,t)^2 + \Gamma d\, R(x=0,t), \tag{16}$$

where $R(x=0,t)$ is obtained from the full analytical solution of Eqs. 6 and 7. Figure 5 shows the predicted shape of $Q(x=0,t)$ as a function of time (solid curve). For comparison, we also plotted, on the same figure, the curve obtained with the assumption of a constant downhill

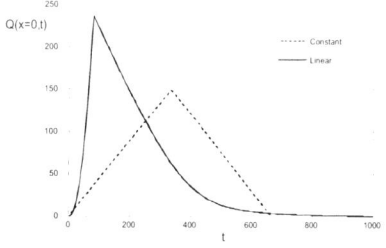

Figure 5. *Loss of material Q(x=0,t) at the bottom edge of the cell as a function of time. Solid line: predicted shape with a linear velocity profile in the flow. Dashed line: predicted shape in the case of a constant velocity (from Ref. [8]). Q is in units of* Γd^2, *t in units of* $1/\Gamma$.

that the flow takes place at the surface of a heap, might therefore not be appropriate to describe the experiments of Refs. [15] and [16].

velocity (dashed line) [8]. We observe that the solid curve displays a maximum corresponding to the moment when the maximum amplitude R_{max} rolls out of the cell. This maximum flow rate is obtained by replacing $R(x=0,t)$ by R_{max} in Eq.16.

CONCLUDING REMARKS

Effects of polydispersity

It is of common knowledge that real granular materials are generally intrinsically polydisperse. This may have drastic effects on the behavior of the flow, and capturing more precisely the physics of real avalanches would certainly suppose to take polydispersity into account. However, the treatment of full polydispersity is a difficult task. Yet, the BCRE equations have been extended to the case of binary mixtures [18,19], and it could be interesting to study the changes brought up in this case by a velocity gradient in the flow.

Domain of validity of the BCRE approach

The general approach introduced by Bouchaud *et al.* to describe surface flows is rather phenomenological. In a recent work, Douady *et al.* [13] have extended the classical approach of Savage and Hutter [20] to physical situations, like the one considered in the present paper, where the boundary between the static part of the pile and the flowing one is unknown. For the particular case of a flow velocity linear in R, the model of Douady *et al.* allows one to recover the governing equations 6 and 7. For other cases, a supplementary term coupling the quantities R and h should be added to Eq. 6.

Other directions of study

Very recently, Dorogovtsev and Mendes [21] have analysed the progressive build-up of a sanpile using the thick flow equations 2 and 3. They have shown that space periodicity takes place while the sandpile evolves (the grains are piling layer by layer), even for a pile made of only one type of particles.

All our study was based on a thick flow assumption. The features generated by the nonlinear thin flow equation 1, such as conservation equations, or shock fronts propagation, have also been recently studied by Mahadevan and Pomeau [22], and by Emig, Claudin and Bouchaud [23].

ACKNOWLEDGEMENTS

We thank T. Boutreux, F. Chevoir, A. Daerr, S. Douady, J. Duran and O. Pouliquen for oral and/or written exchanges.

REFERENCES

[1] For a detailed description of the Coulomb approach and its extensions, see R.M. Nedderman, *Statics and Kinematics of Granular Materials* (Cambridge University Press, Cambridge, 1992).

[2] J. Rajchenbach, in *Physics of Dry Granular Media*, H.J. Hermann, J.P. Hovi, and S. Luding, eds. (Kluwer Academic Publishers, Dordrecht,1998), p. 421.

[3] P.-G. de Gennes, in *From Rice to Snow*, Lecture at the Nishina Memorial Foundation, Publication Nr 38 (Nishina Memorial Foundation, April 1998). See also Ref. [8].

[4] F. Radjai, D. Wolf, and S. Roux, *Phys. Rev. Lett.*, **77**, 274 (1996); C.H. Liu, S. Nagel, D. Shechter, S. Coppersmith, S. Majumdar, O. Narayan, and T. Witten, *Science*, **269**, 513 (1995); F. Radjai, D.E. Wolf, S. Roux, M. Jean, and J.-J. Moreau, in *Powders and Grains 97*, R. Behringer and J. Jenkins, eds., Balkema Publishers, Rotterdam, 1997), p. 211.

[5] J.-P. Bouchaud, M.E. Cates, J. Ravi Prakash, and S.F. Edwards, *J. Phys. France I*, **4**, 1383 (1994); J.-P. Bouchaud, M.E. Cates, J. Ravi Prakash, and S.F. Edwards, *Phys. Rev. Lett.*, **74**, 1982 (1995). See also A. Mehta, in *Granular Matter*, A. Mehta, ed. (Springer Verlag, Heidelberg, 1994).

[6] P.-G. de Gennes, *C. R. Acad. Sci. Paris*, *Ser. IIb,* **321**, 501 (1995).

[7] P.-G. de Gennes, *Cours du Collège de France*, unpublished (Collège de France, Paris, 1997).

[8] T. Boutreux, E. Raphaël and P.-G. de Gennes, *Phys. Rev. E*, **58**, 4692 (1998).

[9] T. Boutreux and E. Raphaël, *Phys. Rev. E*, **58**, 7645 (1998).

[10] J.-P. Bouchaud and M.E. Cates, in *Physics of Dry Granular Media*, H.J. Hermann, J.P. Hovi, and S. Luding eds., (Kluwer Academic Publishers, Dordrecht,1998), p. 465.

[11] L. Samson, I. Ippolito, S. Dippel, G.G. Batrouni, in *Powders and Grains 97*, R. Behringer and J. Jenkins, eds. (Balkema Publishers, Rotterdam, 1997), p. 503.

[12] S. Dippel, G.G. Batrouni, D.E. Wolf, in *Powders and Grains 97*, R. Behringer and J. Jenkins, eds. (Balkema Publishers, Rotterdam, 1997), p. 559.

[13] S. Douady, B. Andreotti, and A. Daerr, *Eur. Phys. J. B*, **11**, 131 (1999).

[14] J. Rajchenbach, E. Clément and J. Duran, in *Fractal Aspects of Materials*, F. Family, P. Meakin, B. Sapoval and R. Wool, eds., M.R.S. Symposia Proceedings No. 367 (Materials Research Society, Pittsburgh, 1995), p. 525.

[15] E. Azanza, P. Chevoir and P. Moucheron, in *Powders and Grains 97*, R. Behringer and J. Jenkins, eds. (Balkema Publishers, Rotterdam, 1997), p. 455.

[16] O. Pouliquen, *Phys. Fluid*, **11**, 542 (1999).

[17] A. Aradian, E. Raphaël, P.-G. de Gennes, *Phys. Rev. E*, **60**, 2009 (1999).

[18] T. Boutreux and P.-G. de Gennes, *J. Phys. France I*, **6**, 1295 (1996).

[19] H.A. Makse, *Phys. Rev. E*, **56**, 7008 (1998); H.A. Makse, P. Cizeau and H.E. Stanley, *Phys. Rev. Lett.*, **78**, 3298 (1997).

[20] S.B. Savage and K. Hutter, *J. Fluid Mech.*, **199**, 177 (1989).

[21] S.N. Dorogovtsev and J.F.F. Mendes, *Phys. Rev. Lett.*, **83**, 2946 (1999).

[22] L. Mahadevan and Y. Pomeau, *Europhys. Lett.*, **46**, 595 (1999).

[23] T. Emig, P. Claudin and J.-P. Bouchaud, *Europhys. Lett.*, **50**, 594 (2000).

Nonlinear Waves in
Granular Media

Mat. Res. Soc. Symp. Proc. Vol 627 © 2000 Materials Research Society

New Wave Dynamics in Granular State

Vitali F. Nesterenko
Department of Mechanical and Aerospace Engineering,
University of California, San Diego,
La Jolla, CA 92093, U.S.A.

ABSTRACT

The unusual feature of granular state is the negligible linear range of the interaction force between neighboring particles resulting in zero sound speed in uncompressed material. This makes linear and weakly nonlinear continuum approach based on Korteveg - de Vries equation invalid. This paper describes a results based on more general approach.

INTRODUCTION

Granular materials are highly nonlinear according to a few physically different reasons. For example, Hertz law for compression of two elastic granules (or more general power law) has no linear part even for relatively small displacements in the vicinity of zero compression force. Nonlinearity can be caused by another important reason - structural rearrangements under applied load. This paper addresses nonlinearities connected with interaction law between particles. Even this part of nonlinear behavior is able to produce a qualitatively different mode of wave propagation and places granular materials in a special class according to wave dynamics. This was a reason for the introduction of the concept of "sonic vacuum" - the medium where the traditional wave equation is not a basic equation for wave dynamics [1]. The synthesis of all components of highly nonlinear behavior is a very exciting area for future research. Hopefully this development will also extend to other newly designed materials, which will possess highly nonlinear properties desirable for applications.

LONG-WAVE EQUATION FOR "STRONGLY COMPRESSED" HERTZ CHAIN

Wave propagation in one-dimensional granular material is considered taking into account that particles interact with each other according to the Hertz law [2]. It is assumed that a one-dimensional chain of identical spherical granules with diameter a is subjected to constant compression forces F applied to both ends and securing the initial displacement δ_0 between particle centers. The particle equation of motion becomes [1]:

$$\frac{d^2 u_i}{dt^2} = A\left(\delta_0 - u_i + u_{i-1}\right)^{\frac{3}{2}} - A\left(\delta_0 - u_{i+1} + u_i\right)^{\frac{3}{2}}, \quad N-1 \geq i \geq 2 \tag{1}$$

$$m = \frac{4}{3}\pi R^3 \rho_0, \quad A = \frac{E(2R)^{\frac{1}{2}}}{3(1-\nu^2)m},$$

here m is the mass of the particle, E and ρ_0 are the Young's modules and density of particle material, R is the granule radius, ν is the Poisson coefficient. It is assumed that the distance between the particle centers does not exceed $a = 2R$ if particles are spherical. After anharmonic and long-wave approximations, Eq. 1 can be transformed into nonlinear Boussinesq equation [1].

It can be further transformed [1,3] inside the same approximation into Korteweg-de Vries equation [4] which describes propagation of disturbances in one direction

$$\xi_t + c_0\xi_x + \gamma\xi_{xxx} + \frac{\sigma}{2c_0}\xi\xi_x = 0, \qquad \xi = -u_x. \tag{2}$$

From the well known properties of Korteweg-de Vries equation it can be concluded that in the granular material, initially compressed by a static load, weakly nonlinear periodic, solitary and shock (the latter if dissipation is included) waves can propagate [5]. The qualitative peculiarities of waves are determined by initial characteristics of disturbances.

EQUATION FOR "WEAKLY COMPRESSED" HERTZ CHAIN

A very interesting, non-classical wave behavior appears if granular material is weakly compressed. It means that the change of the neighboring particle displacements in a wave is much larger than the initial one δ_0, resulting from the static compression. The principal difference between this case and the "strongly" compressed chain is due to the lack of a small parameter with respect to the wave amplitude in the former case. For long wave disturbance the displacements u_{i-1}, u_{i+1} in Eq. 1(modified in a such way that initial relative displacement δ_0 is included into u_i) can be expanded in a power series according to small parameter $\varepsilon = a/L$ up to the fourth order. As a result the new wave equations for displacement is obtained [1]:

$$u_{tt} = -c^2 \left\{ \left(-u_x\right)^{\frac{3}{2}} + \frac{a^2}{10}\left[\left(-u_x\right)^{\frac{1}{4}}\left(\left(-u_x\right)^{\frac{5}{4}}\right)_{xx}\right]\right\}_x, \qquad -u_x > 0, \qquad c^2 = \frac{2E}{\pi\rho_0\left(1-v^2\right)} = Aa^{\frac{5}{2}}. \tag{3}$$

Partial differential equations with the high spatial derivative such as Equations 2, 3 in the linear approximation have some undesirable properties [6,7]. One can overcome this problem by the regularization procedure which is based on replacement of spatial derivative by time derivative using the first order approximation [8]. The analogous regularization can be easily made for "sonic vacuum" Eq. 3 [9]:

$$u_{tt} = -\left\{ c^2\left(-u_x\right)^{\frac{3}{2}} - \frac{a^2}{12}\left[u_{ttx} + \frac{3}{8}c^2\left(-u_x\right)^{-\frac{1}{2}}u_{xx}^2\right]\right\}_x. \tag{4}$$

Eq. 3 and Eq. 4 are formally equivalent according to the same order of the error terms. The stationary solutions of Eq. 3 can be found in the form $u(x - Vt)$. Introducing the strain as a new variable $\xi = -u_x$, obtain from Eq. 3 (C_3 is constant) [1,10]:

$$y_{\eta\eta} = -\frac{\partial}{\partial y}W(y),$$

$$W(y) = -\frac{5}{8}y^{\frac{8}{5}} + \frac{1}{2}y^2 + C_3 y^{\frac{4}{5}}, \tag{5}$$

$$y = \left(\frac{c}{V}\right)^{\frac{1}{5}}\xi^{\frac{5}{4}}, \quad \eta = \frac{\sqrt{10}}{a}x.$$

This form prompts use of the analogy with particle motion in the "potential field" W(y) [3]. In this analogy, η is the "time" and y is the "coordinate". "Potential energy" W(y) is an effective quantity and is different from real potential energy of particle's interaction.

Dependence of phase speed of periodic wave V_p on minimal and maximum strains ξ_{min} and ξ_{max} can be obtained from equations $W(y_{min})=W(y_{max})$ and $W_y(y_2)=0$:

$$V_p = c \left\{ \frac{2}{5\left(\xi_{max} - \xi_{min}\right)} \left[\frac{2\left(\xi_{max}^{\frac{5}{2}} - \xi_{min}^{\frac{5}{2}}\right) - 5\xi_2^{\frac{3}{2}}\left(\xi_{max} - \xi_{min}\right)}{\xi_{max} + \xi_{min} - 2\xi_2} \right] \right\}^{\frac{1}{2}} \tag{6}$$

where ξ_2 is connected with variable y_2 - coordinate of the minimum of function $W(y)$.

Solitary phase speed V_s is a function of the minimum ξ_0 and maximum ξ_m strains, these strains correspond to y_1 and y_m and can be determined from the equation $W(y_1) = W(y_m)$, taking into account that $W_y(y_1)=0$:

$$V_s = \frac{c}{\left(\xi_m - \xi_0\right)} \left\{ \frac{2}{5}\left[3\xi_0^{\frac{5}{2}} + 2\xi_m^{\frac{5}{2}} - 5\xi_0^{\frac{3}{2}}\xi_m \right] \right\}^{\frac{1}{2}} \tag{7}$$

It is important that "potential energy" $W(y)$ at $0 < C_3 < 5/27$ ensures the existence of stationary solutions only like compression solitary waves, it is clear from relative positions of maximum at y_1 and minimum at y_2 ($y_2 > y_1$) [1,10].

Thus the strongly nonlinear periodic and solitary waves are the stationary solutions of Eq. 3. In the case when $C_3 = 0$ each hump of the periodic solution at the same time may be considered as a representative of a solitary wave of Eq. 3 under infinitesimaly small prestrain in a sense that the difference between them can be made as small as desired by managing initial strain.

The existence of a solitary wave as a stationary solution of strongly nonlinear wave Eq. 3 is very interesting fact because this equation is more general than the KdV equation. This solitary wave is a supersonic one as well as KdV soliton. In fact, restriction of $C_3 < 5/27$ guarantees a value of the phase velocity of a solitary compression wave larger than the initial velocity of sound [1,10]. The constant C_3 determines the ratio of the maximum wave amplitude to the initial strain ξ_0. For $C_3 = 0$ a solution of Eq. 3 is a sequence of positive humps connected at the points with zero strains. In a system moving with velocity V_p this solution is

$$\xi = \left(\frac{5V_p^2}{4c^2}\right)^2 \cos^4\left(\frac{\sqrt{10}}{5a}x\right). \tag{8}$$

At some local points ξ vanishes. This contradicts the condition $\xi > 0$, under which Eq. 3 was derived. Nevertheless it is easy to verify that the solution described by Eq. 8 for strain satisfies Eq. 3 at these points as well.

The fact that partial exact periodic solution of the Eq.3 was found in explicit form allows straightforward application of the Whitham method [9] for the investigation of the modulation stability of this solution [11].

For C_3 approaching 0 the value of the maximum $W(y_1)$ tends to zero, as does the value of its coordinate y_1. The important point here is that maximum at y_1 and corresponding solitary solution exists at any infinitesimally small value of constant C_3, y_1 corresponds to the "initial" compression strain of the chain ξ_0 (strain at infinity).

If the value of the total "energy" in this case is $W(y_1)$, the behavior of the solution for $y \gg y_1$ coincides with the behavior of one hump described by Eq. 8 corresponding to the case $C_3 = 0$. Physically it means that solitary shape can be taken as one hump of periodic solution (Eq. 8) with finite wave length. Consequently, in this case the phase speed of wave V_s has a nonlinear dependence on maximum strain

$$V_s = \frac{2}{\sqrt{5}}\, c\, \xi_m^{\frac{1}{4}} = \left(\frac{16}{25}\right)^{\frac{1}{5}} c^{\frac{4}{5}}\, \upsilon_m^{\frac{1}{5}}\,. \tag{9}$$

The characteristic spatial size of a solitary wave L_s is determined by the period of the solution described by Eq. 8, which equals to

$$L_s = \left(\frac{5a}{\sqrt{10}}\right)\pi \approx 5a\,. \tag{10}$$

This solution can be expected to describe the solitary impulse in uncompressed chain, as it was verified by computer calculations and in experiments [1, 10, 12-15, 27]. The finite spatial size of the solitary wave in "sonic vacuum" is a unique feature in comparison with the KdV soliton. The most important properties of the former wave are independence of spatial size on amplitude and strongly nonliniear dependence of speed on particle velocity (or strain) in the crest [1, 12]. Wright [16] obtained analytically the finite length solitary wave (solitary wave with compact support) for nonlinear waves in a nonlinear elastic rod and pointed out the analogy between steady waves in elastic rods and granular materials [17]. There are also another examples of the soliton with compact support [18-20] in very different media. Recent results on solitons with compact support can be found in [5].

Nonlinearity due to Convective Derivative Can be Weak for the Stationary Wave

Convective derivative was dropped during the derivation of Equation 3 without any justification. The physical reason for this could only be the small particle velocity in comparison with the phase speed of the wave. It can not be done a priori because there is no characteristic wave speed in Eq. 3 for strongly nonlinear case. Having introduced the phase speed of stationary wave V, it may be proved that the previously dropped nonlinear term, connected with convective derivative, is really small in comparison with kept terms. The ratio of kept terms in Eq. 3 to the neglected term representing a convective derivative have the orders of magnitude $\sim \xi$ and $\xi\, L/a$ correspondingly. That is why relatively small strains (but not in the connection with initial one ξ_0) ensures the neglect of convective derivative in comparison with other nonlinear terms [21].

Stationary Shock Waves

In the presence of dissipation in the medium, viscous-like terms should be added to Eq. 3. The exact type of dissipation does not effect the equilibrium state behind shock. For C_3 approaching zero the particle velocity (υ_c) behind the front of the stationary shock corresponds to the minimum $W(y_2)$ and equal the value which depends on shock speed V_{sh}:

$$V_{sh} = c^{\frac{4}{5}}\, \upsilon_c^{\frac{1}{5}}\,. \tag{11}$$

This equation is quite different in comparison with traditional one ($V_{sh} = c_0 + const$ υ_c). Comparison of Eq. 11 with equation for solitary wave (Eq. 9) shows that at the same phase speeds the final equilibrium strain behind shock is always smaller than in the solitary wave.

WAVE DYNAMICS OF UNSTRESSED 1-D GRANULAR MATERIALS – NUMERICAL CALCULATIONS

Very important conclusions useful for validation of numerical calculations can be made using dimensionless analysis of "exact" equations for discrete chain [1]. For $\delta_0=0$ and identical R_i Eq.1 with variable replacement is transformed to dimensionless form

$$\ddot{\zeta}_i = \left[\int_0^\tau \left(\zeta_{i-1} - \zeta_i\right)d\tau\right]^{\frac{3}{2}} - \left[\int_0^\tau \left(\zeta_i - \zeta_{i+1}\right)d\tau\right]^{\frac{3}{2}},$$

$$\zeta_i = \frac{\dot{u}_i}{\upsilon_0}, \qquad \tau = t\left(A^2\upsilon_0\right)^{\frac{1}{5}}. \tag{12}$$

In Eq. 12 dot means derivative with respect to dimensionless time τ. Clearly, for a chain with free ends identical τ correspond to identical ζ_i if the initial value problem is solved where particle velocities are prescribed for certain number of particles. It is also valid for stationary pulse. This comment is very important when finding temporal width and wave speed dependencies on amplitude of stationary solitary wave as soon as its existence is established. It predicts independence of normalized particle velocity profiles in solitary waves (and of course their space width) on amplitude: according to Eq. 12 the dimensionless particle's velocities normalized by velocity amplitude in solitary wave are just repeated in corresponding dimensionless times. A similar conclusion is also valid for solving the problem of a piston moving one of the ends of the chain with constant velocity when the overall motion is not stationary one. Therefore, an additional control method was verification of the observed equality of dimensionless particle velocity at identical moments τ for problems with different values of initial velocities, which included soliton formation and nonstationary "shocks" [1,10,21].

Solitary Waves in a Chain of Granules with Free Ends, Chain Fragmentation

A chain with free ends has zero sound speed. Due to the impact caused by the first two, four, five, six particles with the same initial velocity on such chain (all particle radius R_i are identical) the perturbation was decomposed correspondingly into two, four, five, six solitary waves [1,22]. If only one first particle impacted a system, a single solitary wave was formed in the system [1,15,22, 27]. Numerical results demonstrating solitary wave in this chain are consistent with the existence theorem for solitary waves on lattices [23, 24].

To create solitary waves with different amplitudes moving in the opposite directions the corresponding initial velocity conditions were used resulting into 4 solitons - 2 pairs moving toward each other in opposite directions [1]. Only phase changes occur after the collision, each of the pairs kept their amplitudes and shape. A similar solitons interaction without a change in their shape was observed during reflection from a rigid wall of a sequence of 6 solitary waves, generated by impact of a piston [10]. Small amplitude secondary solitons spawn from the collision of two solitons and containing less than 0.5 % of total energy were observed in [27]. This means that these solitary waves can be called solitons only within certain approximation.

Solitary waves with different amplitudes moving in the same direction were created using the appropriate initial conditions [1]. They correspond to the simultaneous impact of two particles onto the left end of the chain and "inner" impact by pair of particles inside the chain with smaller velocity. It resulted in formation of the leading group composed from 3 solitary waves with different amplitudes moving into the same direction. The amplitude and the shape of the solitary waves did not change during these interactions.

Thus, compression solitary waves originating at the decomposition of the initial disturbances in the discrete, strongly nonlinear chain of particles possess the main properties of solitons. They are stationary waves, which do not change their properties at the interaction with each other. Nevertheless, this statement was not analytically proved yet according to the difficulties with solutions of Eq. 3 for nonstationary perturbations.

Numerical calculations [1,15,22] provide a detailed picture of soliton propagation and system "fracture" due to the impact by one particle. For example very fast drop of rebounding velocity magnitudes with ball number starting from the impacted end is observed. The practically zero velocity of balls is maintained behind the propagated soliton being twelve orders of magnitude smaller than velocity in soliton maximum [15,22].

Solitons in Numerical Calculations for 1-D Lattice Versus Solitary Waves in Continuum

It is interesting to compare solitary waves obtained as stationary solutions of sonic vacuum wave equation (Eq. 3) and solitary waves found in computer calculations, which are stationary solutions for system of discrete particle (Eq.1).

The comparison of the solitary solutions obtained for the continuum (Eq. 3) with those obtained in a discrete system when $\xi_0 = 0$ clearly demonstrated that these profiles are very similar and "close enough" to each other [1]. They have the same space and temporal width, similar dependence of phase speed on maximum velocity and strain.

An analytical solution of the form $Atanh(f_n)$ for stationary wave in discrete chain, where f_n is represented by series is presented in [25]. An asymptotic description of the tail of the soliton in discrete chain and a new asymptotic solution for the full solitary wave is found in [15, 25].

Impulsive External Force, Fast Decomposition into Solitons

In computer calculations the corresponding force with duration 100 microsecond acting on the left end of the chain of steel spheres with radius 4.75 mm resulted in the fast decomposition of perturbation into 7 solitons. If the duration of force was decreased to 10 microsecond the number of solitons decreased to one. Piston impact with velocity 1 m/s and masses equal 5 - 10 mass of particle in the chain results in the proportional increase of number of solitons at the same distance from impact end with increasing of piston mass. It is worth mentioning that quick decomposition of the initial impulse into soliton train close to the entrance is one of the main properties of given strongly nonlinear system with nonlinear dispersion [1,10].

There are no analytical results on nonstationary wave behavior in strongly nonlinear case which are similar to weakly nonlinear case, for example the relation between number of emerging solitons and initial impulse duration. Nevertheless, the general pictures in these two cases are similar - the longer pulse creates larger number of solitons. This number is proportional to the mass of the striker [10].

The Piston Moving with Constant Velocity into Uncompressed Chain of Identical Particles

If the system of equations for discrete chain does not contain dissipative terms then piston impact should not result in the stationary shock wave. At the same time formation of leading soliton was observed with the first maximum particle velocity υ_m approaches twice the value of the piston velocity υ_o. This result is similar to those obtained for a chains with different interaction law. This property seems to be a very common property of any discrete system and does not depend on interaction law between particles and similar to the situation with collision of rigid wall moving with constant speed υ_o and particle at rest.

Another interesting feature is the separation of particles into two groups under piston impact - the leading group has relative velocities comparable to the velocity of the piston, and the tail is composed from particles with practically no relative motion. Both numbers are increasing with propagation distance (this property is not preserved in case of random chain of particles) [1,10]. Piston problem has some analogy with the case where 1-D chain collides rigid wall [26].

Influence of Initial Compression on Wave Transformation in Strongly Nonlinear Chains

The dependence of phase speed of solitary wave on initial compression (ξ_0 is a strain in infinity) is given by Eq. 7. Introducing initial sound speed c_0 in compressed chain, obtain the following equation for the phase speed V_s for wave with strain amplitude ξ_m:

$$\frac{V_s}{c_0} = \frac{1}{(\xi_r - 1)} \left\{ \frac{4}{15} \left[3 + 2 \xi_r^{\frac{5}{2}} - 5 \xi_r \right] \right\}^{\frac{1}{2}} \tag{13}$$

where $\xi_r = \xi_m / \xi_0$ is the relative strain amplitude of solitary wave. The distinguished feature of speed V_s is its very weak dependence on relative amplitude ξ_r. To obtain a few times difference between solitary wave speed and sound speed c_0, the relative strains above 100 are necessary.

Nonstationary Waves in Precompressed Chain

Wave evolution in a precompressed chain depends on the ratio of wave amplitude and initial precompression [10]. If wave amplitude is relatively large ($\Delta\xi_m = \xi_m - \xi_0 \gg \xi_0$) then impulse transformation is very similar to the case when $\xi_0 = 0$ at the same type of loading.

When wave amplitude is comparable with initial prestrain ($\Delta\xi_m \sim \xi_0$) the process of impulse transformation at the same length of the chain is qualitatively different with cases $\xi_0 = 0$, and $\Delta\xi_m = \xi_m - \xi_0 \gg \xi_0$. Two new features can be mentioned: the process of impulse decomposition into soliton train is retarded, and periodic wave tail was created following the main impulse.

In case of weak wave ($\Delta\xi_m \ll \xi_0$) quasistationary impulse (later modified by dispersion) accompanied by periodic tail with very small amplitude, propagates in the system [10].

Wave propagation in gravitationally loaded chain is considered in [27].

Soliton Interaction With the Boundary of Two "Sonic Vacuums"

The behavior of compression pulses at the contact of two "sonic vacuums" is of particular interest and was studied experimentally and numerically [28]. Indeed, in this case there is no concept of sonic impedance determining the reflected and passed impulse amplitudes, as in traditional acoustic approach. The observed transformation of the impulses at contact allows such systems to transform external actions into the pulse sequence required. "Soliton pulse spectroscopy" as potential method for probing buried objects in granular beds was proposed [29,30].

Waves in Two-Particle Periodical Chains

The behavior of two-particle 1-D periodical chain is qualitatively different from the behavior of the chain with equal particle masses, even in case of linear interaction law. For example, in the former system, for every wave number there are two characteristic frequencies corresponding to the two branches of vibration specter - acoustical and optical. The long-wave approximation analogous to Eq. 3 can be derived [10] using the approach proposed in [31] for a weakly nonlinear case. In an extreme case when mass of one particle is much larger than the mass of another one ($k = m_1/m_2 >> 1$) and both have the same diameter a, this equation has solitary solution similar to Eq.8 only with characteristic space scale of solitary wave $L=10a$. So, mere redistribution of mass between neighboring particles can result in wider soliton. Computer calculations were used to investigate the behavior of periodical uncompressed diaparticle chains with mass ratio $k = 2, 4, 16, 24, 64$ [10].

Random Uncompressed Chain

The random strongly nonlinear chain of particles can have properties quite different from a periodically ordered system. It does not allow the long wave analytical approach used for periodical chain. The very important differences between ordered and chaotic system of particles were found in numerical calculations [1,10].

For example, in the system with chaotic particle sizes under piston impact there is no tendency toward the uniform steady state of velocities profile even near the piston, unlike in the case of identical sizes. In disordered nondissipative chains there is no possibility of identifying two groups of particles with different values of relative velocities, as was the case for uniform chain. The velocity amplitude at the front does not any more represent the maximum velocity in the system like in periodic lattice.

Short impulse loading of random chain was created by the impact of two particles. In comparison with the case of identical particles, the perturbation does not decompose into 2 solitons, but has a significantly random character. At the same time it is characteristic that the leading perturbation still resembles the soliton shape, demonstrating the robustness of localization even in highly randomized chains. Of course the leading pulse here is not a stationary one. The important feature is the decay in amplitude even in the absence of dissipative losses. A significant increase in the amount of particle chaotization does not lead to an enhanced damping of the velocity amplitude. On the contrary, damping was smaller in the investigated most random case[1,10].

EXPERIMENTAL OBSERVATION OF A NEW TYPE OF SOLITARY WAVES

In experiments was observed the qualitative and quantitative agreement with analytical and numerical results between the amplitudes of the solitary waves, their number, and characteristic time parameters for short system with $N = 20$ [10]. One of the remarkable features of "sonic vacuum" was evident - very rapid decomposition of initial impulse on the distances comparable with the soliton width. In fact, under the given conditions the impulse is split after traveling only through 10 (!) first particles. This peculiarity can be used for controlled impulse transformation in very short transmission lines of different nature. This property can not be obtained through stationary analysis of Eq. 3. It makes nonstationary analysis and establishing some scaling laws very desirable. In experiments [14, 32] a quantitative study of the velocity and shape of solitary waves was performed. The negligible decay of the soliton after traveling distance equal to 50 particle diameters was found. It was concluded that experimentally observed solitons in chains of spheres of different materials are in very good agreement with solitary solutions of Eq. 3.

The elastic nonlinear interaction, described by the Hertz law and discreteness of chain are the main reasons for observed wave phenomena. Nevertheless, this law has a natural limit resulting from the possible plastic flow of material in the vicinity of the contact. The same method of loading - piston impact with the velocity 1 m/s and mass equal to 30 masses of steel particles was applied to the chains composed from steel and lead particles [9]. The plastic deformation of particle contacts for lead particles resulted in the qualitatively different shock structures: monotonic for lead particles and the oscillating for steel particles. Exponential attenuation of soliton amplitude is found in discrete chain for various coefficient of restitution [30].

NONLINEAR WAVES IN DISCRETE 1-D POWER LAW MATERIALS

The general example of strongly nonlinear elastic media are systems with power-law dependence of force F on the displacement δ ($F \sim \delta^n$) like the Hertz law for contact interaction of two particles. Goddard [33] demonstrated that for the conical geometry of particle contact n should be taken 2 instead of 3/2. In a general case of power law material the equation of motion is similar to Eq.1 with exponent n instead of 3/2. In long-wave approximation it can be transformed to strongly nonlinear wave equation similar to Eq. 3 [9]:

$$
u_{tt} = -c_n^2 \left\{ \left(-u_x\right)^n + \frac{n\, a^2}{6\,(n+1)} \left[\left(-u_x\right)^{\frac{n-1}{2}} \left(\left(-u_x\right)^{\frac{n+1}{2}} \right)_{xx} \right] \right\}_x .
$$

(14)

Stationary solutions $\xi(x - Vt)$ of Eq. 14 may be obtained by similar approach used for Hertz law. Cases $n>1$ and $n<1$ represent qualitatively different behavior of materials. The former represents normal behavior - hardening under increasing of the load and latter represents "abnormal" softening under the load. It is possible that in the different range of the load, the behavior of real material can be essentially different with switching from $n>1$ to $n<1$.

A material with $n>1$ represents a general example of a "sonic vacuum". In this case expression for solitary phase speed V_s is:

$$
V_s = \frac{c_n}{\left(\xi_m - \xi_0\right)} \left\{ \frac{2 \left[n\, \xi_0^{n+1} + \xi_m^{n+1} - (n+1)\, \xi_0^n\, \xi_m \right]}{(n+1)} \right\}^{\frac{1}{2}} .
$$

(15)

The solitary wave phase speed V_s approaches sound speed c_0 in case if ξ_m approaches ξ_0 and in the next approximation the relation for the KdV soliton between V_s and strain amplitude can be recovered. If $\xi_m \gg \xi_0$ the compression solitary waves qualitatively differ from the KdV solitons. In this case they are the basic excitations of strongly nonlinear system characterized by the following dependence of phase velocity V_s on ξ_m, or on maximum particle velocity υ_{max} and characteristic spatial length L_n [9,10]:

$$V_s = c_n \sqrt{\frac{2}{n+1}} \left(\xi_m\right)^{\frac{n-1}{2}} = \left(\frac{2c_n^2}{n+1}\right)^{\frac{1}{n+1}} \left(\upsilon_{max}\right)^{\frac{n-1}{n+1}}, \tag{16a}$$

$$L_n = \frac{\pi a}{n-1} \sqrt{\frac{n(n+1)}{6}}. \tag{16b}$$

The phase velocity and the width of these solitary waves are independent of the sound velocity c_0, as opposed to the KdV solitons (where the medium properties in linear description is inherited). Soliton width in numerical calculations for chain shows very close agreement with results of continuum approximation especially for relatively small values of exponent n [27].

Materials with $n<1$ represent different, "abnormal" type of behavior in comparison with $n>1$. In the former case, sound speed is equal to infinity if $\xi_0=0$. The value $n<1$ stipulates anomalous behavior under compression, i.e. the decrease of its elasticity modules as the deformation grows. The speed of the solitary rarefaction wave as a function of the strain in its minimum at total "energy" of nonlinear "oscillator" $W_0 = W(y_1)$ also can be obtained in a similar manner [9]. Thus in discrete systems with abnormal compressibility, the stationary compression solitary waves are prohibited and instead the rarefaction solitary waves are allowed.

GENERAL CASE OF NONLINEAR DISCRETE SYSTEM

A continuous approximation for general interaction law between particles is given by Eq.17:

$$\rho\xi_{tt} = \left\{ f + \frac{a^2}{24}\left[2f'\xi_{xx} + f''\xi_x^2\right]\right\}_{xx}, \tag{17}$$

where $f(\xi)$ is the interactive force between particles. The corresponding regularized equation with mixed derivatives can be obtained from Eq. 17 [34]. Equation 17 and its regularized form can be considered as a general wave equation for various physical phenomena at large gradients in media with structure, where strong nonlinearity results in nonlinear dispersion.

The corresponding variable transformation helps to reduce Eq. 17 to the form which is easy to analyze at least for stationary case [34, 35]

$$\rho\left[M(z)\right]_{tt} = \left[K(z) + \frac{a^2}{12}P(z)z_{xx}\right]_{xx}. \tag{18}$$

Here the functions $M(z)$, $K(z)$, and $P(z)$ are determined by interaction force. It is supposed that $f(a\xi)$ is always the increasing function of its argument.

The conditions for the existence of solitary waves, periodic waves, and their parameters in the continuous approximation for an arbitrary interaction law between the particles can thus be determined by analyzing the extremes of the corresponding effective potential energy $W(z)$ and their relative positions. It is possible to make some general conclusions about the properties of stationary solutions (if they exist) without finding their concrete form based on Eq. 18 [35].

Particularly the strongly nonlinear compression solitary waves and shock waves (if dissipation is introduced) exist for "normal" materials (($f''(a\xi)>0$) and rarefaction solitary waves and shocks exist for "abnormal" materials (($f''(a\xi)<0$). For chain with "normal" interaction this result is similar to the condition for solitary wave existence in discrete chain [23]. It explains why continuum approximation being truncated version of the equations for discrete chain is able to support the same solitary solutions.

CONCLUSIONS

Periodic waves, compression solitary and shock waves in granular materials ("sonic vacuum" if weakly compressed) are qualitatively different from weakly nonlinear case and can not be obtained in the frame of the latter approach. The existence of these new type wave disturbances is confirmed by different authors through analytical approach, numerical calculations and in experiments. They have such unique features, i.e. spatial extent of compression solitons does not depend on amplitude, initial sound speed does not determine the soliton parameters if strain in the wave is much greater than its initial value, and initial impulse is split into soliton train quickly on very short distances from entrance.

In a case opposite to the "sonic vacuum" - strongly nonlinear materials with abnormal behavior, there are periodic waves, rarefaction shock waves and solitary waves. They are supersonic, relative to the initial state. No stationary compression waves are allowed in such materials.

Strongly nonlinear waves due to their specific nature may find applications not only in areas traditional for granular materials but also in other areas like delay lines or signal processing (generating) devices which allow a strong modification of properties by external electrical fields or in other areas. They also represent a very compact way for transmitting lines where solitons with minimal possible space width may be effective information carriers. The very sharp shock front in these strongly nonlinear dispersive systems may be also useful for development of fast-rise-time electromagnetic nonlinear transmission lines. Another possible application may be in neurophysiology where the ultimate response in a given point is affected by the stimulation of other neighboring points, and where corresponding wave equations may be strongly nonlinear at least for intensive excitations.

REFERENCES

1. V.F. Nesterenko, *Journal of Applied Mechanics and Technical Physics*, **5**, 733 (1983).
2. L. D. Landau, E.M. Lifshitz, *Theory of Elasticity,* (Pergamon Press, 1986), pp.26-31.
3. I. A. Kunin, *Theory of Elastic Media with Microstructure,* (Nauka, 1975).
4. D.J. Korteweg and de G. Vries, *London, Edinburgh and Dublin Philosophical Magazine and Journal of Science*, **39**, 422, (1895).
5. M. Remoissenet, *Waves Called Solitons* (Concepts and Experiments), 3-rd revised and enlarged edition, (Springer-Verlag, 1999).
6. T.B. Benjamin, J.L. Bona, and J.J. Mahony, *Philosophical Transactions of the Royal Society of London A. Mathematical and Physical Sciences*, **272**, 47 (1972).
7. M. Kruskal, in: *Lecture Notes in Physics*, **38** (Dynamical Systems, Theory and Applications), (Springer-Verlag, 1975) pp. 310-354.
8. G.B. Whitham, *Linear and Nonlinear Waves*, (John Wiley & Sons Inc., 1974).

9. V.F. Nesterenko, *Journal de Physique IV,* Colloque C8, **4**, C8-729 (1994).

10. V.F. Nesterenko, *High-Rate Deformation of Heterogeneous Materials.* (Nauka, 1992), Chapter **2**, pp. 51-80 (in Russian).

11. S.L Gavrilyuk, and V.F. Nesterenko, *J. of Appl. Mechanics and Tech. Physics* **6**, 784 (1993).

12. A.N. Lazaridi, and V.F. Nesterenko, *J.of Appl. Mechanics and Tech. Physics,* **3**, 405 (1985).

13. R.S. Sinkovits, and S. Sen, *Phys. Rev., E,* **54**, 6857 (1996).

14. C. Coste, E. Falcon, and S. Fauve, *Physical Review E*, **56**, 6104 (1997).

15. A. Chatterjee, *Phys. Rev. E,* **59**, 5912 (1999).

16. T.W. Wright, *Studies in Applied Mechanics*, **72**, 149 (1985).

17. T.W. Wright, *Int. J. Solids Structures*, **20**, 911 (1984).

18. D.R. Scott, and D.J. Stevenson, *Geophysical Research Letters,* **11**, 1161 (1984).

19. D. Takahashi, and J. Satsuma, *J. Phys. Soc. Japan*, **57**, 417(1988).

20. D. Takahashi, J.R. Sachs, and J. Satsuma, in *Research Reports in Physics, Nonlinear Physics*, Eds. Gu Chaohao, Li Yishen, and Tu Guizhang (Springer-Verlag,1990), pp. 214-220.

21. V.F. Nesterenko, *Dynamics of Heterogeneous Materials* (Springer, in press).

22. E.J. Hinch and S. Saint-Jean, *Proc. R. Soc. Lond. A*, **455**, 3201 (1999).

23. G. Friesecke, and J.A.D. Wattis, *Communications in Mathematical Physics*, **161**, 391(1994).

24. R.S. MacKay, *Physics Letters A*, **251**, 191 (1999).

25. S. Sen, M. Manciu, *Physica A,* **268**, 644 (1999).

26. E. Falcon, C. Laroche, S. Fauve, and C. Coste, *The EuropPhys. Journal B*, **5**, 111 (1998).

27. M. Manciu, S. Sen, and A.J. Hurd, *Physica A,* **274**, 588 (1999).

28. V.F. Nesterenko, A.N. Lazaridi, and E.B Sibiryakov, *Journal of Applied Mechanics and Technical Physics,* **4**, 166 (1995).

29. S. Sen, M. Manciu, and J.D. Wright, *Phys. Rev., E,* **57**, 2386 (1998).

30. M. Manciu, S. Sen, and A.J. Hurd, *Physica A,* **274**, 607 (1999).

31. P. C. Dash, and K. Patnaik, *Progress in Theoretical Physics,* **65**, 526 (1981).

32. C. Coste, B. Gilles, *The European Physical Journal B*, **7**, 155 (1999).

33. J.D. Goddard, *Proc. R. Soc. London A*, **430**, 105 (1990).

34. V.F. Nesterenko, *Combustion, Explosion and Shock Waves,* **July**, 116 (1995).

35. V.F. Nesterenko, Solitons, Shock Waves in Strongly Nonlinear Particulate Media, in *Shock Compression of Condensed Matter-1999*, edited by M.D. Furnish, L.C. Chabildas, and R.S. Hixson, AIP, 2000, pp. 177-180.

Mat. Res. Soc. Symp. Proc. Vol. 627 © 2000 Materials Research Society

Signal Propagation in Nonlinear Granular Chain

Jongbae Hong
Department of Physics and Center for Strongly Correlated Materials Research, Seoul National University, Seoul 151-742, Korea

ABSTRACT

The vertical granular chain changes the elastic property of the medium due to gravity. Therefore, the solitary propagating mode in the horizontal chain disperses and damps in the vertical chain. We show that there are two different types of propagating modes, i.e., quasi-solitary and oscillatory, depending on the strength of impulse and the dispersion and damping of the signal follow power-laws. The power-law behavior in the weakly nonlinear oscillatory regime is explained analytically in the limit of linear approximation. However, the nonlinear solitary regime in which soliton damps and disperses due to gravity is discussed briefly in the strong impulse limit.

INTRODUCTION

The study of the propagation of elastic impulse in the granular medium is useful and interesting in connection with finding the information inside the granular medium [1,2] and with the nonlinear behavior originated by the nonlinear contact force of grains. Some years ago Nesterenko [3] has shown that the propagating mode of a strong impulse is a soliton. Recently, MacKay [4] has also proved the existence of solitary waves in the horizontal Hertzian chain using a rather general mathematical theorem given by Friesecke and Wattis [5] and Ji and Hong [6] extended the proof to the more general case. Coste et al. [7] have presented experimental measurements on the properties of the soliton in the horizontal Hertzian chain. Even though the geometrical effect [8] is important in real situation, the one-dimensional chain with arbitrary power-law type contact force may simulate real situation.

In this work we study the gravity effect in the signal propagation in the granular chain. Gravity changes the elastic property of medium, and the signal changes as it goes down in the vertical granular chain. An interesting discovery we find is that gravity makes the signal damp and disperse, and the type of damping and dispersion is power-law. We also find that the depth-dependent power-law behavior of the propagating signal is generic for the whole range of strength of initial impulse.

We also found that there are two regimes which discriminate weakly nonlinear oscillatory regime and strongly nonlinear solitary regime. The power-law exponent for a given contact force is independent of initial impulse in the former regime, while it changes in the latter regime depending on the strength of impulse. Analytical studies are possible for the power-law type behavior of the signal for the weakly nonlinear oscillatory regime [9]. The strongly nonlinear soliton regime, however, may be studied mostly numerically [10]. We focus on the motion of grains in a vertical granular chain whose contact force is a power-law type of arbitrary index.

The equation of motion of a grain at z_n from the top of the chain is written as

$$m\ddot{z}_n = \eta \left[\{\Delta_0 - (z_n - z_{n-1})\}^p - \{\Delta_0 - (z_{n+1} - z_n)\}^p \right] + mg \qquad (1)$$

where m is the mass of the grain, Δ_0 is the distance between adjacent centers of the spherical grain, p is the exponent of the power-law type contact force, and η is the elastic constant of the grain under consideration. We do not consider the plastic deformation.

ANALYSIS

To analyze Eq. (1), we introduce a new variable φ_n, denoting the displacement of nth grain from equilibrium, defined by

$$\varphi_n = z_n - n\Delta_0 + \sum_{l=1}^{n}\left(\frac{mgl}{\eta}\right)^{1/p} \qquad (2)$$

where the last term is the sum of overlaps up to nth contact and we set $z_0 = \varphi_0 = 0$. Eq. (1) can be transformed into

$$m\ddot{\varphi}_n = \eta\left[\left(\frac{mgn}{\eta}\right)^{1/p} + (\varphi_{n-1} - \varphi_n)\right]^p - \eta\left[\left(\frac{mg(n+1)}{\eta}\right)^{1/p} + (\varphi_n - \varphi_{n+1})\right]^p + mg \qquad (3)$$

using Eq. (2).

For the weak impulse case, the condition

$$(\varphi_{n-1} - \varphi_n) >> \left(\frac{mgn}{\eta}\right)^{1/p} \qquad (4)$$

is valid and the leading terms of the expansion of Eq. (3) gives rise to

$$m\ddot{\varphi}_n = -\mu_n(\varphi_n - \varphi_{n-1}) + \mu_{n+1}(\varphi_{n+1} - \varphi_n) \qquad (5)$$

where $\mu_n = mpg(\eta / mg)^{1/p} n^{1-(1/p)}$ is the force constant of nth contact. This equation of motion corresponds to that of the horizontal chain with varying force constant. Therefore, the effect of gravity has been used to change the force constant. Both left and right of Eq. (5) are linear in φ_n. The scaling analysis tells us that the motion is independent of the strength of initial impulse v_i.

To study analytically, Eq. (5) is written in a continuum form, i.e.

$$\rho \frac{\partial^2}{\partial t^2}\varphi(h,t) = \frac{\partial}{\partial h}\left[\tau(h)\frac{\partial}{\partial h}\varphi(h,t)\right] \qquad (6)$$

when the lattice constant $\Delta_0 \to 0$, where $\tau(h) = \tau_1(h/\Delta_0)^{1-(1/p)}$ denotes the depth-dependent tension, and $\rho = m/\Delta_0$ and $\tau_1 = \mu_1\Delta_0$ are the density and the tension at the first contact, respectively. We set $c_1 = \sqrt{\tau_1/\rho}$ which is the well-known speed of wave in the string of tension τ_1 and density ρ.

Since Eq. (6) is the linear differential equation, we can apply Fourier analysis to this equation. Then we have the following dispersion relation for the complex wave number $k(\omega) = k_r + ik_i$:

$$\omega(k) = k_r \sqrt{\frac{\tau(h)}{\rho}} \left(1 + \frac{\tau'^2}{4\tau^2 k_r^2}\right)^{1/2} \tag{7}$$

and $k_i = -\tau'/2\tau$. From Eq. (7), we obtain the phase and group velocity as follows:

$$v_p = \frac{\omega}{k} \approx \sqrt{\frac{\tau(h)}{\rho}} \left(1 + \frac{\tau'^2}{8\tau^2 k_r^2}\right) \tag{8}$$

and

$$v_g = \frac{d\omega}{dk} \approx \sqrt{\frac{\tau(h)}{\rho}} \left(1 - \frac{\tau'^2}{8\tau^2 k_r^2}\right) \tag{9}$$

Here and in what follows k means k_r. The difference between v_p and v_g denotes wave is dispersive, and $k_i \neq 0$ means wave is normally diffusive.

Treating dispersive and diffusive wave is not simple. Since $\tau'/\tau \propto h^{-1}$ and therefore $e^{-k_i h}$ is constant in h, the envelope function of the wave is not exponential and not diffusive. Therefore, the general solution of $\varphi(h,t)$ is written as

$$\varphi(h,t) = \sum_{\omega} A(\omega) e^{i(k_r h - \omega t)} \tag{10}$$

and the h-dependence of the envelop of the function $\varphi(h,t)$ is solely given by the coefficient $A(\omega)$. Now we solve Eq. (6) again using $\varphi_\varsigma(h,t) = u_\varsigma(h)e^{-i\varsigma}$ as a normal mode solution, where $u_\varsigma(h) = A(\omega)e^{ikh}$ and $\omega \propto \varsigma$. Then $u_\varsigma(h)$ satisfies

$$\frac{d^2}{dh^2} u_\varsigma(h) + \frac{1-(1/p)}{h} \frac{d}{dh} u_\varsigma(h) + \frac{\varsigma^2}{h^{1-(1/p)}} u_\varsigma(h) = 0 \tag{11}$$

which is a type of Bessel's differential equations [11]. A solution of Eq. (11) propagating to the positive h-direction is given by the Hankel function [11],

$$u_\varsigma(h) = h^\xi H_\nu^{(1)}(\theta h^\gamma) \tag{12}$$

where $\xi = 1/2p, \gamma = (1/2) + \xi = (1/2)[1 + 1/p],\ \theta = \varsigma/\gamma, \nu = \xi/\gamma = 1/(1+p)$.

The asymptotic form of Eq. (12) at large h for a fixed v is

$$u_\varsigma(h) \approx \sqrt{\frac{2}{\pi\theta}}h^{\varsigma-\frac{\gamma}{2}}e^{i\left[\theta h^\gamma-\frac{\pi}{2}v-\frac{\pi}{4}\right]} \tag{13}$$

and the displacement is written as

$$\varphi_\varsigma(h,t) \approx h^{\varsigma-\frac{\gamma}{2}}e^{i\left[\frac{\varsigma}{\gamma}h^\gamma-g\right]} \tag{14}$$

Therefore, the depth-dependence of the coefficient $A(\omega)$ of the displacement signal is

$A[\omega(h)]=h^{\varsigma-\frac{\gamma}{2}}=h^{-\frac{1}{4}\left(1-\frac{1}{p}\right)}$ for all ω.

The asymptotic form of the solution of the general linear equation is given by the saddle-point method or the steepest descent method. The result is written as [12]

$$\varphi(h,t) \cong \frac{\sqrt{2\pi}\,A(\omega_s)\exp\left[i\{k_s h-\omega(k_s)t\}-\alpha\right]}{\{t\,|\,\omega''(k_s)\,|\}^{1/2}} \tag{15}$$

when $\omega''(k) \neq 0$, where k_s means the wave number at the saddle-point and $\alpha = (i\pi/4)\mathrm{sgn}\,\omega''(k_s)$. When there are many saddle-points, the asymptotic solution must be the sum over all saddle-points. This work is the case of single saddle-point, therefore, the amplitude of the general solution of the linear wave equation in the asymptotic regime is given by $A(\omega_s)\{t\,|\,\omega''(k_s)\,|\}^{-1/2}$ where $t = h/v_g$. Since we showed that $A(\omega_s)$ exhausts depth-dependence of the amplitude of $\varphi(h,t)$, $\{t\,|\,\omega''(k_s)\,|\}^{1/2}$ must be depth-independent. Differentiating Eq. (9) once more, we get

$$t\,|\,\omega''(k_s)\,| \propto \frac{h\tau'^2}{\tau^2}k^{-3} \propto h^0 \tag{16}$$

This relation gives rise to $k \propto h^{-1/3}$. Using this and the amplitude function $A[\omega(h)]=$ $h^{\varsigma-\frac{\gamma}{2}}=h^{-\frac{1}{4}\left(1-\frac{1}{p}\right)}$, the phase velocity $v_p(h) \propto h^{(1/2)\left(1-\frac{1}{p}\right)}$, which is the leading h-dependent behavior of the phase velocity of Eq. (8), and the relation $v(h) \propto A(h)/\omega(h)$ and $\omega(h) = k(h)v_p(h)$, we get the depth-dependence of velocity and frequency as

$$v(h) \propto h^{-\frac{1}{4}\left(\frac{1}{3}+\frac{1}{p}\right)} \tag{17}$$

and

$$\omega(h) \propto h^{\frac{1}{6}-\frac{1}{2p}} \tag{18}$$

This analysis explains the damping and dispersive behavior due to gravity in the vertical granular chain. Signal becomes solider as the strength of impulse increases.

We will discuss briefly the case of strong impulse at which soliton is the propagating mode of signal. The numerical simulation of Eq. (1) for a rather strong impulsive initial speed shows that a solitary signal is created and propagated. But this soliton-like signal is still damping and dispersing, because medium changes gradually due to gravity. One can observe power-type damping and dispersion like the above weak impulse case [10]. It is uneasy to analyze strongly nonlinear phenomena analytically. We present here just a simple analogy with above analytical work to understand the phenomena of strongly nonlinear regime rather conceptually.

For the regime where the strong impulse condition

$$| \varphi_{n-1} - \varphi_n | >> \left(\frac{mgn}{\eta} \right)^{1/p} \tag{19}$$

is satisfied, the signal becomes much solider and we call it the nonlinear soliton regime. The equation corresponding to Eq. (5) is written as

$$m\ddot{\varphi}_n = \eta\left[(\varphi_{n-1} - \varphi_n)^p - (\varphi_n - \varphi_{n+1})^p \right] - \eta p\left[\delta_{n+1}(\varphi_n - \varphi_{n+1})^{p-1} - \delta_n(\varphi_{n-1} - \varphi_n)^{p-1} \right] + mg \tag{20}$$

where $\delta_n = (mgn/\eta)^{1/p}$ denotes the overlap distance at nth contact. The first term on the right side of Eq. (20), which is the leading term of the expansion, is the same as the contact force of the horizontal chain in which soliton is created under strong enough impulse if $p > 1$ [3,4,6]. The second term of Eq. (20), on the other hand, is the nonlinear contact forces varying at each contact of the horizontal chain. This term is quite similar to that of Eq. (5). Therefore, one can expect that this term is in charge of playing the role of damping and dispersing of the soliton signal. Since the effect of gravity is already immersed in the second term, the constant mg of the last term does not play an important role.

CONCLUSION

Numerical simulations for Eq. (1) [10] have shown that the signal created by an impulse may be classified into two regimes, i.e., one is weakly nonlinear oscillatory regime and the other is strongly nonlinear solitary regime. We were able to analyze the former in the limit of weakly nonlinear, i.e., in the linear approximation, where Fourier analysis, normal mode analysis, and saddle-point method are possible. We derived power-law behaviors in most respects of signal propagation under gravity.

The strongly nonlinear solitary regime, however, requires rather different approach to analyze the phenomena. We just found that the exerting force in the equation of motion of the strongly nonlinear solitary regime may be divided into non-gravity part and gravity part as shown in Eq. (20). Therefore, one can understand that soliton is created by the non-gravity part and damping and dispersion are caused by the gravity part. Analyzing strongly nonlinear solitary regime will be our future work in this field.

ACKNOWLEDGMENTS

This was supported by the Brain Korea 21 Project.

REFERENCES

1. S. Sen and R. S. Sinkovits, Phys. Rev. E **54**, 6857 (1996), R. S. Sinkovits and S. Sen, Phys. Rev. Lett. **74**, 2686 (1995),
2. S. Sen, M. Manciu, and J. D. Wright, Phys. Rev. E **57**, 2386 (1998).
3. V. F. Nesterenko, J. Appl. Mech. Tech. Phys. (USSR) **5**, 733 (1983).
4. R. S. MacKay, Phys. Lett. A **251**, 8589 (1999).
5. G. Friesecke and J. A. D. Wattis, Commun. Math. Phys. **161**, 391 (1994).
6. J.-Y. Ji and J. Hong, Phys. Lett. A **260**, 60 (1999).
7. C. Coste, E. Falcon, and S. Fauve, Phys. Rev. E **56**, 6104 (1997).
8. J. D. Goddard, Proc. R. Soc. London, Ser. A **430**, 105 (1990).
9. J. Hong, J.-Y. Ji, and H. Kim, Phys. Rev. Lett. **82**, 3058 (1999).
10. J. Hong, H. Kim, and J.-P. Hwang, Phys. Rev. E **61**, 964 (2000).
11. M. Abramowitz and I. A. Stegun, *Handbook of Mathematical Functions*, AMS 55 (National Bureau of Standards, 1972).
12. P. L. Bhatnagar, *Nonlinear waves in one-dimensional dispersive systems* (Clarendon, Oxford, 1979).

Mat. Res. Soc. Symp. Proc. Vol. 627 © 2000 Materials Research Society

Impulse and Low Frequency Acoustic Wave Propagation in Granular Beds

Surajit Sen, Marian Manciu, Victoria Tehan
Department of Physics, State University of New York at Buffalo, Buffalo, NY 14260-1500, USA
Alan J. Hurd
Department 1841, Sandia National Laboratories, Albuquerque, NM 87185, USA

Abstract

The study of sound propagation in granular beds at frequencies exceeding a MHz has been a subject of study for many years. Much remains to be learnt about sound propagation at lower frequencies. We shall present our studies on the problem of impulse propagation and briefly comment on low frequency acoustic propagation in model granular beds using particle dynamical simulations. The following results will be briefly discussed. (i) Impulses propagating as solitary waves in 1-D granular chains with Hertz contacts in the absence of precompression. (ii) The effects of uniform precompression and gravitational loading on wave propagation. (iii) Impulse propagation in 3-D granular beds. The research presented shall highlight the intrinsically nonlinear nature of wave propagation in granular beds.

I. Introduction

The study of sound propagation in granular media dates back to the work of Janssen [1], Biot [2] and others. It is well known that a pressure impulse that is propagating through a grain silo suffers amplitude attenuation and develops a long elongated oscillatory tail as it travels into the bed [3]. Sound waves at various frequencies are also known to propagate reasonably well through wet and saturated soil [2, 4]. Unfortunately, acoustic attenuation typically becomes a serious problem with increasing frequency of an acoustic signal [5].

In the present study we address the problem of propagation of an acoustic impulse [6, 7] through a granular bed and comment briefly on the propagation of low frequency acoustic signals through a granular bed. The studies reported here are simulational in nature. The connection of our work with available experimental results is briefly discussed.

The article is arranged as follows. In Section II we present a discussion of the model we study and of the details of our calculations. Section III presents our main results and Section IV summarizes our work.

II. Model, Analytical Results and Calculational Methodology

We shall consider two granular systems in the present study, (i) a chain of spherical grains in which the grains touch one another, and (ii) a 3D body centered cubic lattice of spherical grains. The grains repel upon compression according to the Hertz potential as follows:

$$V(\delta) = a\delta^{5/2}, \tag{1}$$

where

$$a = (2/5D(Y,Y',\sigma,\sigma'))[RR'/(R+R')]^{1/2}, \tag{2}$$

and

$$D(Y,Y',\sigma,\sigma') = (3/4)[(1-\sigma^2)/Y + (1-\sigma'^2)/Y']. \tag{3}$$

The quantity $\delta \equiv R + R' - (z_i - z_{i+1}) > 0$ is called the overlap function between the grains and Y, Y', σ, σ' denote the Young's moduli and Poisson's ratio for the two spherical grains in contact.

For granular contacts that are not between spherical grains, the index of δ is different. Typically, the index is more than 2 and lies between 5/2 and 3 [8]. Hence, the potential associated with the repulsion between overlapping grains is nonlinear. Given Eqs. (1-3), it is possible to construct the equation of motion of a single grain, which is in contact with its neighbors, in a chain [6]. The equation of motion for a grain of mass m in location z_i with neighbors at z_{i-1} and z_{i+1} reads,

$$m d^2 z_i / dt^2 = 5/2 \, a \, [(\Delta_{i,i-1} - z_i + z_{i-1})^{3/2} - (\Delta_{i,i+1} - z_{i+1} + z_i)^{3/2}] + mg, \qquad (4)$$

The quantity $\Delta_{i,i-1}$ represents the initial distance between the grain i and i-1 and is equal the grain diameter in a chain without precompression (g=0) and is less than the diameter in the presence of precompression (g≠0). We ignore restitution for now but will comment upon the effects of restitution below.

When g=0, the position of the grain i can be written as $z_i(t)=z_i(0)+u_i(t)$, where u_i represent the displacement of grain i from initial equilibrium position. The existence of a wave solution implies that one can write $u_i(t)=u(z_i,t)=u(z_i-ct)$, with c being the propagation velocity of the wave [17]. In terms of the function u, Eq.(4) becomes [11]:

$$[2mc^2 / (5a)] \, d^2u(z) / dz^2 = (u_{i-1}(z) - u_i(z))^{3/2} - (u_i(z) - u_{i+1}(z))^{3/2}. \qquad (5)$$

One can show [11,12,17] that a generic solution to Eq. (5) can be written as :

$$u(z) = (A/2)[1 - \tanh(f(z)/2)], \qquad (6a)$$

with $f(z) = \Sigma_{q=0}^{\infty} C_{2q+1} z^{2q+1}$. $\qquad$ (6b)

Eqs.(6a,b) describes the shape of the propagating solitary wave with amplitude A [6, 10, 11]. The coefficients C_{2q+1} are independent of material parameters (a,m) or solitary wave amplitude (A), and can be obtained in several ways [11,17].

In general, in the presence of finite precompression, there is no solitary wave propagation. However, for vanishing small precompression, it was shown by Nesterenko [6], that, in the long wavelength limit, Eq.(4) is well approximated by a KdV type nonlinear equation which support solitons.

In the case g≠0, there is always a depth at which the precompression cannot be neglected. In this case, following Hong [9], we denote $\psi_i = z_i - i\Delta_0 + \Sigma_{l=1}^{i} (mgl/A)^{2/3}$. For $(\psi_i - \psi_{i-1}) \ll (mgi/A)^{2/3}$ (we define this statement as the "acoustic limit"), Equation (4) simplifies to

$$m \partial^2 \psi_i / \partial t^2 = -\mu_i(\psi_i - \psi_{i-1}) + \mu_{i+1}(\psi_{i+1} - \psi_i), \qquad (7)$$

where $\mu_i = \mu_1 i^{[1-1/p]}$ is the force constant of the ith contact and $\mu_1 = mpg(A/mg)^{1/p}$ is the force constant of the first contact [9]. The displacement function has strongly oscillatory components in space and time and is dispersive as the wave propagates [9, 12]. The primary limitation of the above treatment is that it is not valid near that end of the chain where the impulse is generated. This is because of the fact that gravitational loading is vanishingly small at the surface and the acoustic approximation loses its validity. Our numerically calculated behavior of displacement [12] suggests that at finite depths no perturbation can be regarded as "weak" in a meaningful way. Dynamical simulations therefore play an important role in studying impulse propagation in granular media.

We use velocity Verlet algorithm for carrying out dynamical simulations of impulse propagation in 1D and in 3D granular systems. The details of the algorithm may be found elsewhere [13, 10, 14-15]. The program units of length and time are millimeters and microseconds, respectively. In these units, a = 18.95 in Equation (2). The time units used in the numerical integration is 0.1 microsecond. The initial velocity imparted to the system is chosen to be between 0.02 and 0.05 m/s. The typical velocities of the traveling impulses are ~ 500 m/s.

III Results
1D Results

Figure 1 presents our results on the propagation of an impulse as a solitary wave, more than ~50 grains away from the boundary in a 1D chain without precompression and gravitational loading. As seen in Figure 1, the solitary wave is about 5 grain diameters wide. When the grain compression is small compared to the compression due to constant loading, one obtains an acoustic impulse propagation through the chain [10]. We have recently studied this problem [14] and reported that

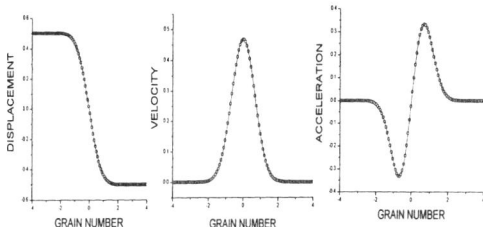

Figure 1: Plots of normalized displacement, velocity and acceleration versus space of a propagating solitary wave in a 1D chain of Hertzian grains.

the amplitude of the maximum velocity of the propagating impulse decays with depth z (measured from the surface) as $z^{-1/2}$. The envelope of the velocity function associated with the propagating perturbation as experienced by a typical grain decays at large times as $t^{-1.4}$. The tail is measured in such a way that the behavior of the leading edge of the impulse does not affect the measurement of decay of the tail [14]. To our knowledge, there is no analytic understanding of the relaxation processes associated with the response of grains to a perturbation in a uniformly loaded chain.

We now consider the propagation of an impulse in a gravitationally loaded chain of grains. The impulse propagates as a dispersive bundle of energy. The leading edge of the impulse decays in space and time. The tail of the impulse elongates in space and time. The spatio-temporal stretching of the tail is clearly related to the spatio-temporal reduction in the

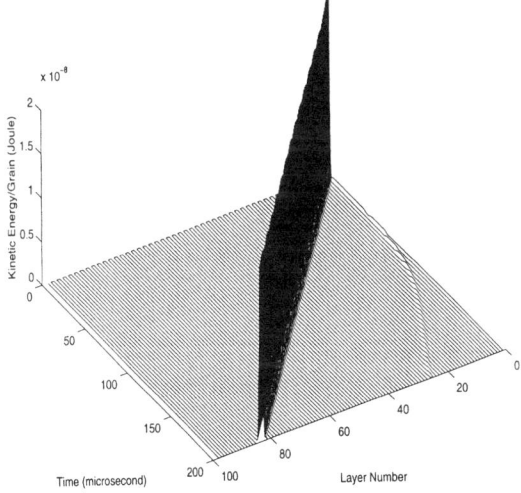

Figure 2: Propagation of a solitary wave generated at the entire surface of an infinite 3D granular bed. The weak secondary pulse forms naturally from the initial impulse.

leading amplitude of the velocity function. Our calculations indicate that the leading amplitude of the velocity function decays with depth algebraically with slope -0.250 [12]. We have also carried out the corresponding check for the algebraic decay of the leading amplitude of displacement function with depth and find a value of -0.0835. Both of these results are consistent with the calculations of Hong et al. [9].

3D Results

We next consider impulse propagation in pristine, periodic (and hence infinite) 3D granular beds. We start with a periodic body centered cubic lattice and initiate an impulse across an infinite surface of the bed. The grains at the bed surface are constrained to move along the direction of the impulse. The calculations are set up as follows. A surface grain can be thought of as located at the top of a regular tetrahedron with the three vertices at the bottom being the locations of three nearest neighbor grains immediately below the top grain. The force experienced by the top grain as it compresses by amount x against the 3 lower grains can be shown to be

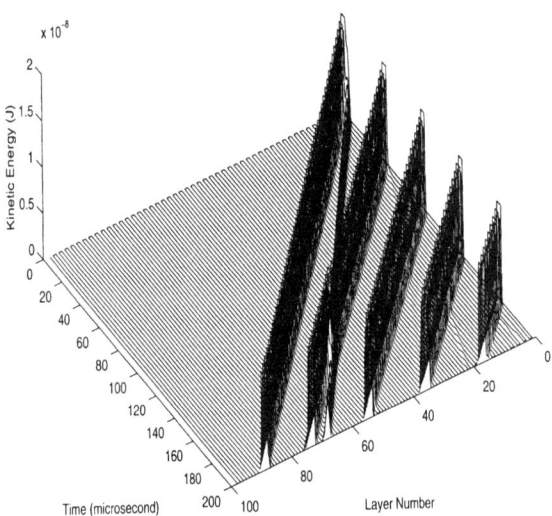

Figure 3: Impulses are generated at a frequency of 20 kHz. The amplitudes of the impulses are gradually increased in time to generate trains of solitary waves. Some interaction between trains is typically unavoidable.

$$F = (5a/2)x^{3/2} \ [(2\sqrt{3}/3)6^{1/4} - (7\sqrt{2}/12) (6)^{1/4}(x/d) - ...], \tag{8}$$

where d is the distance between the centers of nearest neighbor grains and a is defined by Eq.(2) above. Not surprisingly, an impulse initiated at all the grains of an infinite interface propagates down the infinite bed as a solitary wave as shown in Figure 2.

It is interesting to observe that the impulse velocities are fairly small, being somewhere between 10^2 and 10^3 m/s. Hence, when impulses are sent at sufficiently high frequencies or when one considers continuous signals at MHz frequencies, the individual pushes on each grain interact with each other and the solitary wave-like propagation of a impulse is lost. In this "high-frequency" regime, the signal becomes strongly dissipative as it excites internal vibrational modes in individual grains.

When impulses of sufficiently low frequencies are used, the individual impulses do not interfere with one another. Each impulse travels as a solitary wave and there is minimal

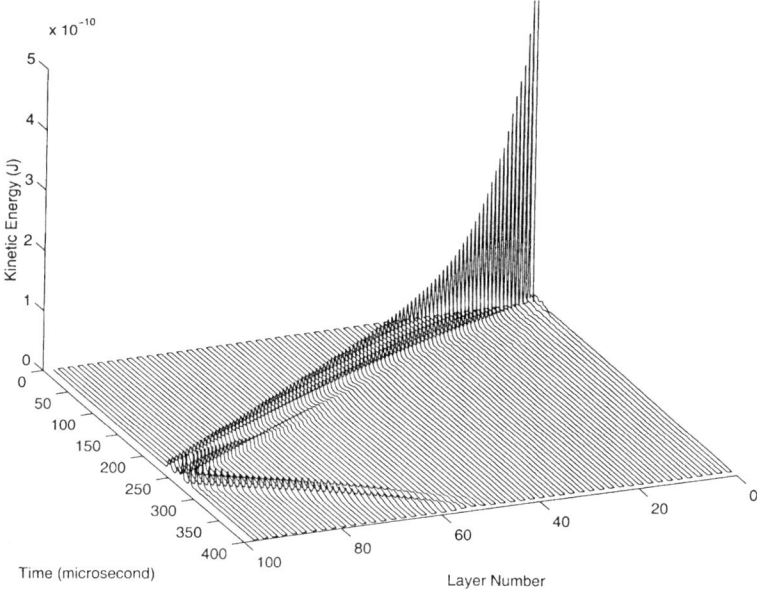

Figure 4: Propagation of an impulse initiated at across an area with 10x10 grains in a bed of dimension 30x30x100. Typical impulses that we have probed have initial velocities of about 0.05 m/s. The impulses travel at velocities of about 500 m/s.

interference between individual waves. Figure 3 shows such an infinite surface being pulsed at about 20 kHz. The solitary wave-like object generated by each pressure impulse travels as an individual object. Lower frequencies reveal propagation of impulses that are far apart in space and time and can hence be easily described in terms of individual impulses.

Given that one can propagate solitary waves in 3D granular beds, is it possible to use such solitary waves for detecting buried objects? To address this matter, we consider a slightly more complicated dynamical problem than the one addressed above, namely the propagation of an impulse initiated across a small region of the surface of a large bed. Clearly, an impulse generated across a small surface area will spread along the x and y axes as it travels down the depth of a bed. Figure 4 shows the propagation of a pulse generated across a region of 10x10 grain diameters in a bed of sides 30x30 grain diameters. As the data shows, the signal propagates without significant corruption in its spatial width except that it spatio-temporally attenuates and breaks up into several weaker pulses. The dynamics of the pulse break-up and attenuation is complex and more work is needed to analyze it. However, it is clear that surface modes play a significant role in attenuating the signal and hence significant part of the attenuation occurs in the first 20 layers or so [16]. To our knowledge, there is no simple decay law to explain the recorded amplitude attenuation. The reflection of the pulse at layer 100 occurs due to the presence of a reflective boundary at the base of layer 100.

Typical dry granular materials possess significant polydispersity. In addition, some energy is always lost as energy propagates through every granular contact due to restitution. We have addressed the role of polydispersity and restitution in a detailed dynamical study [17]. The general role of polydispersity is to introduce a background noise at the wake of a propagating impulse. An important question to address is how much polydispersity can destroy a propagating impulse. We address the effects of polydispersity on propagation and backscattering of impulses in the following article.

We thank Professors V. Nesterenko, J. Hong, M.J. Naughton and Dr. R. Sinkovits for their interest in this work. Partial support of the DoE and SUNY-Buffalo are gratefully acknowledged.

References
[1] H.A. Janssen, Zeits. Ver. Dtsch. Ing. **39**, pp 1045-1049 (1895).
[2] M.A. Biot, J. Acoust. Soc. Am. **28**, pp 168-191 (1956).
[3] T. Boutreux, E. Raphael and P.G. de Gennes, Phys. Rev. E **55**, pp 5759-5773 (1997).
[4] B.O. Hardin and F.E. Richart, Jr., J. of Soil Mech. and Found. Div., Proc. ASCE, **89**, No. SM 1, Feb. pp 33-65 (1963).
[5] F.E. Richart, Jr., R.D. Woods and J.R. Hall, Jr., *Vibrations of Soils and Foundations* (Prentice Hall, Englewood Cliffs, NJ, 1970) and references therein.
[6] V.F. Nesterenko, J. Appl. Mech. Tech. Phys. **5**, pp 733-743 (1983).
[7] A.J. Rogers and C.G. Don, Acoust. Australia **22**, pp 5-9 (1994).
[8] D.A. Spence, Proc. R. Soc. Lond. A **305**, pp 55-80 (1968).
[9] J. Hong, J.-Y. Ji and H. Kim, Phys. Rev. Lett. **82**, pp 3058-3061 (1999).
[10] S. Sen, M. Manciu and J.D. Wright, Phys. Rev. E **57**, pp 2386-2397 (1998).
[11] S. Sen and M. Manciu, Physica A **268**, pp 644-649 (1999).
[12] M. Manciu, V.N. Tehan and S. Sen, Chaos **10** pp 658-669 (2000).
[13] R.S. Sinkovits and S. Sen, Phys. Rev. Lett. **74**, pp 2686-2689 (1995).
[14] M. Manciu, S. Sen and A.J. Hurd, Physica A **274**, pp 588-607 (1999).
[15] M. Manciu, S. Sen and A.J. Hurd, Physica A **274**, pp 607-618 (1999).
[16] S. Sen, M. Manciu, R.S. Sinkovits and A.J. Hurd, Gran. Matt. (to appear, 2000).
[17] M. Manciu and S. Sen, Phys. Rev. Lett. (submitted).

Mat. Res. Soc. Symp. Proc. Vol. 627 © 2000 Materials Research Society

Backscattering of Nonlinear Acoustic Impulses from Buried Inclusions in Granular Beds

Marian Manciu and Surajit Sen
Department of Physics, State University of New York at Buffalo, Buffalo, NY 14260-1500, USA
Alan J. Hurd
Department 1841, Sandia National Laboratories, Albuquerque, NM 87185, USA

Abstract
Nonlinear acoustic impulses travel as weakly dispersive bundles of energy through 3-D, dry, granular assemblies. Such impulses can therefore be exploited to probe for buried objects in 3-D granular assemblies. In this article, we shall discuss the details of propagation of an impulse generated at the surface of a granular bed. The role of polydispersity will be briefly addressed. The calculations suggest that backscattered data, as received at predetermined locations at the surface, may be helpful in reconstructing an approximate image of the buried inclusion, as viewed from above.

I. Introduction

We have discussed the propagation of an impulse in a 3D granular assembly in the preceding article [1] and in Manciu et al. [2]. Our studies show that an appropriate impulse generated at the surface of a granular bed can travel into the bed as a weakly dispersive bundle of energy. The dispersion of the impulse depends upon several factors. These factors are: (i) the surface area associated with the impulse and the magnitude of the impulse, (ii) the nature of the boundary conditions employed in the study, (iii) the

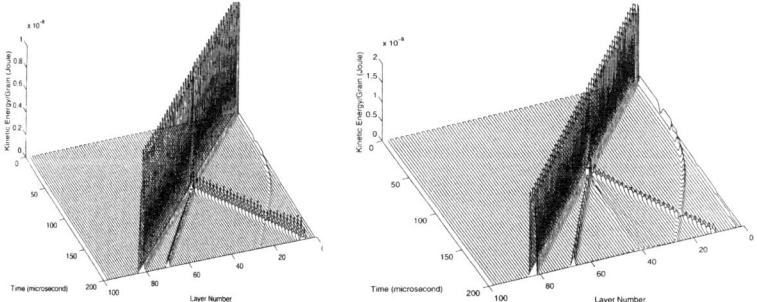

Figure 1: (Left panel) An impulse generated to an infinite layer at the surface of a granular bed propagates into the bed and backscatters off a buried infinite layer of grains that are three-fold more massive. (Right panel) The calculations presented are identical to that in the left panel except that the area of the impurity layer is 4x4. Observe that the magnitude of the backscattered signal is sensitive to the ratio between the area across which the impulse is generated and the surface area of the backscatterer.

degree of polydispersity in the medium, (iv) the magnitude of restitution invoked at the granular contacts and (v) issues such as wetness. In the present study, we consider the role of (i)-(iii) via simple numerical experiments. We comment on the role of (iv) with reference to our work and to available literature. Our objective is to explore whether one can exploit the nonlinear behavior of impulse propagation to detect buried objects in granular beds. It may be noted that to the best of our knowledge there are only two available experiments that have demonstrated that it is possible to use acoustic impulses and low frequency acoustic signals to detect buried objects at shallow depths (of about 30 cm. or so) [3-4].

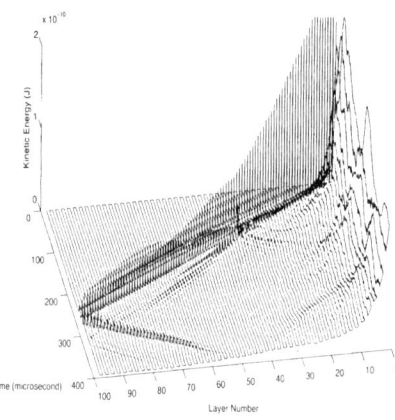

Figure 2: Propagation of an impulse into a three-dimensional granular bed with 30x30x100 grains with perfectly reflecting boundary conditions. Impulse is generated at the entire surface and backscatters off buried layer, which has grains that are three times more massive. Observe the dominant surface excitations.

2. Model Calculations

The details of the model system are presented in the preceding article [1]. We discuss below the approximations used and the conclusions drawn based on our particle dynamical calculations.

Figure 1 describes particle dynamical calculations of impulse propagation in an ideal, hexagonal-close-packed lattice of (monodisperse) grains that constitute a 3D granular bed. Periodic boundary conditions have been employed to generate an infinite 3D lattice. The interactions between the grains are Hertzian [5], and the grain-grain forces can be described as

$$F = a\delta^{3/2}, \tag{1}$$

where δ is the overlap between the grains [1,5].

In the calculations presented in the left panel of Figure 1, the buried object is assumed to occupy the entire infinite layer (layer number 40) and the masses that make up the infinite layer are three times more massive than those in the pristine host. The lattice dimensions are taken to be (10x10x100) in grain diameters and periodic boundary conditions are imposed in the planes that are perpendicular to the direction of propagation. The impulse propagates as a solitary wave in the system [1] and the backscattering of the impulse by the layer resembles our earlier calculations, namely that of impurity backscattering in 1D Hertzian chains reported elsewhere [1,6]. When an impurity layer of smaller area (4x4, also with periodic boundary conditions) is used to carry out the calculations, we find that the amplitude of the backscattered signal, as measured using kinetic energies of the grains, is weaker compared to the data in the left panel of Figure 1 [8]. The calculations presented in Figure 1 may be of interest in the context of geophysical explorations of extended buried layers. However, when detecting buried objects of finite dimensions, it is of interest to know the area across which an impulse should be generated to best detect the object.

In Figure 2, we present dynamical calculations to study the propagation of an impulse that is generated across an entire layer of a 30x30x100 system to detect a finite object in layer number 40. *Perfectly reflecting walls* have been used in all of the calculations reported here. Thus, any perturbation reaching the wall is reflected back into the system without any loss of amplitude. While such boundary conditions may seem highly artificial, as indeed they are, we argue that they are still worthwhile given (a) it is of interest in detecting objects using impulses in confined areas, and (b) that in small system calculations with buried objects at shallow depths (of about 40 grain diameters), where the backscattering is largely normal [6], the interference between reflected wall signals have negligible effects on the backscattered signal from an object.

The results show that one obtains a very weak return signal at the surface. The data in Figure 2 suggests that a significant part of the energy is trapped as surface modes in the shallow reaches of the bed and thus it is difficult to decipher the backscattered signal at the surface.

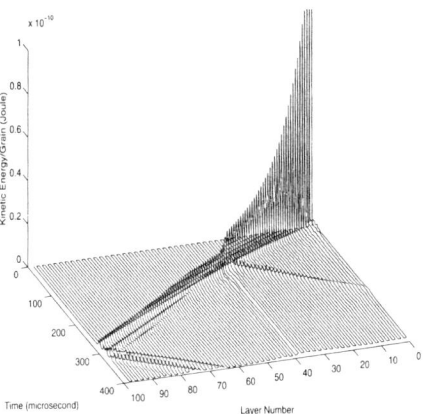

Figure 3: Propagation of an impulse generated across a small area to detect a small massive buried object in a 30x30x100 grain bed with perfectly reflecting boundary conditions. Observe that there are not many surface modes.

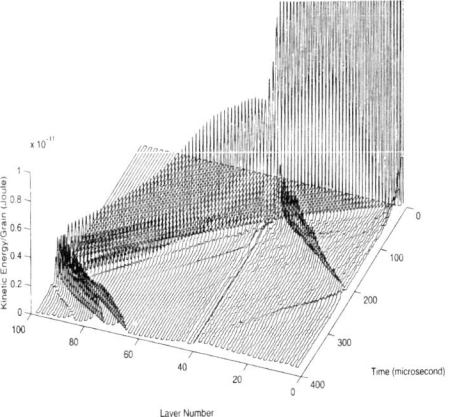

Figure 4: Calculations represent same data as in Figure 3 except that only the vertical component of the kinetic energy (dominant part of the kinetic energy) is being shown.

We next considered the propagation and backscattering of impulses generated across small areas of the surface in a finite, idealized, three-dimensional granular bed. Figure 3 shows the propagation of an impulse generated across an area spanning 10x10 grain diameters in a finite box of 30x30x100 in grain diameters. The impurity is placed at layer number 40 with respect to the surface. The impurity masses are three times the mass of the host system grains. We do not use periodic boundary conditions in this study. However, we retain perfectly reflecting boundaries when carrying out the simulations. The object is composed of grains that are three times more massive than the host grains and is of dimension 1/9 (10x10 in grain diameters) of a layer. The finite area of the region across which the impulse is generated helps to reduce the undesirable effects of dominant surface modes in observing buried objects. Our studies suggest that it is likely that sending impulses across small areas would generate better data to reconstruct images of buried objects.

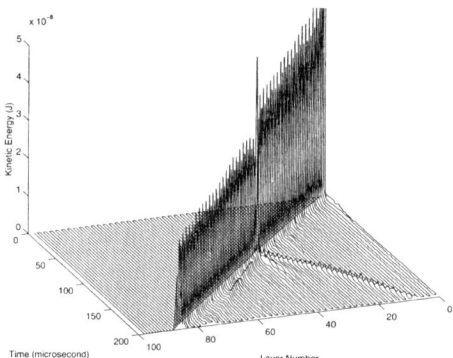

Figure 4 represents the data in Figure 3 except that the normal component of the kinetic energy is shown along the vertical axes.

We shall ignore the effects of restitutional losses during the propagation of the impulse through the systems. We have recently completed a detailed study [7] in which we have probed the role of restitution $w \equiv F_{unloading} / F_{loading}$, in attenuating the amplitude of a propagating impulse in one-dimensional Hertzian chains. Our analyses reveal that restitutional losses can be approximated as exponential attenuations, a well known result and a behavior that has been approximately validated in experiments [8]. However, an important difficulty in addressing the role of restitution lies in the fact that we must better understand the process of energy transfer into the internal degrees of freedom of a grain. The detailed properties of restitutional decay of displacement amplitude of a propagating signal in granular systems therefore warrant sophisticated atomistic level analysis.

Figure 5: We show backscattering data from an entire layer of impurities for a 10x10x100 system with 3.3% of randomly masses that are twice as massive as that in the host system.

In what follows, we briefly address the complications introduced when one introduces polydispersity of grains in the bed. In Figure 5 we show a calculation that has been carried out for a 10x10x100 grain system with periodic boundary conditions. In the calculations, we randomly replace 3.3% of masses by impurities that are twice as massive as the host mass. The reader may observe that in a typical, dry, sand bed, one may not have such a large mass mismatch. In addition, the impurity concentration is quite modest. Thus, the effects of polydispersity are expected to be significant in our model study. The backscattering of an impulse in this case is also reminiscent of normal incidence reflection. We therefore expect that use of periodic boundary conditions will not play a significant role in modifying the backscattering process in model granular beds. The impurity masses make up the entire 40th layer. The entire surface of the bed is initially perturbed by the impulse. The data in Figure 5 demonstrates that polydispersity does not destroy the backscattered signal. Reducing the mass of the impurity layer adversely affects the magnitude of the backscattered signal. However, detectable signals can still be obtained by controlling the amplitude of the initial impulse, by fitering out the planar vibrations of the surface grains and, if necessary, by using multiple impulses. These results are consistent with published studies in Refs. [3,4]. Upon increasing polydispersity to 5% of randomly distributed grains of mass three times that of the host masses, we obtain a dirtier signal (see Figure 6), i.e., a worse signal to noise ratio. It may be noted that the magnitude of the continuing pulse is appreciably lower in Figure 6 when compared with the same in Figure 5. This is because heavier masses backscatter more energy than lighter masses, as discussed in detail in Ref. [2].

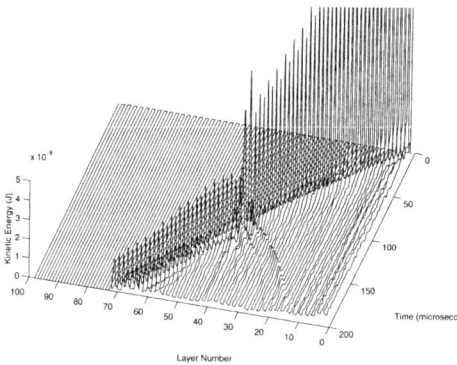

Figure 6: The calculations shown here are almost identical to that in Figure 5 except that the impurity masses are three times as massive as that in the host system. The proportion of randomly distributed masses is 5%. Observe that there is more backscattered signal in the data in Figure 6 than in Figure 5.

The research reported here has been partially supported by Sandia National Laboratories under contract number DE-AC-04-85000, by the National Science Foundation under grant number NSF-CMS-0070055 and by the U.S. Army under contract number DACA-39-97-K0026.

References
[1] M. Manciu, S. Sen and A.J. Hurd, in S. Sen and M.L. Hunt Eds. "The Granular State" MRS Symp. Proc. 627, pp (2000).
[2] M. Manciu, S. Sen and A.J. Hurd, Physica A **274** pp 607-618 (1999).
[3] A.J. Rogers and C.G. Don, Acoust. Australia **22**, pp 5-9 (1994).
[4] M.J. Naughton, R. Shelton, S. Sen and M. Manciu, in MD98, Proceedings of the 2nd International Conference on the detection of Abandoned Landmines, IEE Conf. Proc. **458**, pp 249-251 (IEE, London, 1998).
[5] H. Hertz, J. reine u. angew. Math. **92**, pp 156-171 (1881).
[6] S. Sen, M. Manciu and J.D. Wright, Phys. Rev. E **57**, pp 2386-2397 (1998).

[7] M. Manciu and S. Sen, preprint (2000).
[8] S. Sen and M.J. Naughton, in Research Report to U.S. Army Corps of Engineers, Waterways Experiment Station, Vicksburg, MS for Contract No. DACA-39-97-K0026 (August, 1998).
[9] There is negligible mixing between backscattered impulses from periodically repeated objects.

Mat. Res. Soc. Symp. Proc. Vol. 627 © 2000 Materials Research Society

Ultrasound propagation in disordered granular media

Xiaoping Jia
Groupe de Physique des Solides, Université Paris 7
2 place Jussieu, 75251 Paris Cedex 05, FRANCE

ABSTRACT

We have identified, according to the ratio of the wavelength to the grain size, two distinct types of pulsed ultrasound transmission through a dry bead packing under stress: one corresponds to coherent ballistic waves characterized by the effective medium description, the other to the waves scattered by the inhomogeneous stress field within the granular medium. Over long distances of transport, the multiply scattered waves exhibit a diffusive character. Also we investigate the dynamics of the granular medium during a compaction under cyclic loading-unloading. Both the macroscopic deformation and the microscopic rearrangement have been measured, via an ultrasonic correlation technique using the multiple acoustic scattering very sensitive to the change of the system configuration. It is found that as the packing fraction increases, there is a continuous evolution of the system in response to external loading, from an irreversible behavior towards more elastic one.

INTRODUCTION

The study of granular media is of great current interest both scientifically and practically [1, 2]. In a static assembly of cohesionless grains, experimental observations (e.g., photoelastic visualization) [3, 4] and computer simulations [5] have shown very inhomogeneous spatial distributions of contact forces between particles, organized along force chains which extend over a scale of order 5-10 grains diameters. These force chains, carrying most of the forces in the system, involve only a small fraction of the total number of grains and are only marginally stable ("fragile matter"). Any external (thermal or mechanical) perturbation incompatible with the structure of the force chains or/and of sufficiently high amplitude, can lead to the plastic rearrangement of the system [1, 2]. Some new constitutive laws have been recently proposed to describe the *static* force transmission along the privileged force paths within granular systems and the instability of the force networks [6, 7]. In contrast to the conventional elastic description based on continuum mechanics, the equations governing the transmission of forces are hyperbolic instead elliptical, which lead to certain preferred directions as shown by experiments.

Sound propagation in a disordered granular medium is of fundamental interest (multiple scattering). It also provides a very useful, and sometimes unique, probe of the internal structure such as force networks and the mechanical properties of real 3D granular packings [8, 9], where the coherent photoelastic methods become impracticable. In this paper, we investigate via ultrasound propagation, the *dynamic* force transmission at *small amplitudes* in glass bead packings under stress. Emphasis will be put on relevant length scales of observation for separating the different regimes of wave phenomena. Also the compaction of the granular packings under cyclic loading-unloading will be observed, in particular by measuring the correlation function of multiply scattered waves between successive runs.

EXPERIMENTAL TECHNIQUE

A schematic diagram of the experimental apparatus is shown in Fig. 1. Granular materials consist of random packings of mono- or poly-disperse glass beads. The sizes of the beads used in our experiments range from 400 μm to several millimeters. They are contained in an aluminum cylinder of 30-mm inner diameter with the top and bottom surfaces made of close-fitting pistons. The beads are poured gently into the container being vibrated horizontally, in order to prepare dense packed granular materials. A volume fraction of solid phase is thus obtained to be about 0.63. A normal load P ranged from 0.03 to 3 MPa is applied to the upper piston using a jackscrew

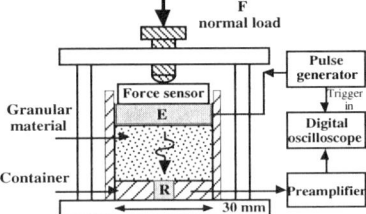

Figure 1. *Schematic diagram of the vertical cross section of the apparatus. An aluminum cylinder contains the granular material. E and R correspond, respectively, to the ultrasonic emitter and receiver. A normal force is applied vertically upon the top piston of the container by using a jackscrew arrangement.*

arrangement (œdometric test). At such load level, stress transmitted to the grains across the hertzian contact force is expected dominant compared with gravity and capillary effect. Compressional and shear waves are excited, by applying input pulses of one or more cycles centered at 500 kHz onto the corresponding piezoelectric transducers. The vibration amplitude of the acoustic source is about 10 nm, measured in air by means of an optical interferometer. Two sizes of ultrasonic receiver are used: a large one, equal to that of the emitter (30 mm in diameter), and a small one (2 mm in diameter).

CHARACTERISTICS OF SOUND PROPAGATION

Typical ultrasonic transmission through the glass packing of bead size $d = 0.4$-0.8 mm under an external load $P = 0.75$ MPa (corresponding to 50 daN) is illustrated in Fig. 2a. It is excited by a *one-cycle* pulse of 2 μs duration and detected by a longitudinal transducer of 30 mm in diameter placed at a distance $L \approx 11$ mm away from the source. The detected signal is basically composed of two parts, as observed in a previous work [9] : a coherent ballistic pulse E traveling through an "effective contact medium" with a velocity $V_{eff} \approx 1000$ m/s, followed closely by speckle-like multiply scattered waves S due to the inhomogeneous distribution of stress within the granular system. Indeed, by performing a separate spectral analysis, we find that E carries a rather narrow band of low frequencies while S has a broadband spectrum dominated by high frequency (Fig. 3a). As the bead size increases ($d = 1.5$ mm), the spectrum of E shifts towards low frequencies while the spectrum of S clearly exhibits a strong filtering out of high frequencies ($f > f_c$ ≈ 500 kHz) (Fig. 3b). This result suggests a frequency cut-off occurring when the wavelength $\lambda_c = V_{eff}/f_c$ is comparable to the bead size d.

Note also that the detected ultrasonic signals (E and S) remain stable against repetitive averaging (~ 100) over the duration of an experimental run (≤ 1 min). This excellent reproducibility of the signals shows that the ultrasound at small amplitudes (~ 10 nm) used here, can propagate in the granular system under stress without producing any significant irreversible rearrangement. The characteristics of acoustic propagation in such a quasielastic medium can therefore be interpreted within the framework used to describe the vibrational properties of amorphous systems in terms of an inhomogeneous stress field [10]. At the scale much larger than the bead size d ($= 0.4$-0.8 mm), the granular medium can thus be considered as homogeneous by a coarse-grained average [2]. Since the coherent wave E can only be self-averaging on a scale of order of its wavelength, this leads us to state that the correlation length of the stress filed is not large and cannot exceed a fraction of λ_{eff}, i.e. a few bead diameters [9] - A result in qualitative agreement with the computer simulation and the photoelastic visualization [3-5].

A fundamental difference between E and S signals lies in their sensitivity to the change of the packing configuration. Fig. 2b illustrate temporal waveforms detected under the same external load $P = 0.75$ MPa after performing an unloading-reloading cycle. As clearly shown, E exhibits a

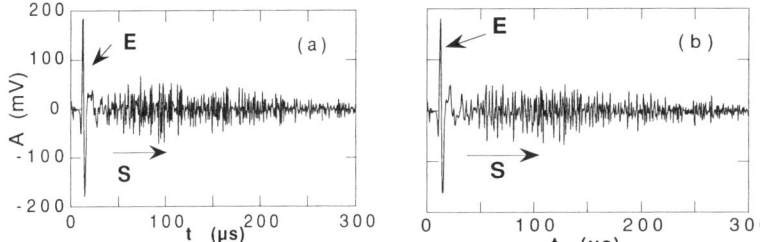

Figure 2. *Transmitted ultrasonic signals through the bead packing of d = 0.4 - 0.8 mm detected by a 30-mm-diam transducer obtained under the same applied stress P = 0.75 MPa , first loading (a) and re-loading (b).*

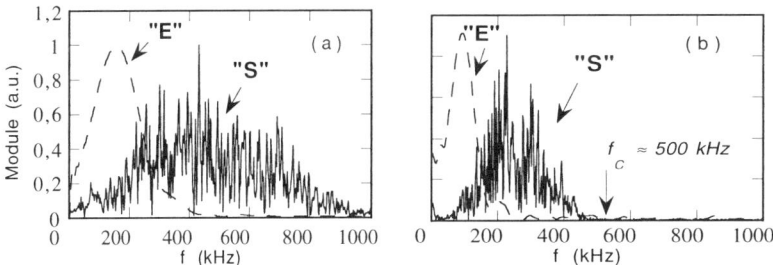

Figure 3. *Spectra of the E and S signals windowed from the total temporal waveform in the bead packing of d = 0.4-0.8 mm (a) and in that of d = 1.5 ± 0.15 mm (b)*

reversible behavior while S is non-reproducible, i.e. configuration specific at the level of individual contacts. It is expected however that the global shape of S, which is determined by the statistical properties of the granular structure, shall be weakly altered by local irreversible events.

In order to make S as a more quantitative probe to configuration variations, we shall study the transport characteristics of *multiply* scattered acoustic waves. Fig. 4 presents the time dependence of the average intensity of the transmitted S signal detected by a small detector of 2 mm in diameter. We used a *ten-cycle* pulse for excitation in order to have a well-defined frequency at $f = 500$ kHz, corresponding to a wavelength $\lambda \approx 2$ mm. This result was *ensemble* averaged over 50 different granular samples prepared in a similar way. Note that the time profile exhibits typical characteristics of a diffusive transport of strongly scattered sound observed in other random media [11, 10]. Using appropriate boundary conditions and resolving the diffusion equation, this will enable us to determine the diffusion coefficient and also the transport mean free path, being of the order of several bead sizes. More details about this study will be given elsewhere [12].

ACOUSTIC PROBING OF GRANULAR MATERIALS

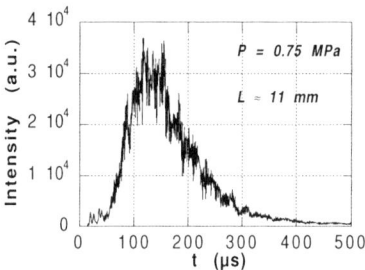

Figure 4. Time profile of the average intensity of multiple acoustic scattering transmitted the randomly close-packed glass beads d = 0.4 - 0.8 mm for sample thickness L ≈ 11 mm.

Determination of elastic moduli

Once the two distinct types of sound propagation in granular media are identified, we can unambiguously determine the compressional or shear wave velocities, V_P and V_S either by the time-of-flight method or by the phase spectrum analysis, if *only* the coherent E signals are windowed from the total detected signal [9]. In the long-wavelength limit, the acoustic velocities are related to the bulk modulus K and shear modulus μ of the granular material $V_P = [(K + 4/3\mu)/\rho]^{1/2}$ and $V_S = (\mu/\rho)^{1/2}$ where ρ is the density of the material's density.

Coherent compressional and shear waves' propagation in the bead packing of $d = 0.2-0.3$ mm, excited and detected by corresponding transducers, are shown in Fig. 5a. The measured shear wave velocity is given as a function of applied normal stress P (Fig. 5b). The data behaves basically as a power law: $V_S \propto P^{1/4.8}$, similar to those observed for the compressional wave velocity [9, 13, 14]. The discrepancy between the experiment and the effective medium theory ($V_S \propto P^{1/6}$), based on the Hertz-Mindlin contact [15], is believed to be associated with the breakdown of the assumptions underlying the effective medium theory, i.e. the co-ordination number is pressure independent, which are not the case in granular media [16, 14]. On the other hand, the ratio of K/μ deduced from the experimental velocities ($V_P/V_S \approx 1.79$) differs significantly from the prediction of the effective medium theory [15]. One interpretation recently proposed points to the failure of the affine (well-bonded) approximation used in the effective medium theory for calculating the shear modulus μ of granular materials [14].

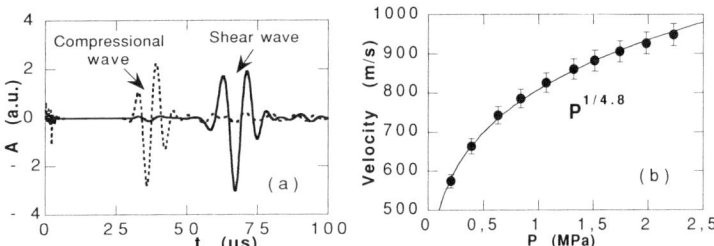

Figure 5. *(a) Propagation of coherent effective waves in the glass bead packings of d = 0.2 - 0.3 mm under external stress P = 2.3 MPa. Solid and dashed curves correspond respectively to compressional and shear ballistic pulses. (b) Shear velocity V_S (data points) measured as a function of the normal stress P. A power-law behavior is fitted approximately.*

Compaction under cyclic loading-unloading

In the experiment a vertical column of 30 mm in diameter filled with polydisperse glass beads d = 0.4 -0.8 mm to a initial height of h_0 = 10 mm was subjected to slowly repeated loading-unloading by the upper piston between 5 daN and 50 daN (Fig. 1). A displacement sensor monitors the resulting height (volume) change of the granular packing. Fig. 6a illustrates the typical height variation of the granular packing measured with a force step of 5 daN for a few first cycles of loading-unloading. A few percent (~ 2%) of the volume compaction ($\Delta h/h_0$), due to plastic rearrangement of grains, is observed after n ~ 10 cycles. Besides, Δh (n) plot follows approximately a logarithmic law: $\Delta h/h_0$ ~ ln (n), similar to those observed in the granular compaction subjected to vertical shakes with small amplitudes [17].

Now we present an original observation in which we make use of configuration-sensitive acoustic speckles S to follow semi-quantitatively the irreversible rearrangement *within* the granular system during a macroscopic plastic deformation (compaction) produced by cyclic loading-unloading. To do so, we define a degree resemblance between two time signals $S_i(t)$ and $S_j(t)$, via their intercorrelation function $C_{i,j}$ (τ) :

$$\Gamma_{i,j} (\tau) \equiv C_{i,j} (\tau) / [C_{i,i} (0) \ C_{j,j} (0)]^{1/2}$$

The value of $\Gamma_{i,j}$ ($\tau = 0$) characterizes their resemblance and it is to be calculated for multiple acoustic scatterings transmitted through the granular material under successive cycles of loading-unloading ($n = i$ and $i + 1$). $\Gamma_{i, i+1}$ (0) = 1 would correspond to an *ideal* solid having a pure elastic response to applied forces.

$\Gamma_{n, n+1}$ (0), deduced from the acoustic speckles between two successive cycles n and $n+1$ (Fig. 6b), increases progressively as the system becomes more and more compacted. Since the acoustic speckle is configuration specific, this result simply indicates that greater is the packing fraction, more reversible and elastic behaves the granular system with respect to applied forces. This is intuitively reasonable because as the packing fraction increases the important arrangement of grains tends to disappear and the force network becomes more tangled, with many shorter intersecting chains as being photoelastically observed in 2D granular systems [4]. The two peaks observed in Fig. 6b are associated with partially unloading, i.e., unloading to 40 daN instead of 5 daN, and then reloading to 50 daN. For such a small perturbation, the granular system exhibits a *quasielastic* character. These results are reminiscent of the pastelike behavior observed in other media such as concentrated suspensions, foams, etc. More quantitative results are expected by using *Diffusing Acoustic Wave Spectroscopy* (DAWS) in development.

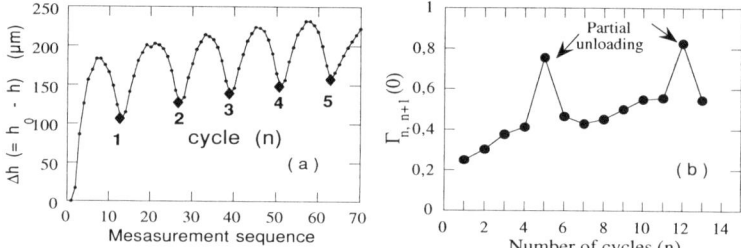

Figure 6. (a) Height (or volume) variation of the granular packing under cyclic loading-unloading. (b) Correlation of two S signals (multiply scattered waves) transmitted through the bead packing between two successive cyclic loading-unloading.

CONCLUSION

Ultrasound propagation in the glass bead packing reveals that for small amplitudes of force transmissions, the static granular system can be considered as a nonlinear quasielastic inhomogeneous medium, given external (œdometric) loading [2]. At scales of coherent waves much larger than the grain and the stress correlation lengths, the effective medium theory based on a coarse-grained average of contact mechanics gives a quite good description. When the wavelength shortened ultrasounds encounter successive scatterings caused by the inhomogeneous stress filed, which leads to the onset of a diffusive transport over long distance propagation. At a scale comparable to the characteristic length of wave transport, of the order of the correlation length of stress, the acoustic force transmission is expected to be along certain preferred directions [6, 7]. We believe that the ultrasonic correlation technique (DAWS) used in the study of the granular compaction will facilitates the connection between the microscopic structure (configurational rearrangements) and the macroscopic behavior of granular materials in response to externally applied forces.

ACKNOWLEDGEMENTS

We thank B. Velicky, C. Caroli and P. Mills for helpful discussions. This work was funded in part by the universities of Paris 7 and Paris 6 under BQR.

REFERENCES

1. H.M. Jaeger, S.R. Nagel, R.P. Behringer, *Rev. Mod. Phys.* **68**, 1259 (1996); *Physics of Dry Granular Media* , edited by H.J. Herrmann et al (Kluwer, Dordrecht, 1998)
2. P.G. de Gennes, *Rev. Mod. Phys.* **71**, S374 (1999); P. Evesque, P.G. de Gennes, *C. R. Acad. Sci.* (Paris), Ser. II **326**, 761 (1998)

3. P. Dantu, in *Proceedings of the 4th International Conference on Soil Mechanics and Foundations Engineering* (Butterworths, London, 1957); T. Travers, M. Ammi, D. Bideau, A. Gervois, J.C. Messager and J.P. Trodec, *Europhys. Lett.* **4**, 329 (1987)
4. D. Howell, R.P. Behringer, C. Veje, *Phys. Rev. Lett.* **82**, 5241 (1999)
5. F. Radjai, M. Jean, J.J. Moreau, S. Roux, *Phys. Rev. Lett.* **77**, 274 (1996)
6. M.E. Cates, J.P. Wittmer, J.P. Bouchaud, P. Claudin, *Phys. Rev. Lett.* **81**, 1841 (1998); J.P. Bouchaud, M. Cates, P. Claudin, J. Phys. I (France) **5**, 639 (1995)
7. A.V. Tkachenko, T. Witten, *Phys. Rev. E* **60**, 687 (1999); S.F. Edwards, D.V. Grinev, *Phys. Rev. Lett.* **82**, 5397 (1999); S.N. Coppersmith, C.H. Liu, S. Majumdar, O. Narayan, T.A. Witten *Phys. Rev. E* **53**, 4673 (1996);
8. C. Liu, S.R. Nagel, *Phys. Rev. Lett.* **68**, 2301 (1992); *Phys. Rev.* **B48**, 15646 (1993)
9   X. Jia, C. Caroli, B. Velicky, *Phys. Rev. Lett.* **82**, 1863 (1999)
10. P. Sheng, *Introduction to Wave Scattering, Localization, and Mesoscopic Phenomena* (Academic Press, san Diego, 1995); S. Alexander, *Phys. Rep.* **296**, 66 (1998)
11. J.H. Page, H.P. Schriemer, A.E. Bailey, D.A. Weitz, *Phys. Rev. E* **52**, 3106 (1995)
12. X. Jia (unpublished)
13. S.N. Domenico, *Geophysics* **42**, 1339 (1977)
14. H.A. Makse, N. Gland, D.L. Johnson, L.M. Schawartz, *Phys. Rev. Lett.* **83**, 5070 (1999)
15. K.W. Winkler, *Geophys. Res. Lett.* **57**, 1073 (1983); P.J. Digby, *J. Appl. Mech.* **48**, 803 (1981)
16. J.D. Goddard, *Proc. R. Soc. London* **A430**, 105 (1990); S. Roux, D. Stauffer, H.J. Herrmann, *J. Phys.* **48**, 341 (1987)
17. E.R. Nowak, J.B. Knight, E. Ben-Naim, H.M. Jaeger, S.R. Nagel, *Phys. Rev. Lett.* **57**, 1971 1 (1998)

Mat. Res. Soc. Symp. Proc. Vol. 627 © 2000 Materials Research Society

Linear and Nonlinear Ultrasonic Properties of Granular Soils

Brian P. Bonner, Patricia A. Berge, Chantel M. Aracne-Ruddle, Hugo Bertete-Aguirre,
Dorthe Wildenschild, Cosette N. Trombino, and Edgar D. Hardy
Experimental Geophysics Group, Lawrence Livermore National Laboratory,
L-201, PO Box 808,
Livermore, CA 94551-9900, U.S.A.

ABSTRACT

The ultrasonic pulse transmission method (100-500 kHz) was adapted to measure
compressional (P) and shear (S) wave velocities for synthetic soils fabricated from quartz-clay
and quartz-peat mixtures. Velocities were determined as samples were loaded by small (up to 0.1
MPa) uniaxial stress to determine how stress at grain contacts affects wave amplitudes,
velocities, and frequency content. Samples were fabricated from quartz sand mixed with either a
swelling clay or peat (natural cellulose). P velocities in these dry synthetic soil samples were
low, ranging from about 230 to 430 m/s for pure sand, about 91 to 420 m/s for sand-peat
mixtures, and about 230 to 470 m/s for dry sand-clay mixtures. S velocities were about half of
the P velocity in most cases, about 130 to 250 m/s for pure sand, about 75-220 m/s for sand-peat
mixtures, and about 88-220 m/s for dry sand-clay mixtures. These experiments demonstrate that
P and S velocities are sensitive to the amount and type of admixed second phase at low
concentrations. We found that dramatic increases in all velocities occur with small uniaxial
loads, indicating strong nonlinearity of the acoustic properties. Composition and grain packing
contribute to the mechanical response at grain contacts and the nonlinear response at low
stresses.

INTRODUCTION

Wave propagation in natural granular media applies to a range of problems in near-surface
geophysics and environmental engineering. Earthquake strong motion, observations of ground
water movement using seismic methods and slope stability problems in civil engineering are
some examples. Interest in these problems has generated an extensive literature on elastic
properties of natural granular media [1], but experimental difficulties have prevented
measurements at the very low pressures corresponding to the shallow subsurface. New
experimental methods were developed to accomplish the measurements reported here [2].
Although these methods can be used for natural materials recovered from in-situ, first
experiments were conducted on synthetic soils so that critical parameters such as composition,
porosity, and packing technique could be controlled and investigated systematically.

EXPERIMENTAL DETAILS

The experimental method is ultrasonic pulse transmission, modified for measurements in
highly attenuating materials [3]. The time-of-flight of an ultrasonic pulse launched from a
transmitting to a receiving transducer and the sample length (44.9 mm) yields the velocity for P

and S waves as appropriate. Changes in the path length produce small changes in travel time compared to concurrent stress-induced changes in the elastic constants and are neglected here in velocity computations. Transducers with center frequencies ranging from 100 to 500 kHz polarized for transverse shear were used for S measurements, although sufficient compressional energy is available for simultaneous P measurements. Further improvements on previous methods [3] were made to allow measurements of highly attenuated signals. The transmitting transducer was excited with 400 V pulses. Received signals were as small as 10^{-5} V, requiring amplification of 60 db for detection and digital averaging of 200-1000 repetitions to pick arrivals. A sample sleeve was constructed to ensure that the signal travels through the soil mixture, suppressing energy traveling on longer, but faster, paths to the receiving transducer [2]. The sample assemblies were closed with latex membranes that transmitted sound from the transducer and contained the soil mixture. Accuracy was limited to 20% at the lowest stresses but improved with signal amplitude to ~3% for compressional waves and 10% for shear waves at higher stresses. Some of the data scatter is caused by differences in subjective picking of arrivals, as well as dramatic changes in the character of the waveforms with stress.

End-load pressures between 0 and 15 psi (0 to 0.11 MPa), simulating up to several meters of overburden, are applied in increments of approximately 1.5 psi to the sample by air-driven, pneumatic pistons that push on the transducer housing. Internal stress in the sample approximates uniaxial strain, modified by edge effects and minor sleeve deformation.

Samples were constructed by mixing pure quartz sand with either clay or peat moss in increasingly larger percentages by weight. The sand (Ottawa, IL) and was sieved to a median grain diameter of 273 micrometers. The clay is Na-montmorillonite from Wyoming, a swelling smectite. The clay is equilibrated in a 100% humid atmosphere for seven days before sample preparation to achieve reproducible water content. The peat is commercial Canadian sphagnum peat moss, composed mainly of natural cellulose fibers. Typical organic content of peat moss is 80 to 95% of the mass fraction. The peat was equilibrated with air at ambient humidity, typically 30%, before sample construction. The samples were layered to include a central section of pure sand to be consistent with earlier experiments that required a high permeability layer to provide access for pore fluid. Measured travel times were corrected to allow for the sand layer, using data from a sand sample prepared using the same methods as used for the mixtures. Velocities, attenuation and load dependence are sensitive to the packing methods (vibration, hand packing) used in sample fabrication for the low stress levels of these experiments. Packing effects contribute to the scatter observed in experimental data.

RESULTS

Representative waveforms for sand-clay mixtures are presented in figure 1 to give a general idea of the quality of the data and to demonstrate the dramatic effects of external loading on transmitted pulses. Complete data sets are archived elsewhere [4,5]. Three waveforms plotted at the same scale (zero gauge units, 7.8 and 15.6 psi) for the 10% clay-sand sample are shown in figure 1. Both the compressional and shear waves arrive earlier with increasing load. The compressional amplitude increases only slightly, if at all. The shear amplitude increases with increasing load and also becomes sharper, consistent with a decrease in shear attenuation and more efficient transmission of high frequencies.

Velocities for sand-clay and sand-peat mixtures are plotted in figure 2 to demonstrate that velocities are low and can increase rapidly with small static loads. Two different pure sand

10% Na-Montmorillonite Clay with F-50 Ottawa Sand - Dry

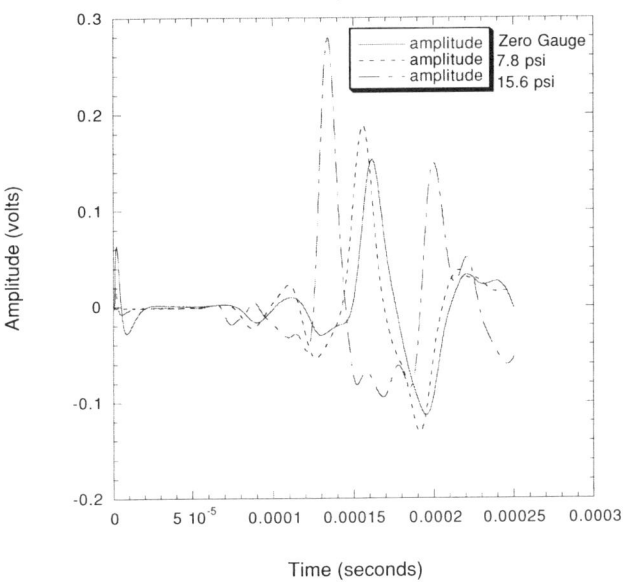

Figure 1. Waveforms for 10% clay, 90% sand sample for three different load values

samples were examined; one after packing by vibration and the other after hand packing. Vibration produces higher velocities and smaller velocity increases with increasing load. All mixtures were hand packed. Adding clay to the sand increases compressional velocity relative to the hand-packed sand sample and eliminates the velocity increase with pressure at low load. Although trends are less clear for the shear velocity, added clay increases velocity and tends to suppress the gradient to the highest loads. Measurements for sand-peat mixtures were generally more difficult and show more scatter because of sample variability and high ultrasonic attenuation. Adding peat to sand decreases compressional and shear velocity, except for the 60% sample, which shows the highest P velocity. Peat mixtures show velocity gradients comparable to hand-packed sand, except for the anomalous 60% peat sample.

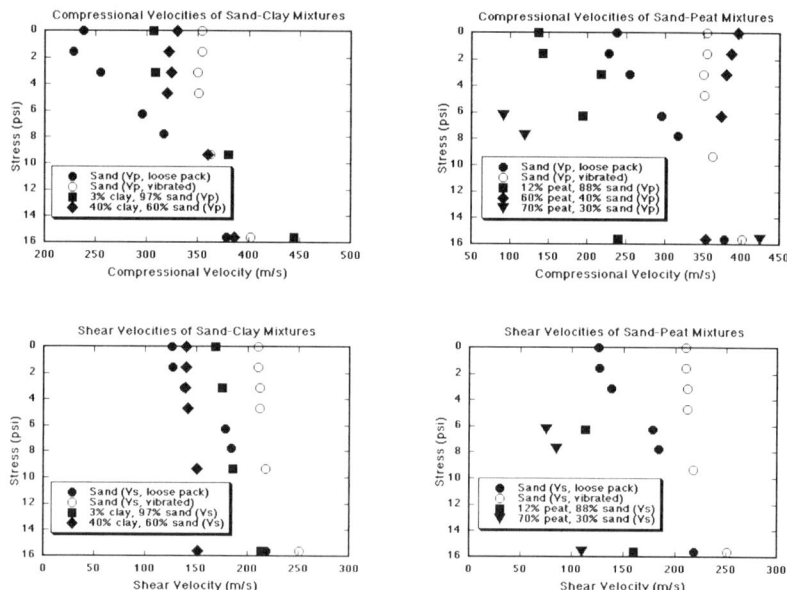

Figure 2. *Compressional and shear velocities for sediments as a function of uniaxial load*

DISCUSSION AND CONCLUSION

Although the behaviors shown in figure 2 are complicated, several dominant trends are evident. Packing technique dominates velocity gradient effects. The stress sensitivity of the sand velocities is altered by the addition of small amounts of clay for loosely packed samples. For these samples, compressional velocity of sand-clay at all concentrations is essentially constant until the stress exceeds 8 psi, at which point P velocity increases at a rate similar to the pure sand. It appears that the clay binds the sand grains until a critical stress is exceeded, and then yields and flows away from regions of local high stress at grain contacts. This behavior appears to persist to the highest clay fraction tested, 40% clay. Compressional velocity gradients of the peat mixtures are variable, and will be investigated further in additional experiments. The anomalous behavior of the 60% peat sample coincides with the elimination of air-filled porosity at that concentration. Discussion of the effects of microstructural parameters and modeling these results with effective medium theories is presented elsewhere [6].

The velocities observed for the synthetic soils tested in this study are low, with a compressional velocity comparable to the sound speed in air for the slowest sand-peat mixtures. The compressional velocities are lower than typical field values [1] and are slightly higher than values for near-surface sand in situ [7]. The admixed second phase can alter seismic attributes even for low mass-fractions. The photomicrograph of 90% sand, 10% clay spread on a glass slide shown in figure 3 suggests that the micromechanics of the small clay particles may explain this strong influence. The clay particles adhere electrostaticly to the quartz grains, with their long axes perpendicular to the surface, and tend to bridge the gaps between quartz grains. The large increase in compressional velocity when clay is first added to sand, accompanied by a decrease in shear attenuation, suggests that the clay alters the grain contacts by acting as an adhesive. The packing effects obscure the dependence of velocities on the presence of the second phase, for our sand-clay mixtures. We do not see monotonic increases in velocities as more clay is added.

In contrast, peat does not increase the velocity at low concentrations; instead, even in small amounts, peat causes both P and S velocities to decrease. Packing is less important for the sand-peat samples than for the sand-clay samples. As peat is added to sand, this soft second phase disrupts the structure of the sand framework, causing a decrease in velocity. As the mass fraction of the second phase continues to increase, porosity reduction dominates, generally producing the highest velocities. Finally, when the free porosity is eliminated, velocities begin to drop as the slow second phase becomes the framework. This behavior is similar to that reported for sand-kaolinite mixtures at high pressures [8].

The pressure dependence of the ultrasonic attributes indicates that wave transmission is extremely sensitive to the details of stress transmission through granular media. The nonlinear behavior of the elastic properties is an effective probe of conditions at grain contacts, which are fundamental to understanding the granular state.

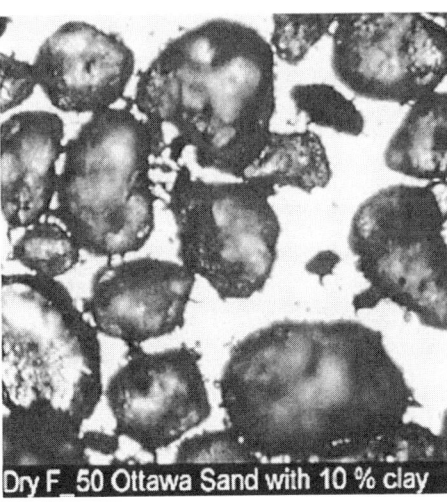

Figure 3. Photomicrograph of 90% Ottawa sand (mean grain diameter 270 microns) and 10% Wyoming bentonite

ACKNOWLEDGMENTS

D. Hart and C. Rowe made initial measurements and helped with experimental design. C. Boro designed and constructed the ultrasonic and loading assembly. This work was performed under the auspices of the U.S. Department of Energy by the University of California Lawrence Livermore National Laboratory under contract number W-7405-ENG-48 and supported specifically by the Environmental Management Science Program of the DOE Office of Environmental Management and the Office of Science.

REFERENCES

1. Bourbie, T., O. Coussy, and B. Zinszner, *Acoustics of Porous Media* (Gulf Publishing, 1987) 334 pp.
2. Bonner, B. P., C. Boro, and D. J. Hart, Anti-waveguide for ultrasonic testing of granular media under elevated stress, LLNL Patent disclosure IL-10607, and patent application, DOE Patent Docket No. S-94182 (1999).
3. Sears, F. M., and B. P. Bonner, Ultrasonic attenuation measurement by spectral ratios utilizing signal processing techniques, *IEEE Trans. On Geoscience and Remote Sensing*, **GE-19**, 95-99 (1981).
4. Aracne-Ruddle, C. M., B. P. Bonner, C. N. Trombino, E. D. Hardy, P. A. Berge, C. O. Boro, D. Wildenschild, C. D. Rowe, and D. J. Hart, Ultrasonic velocities in unconsolidated sand/clay mixtures at low pressures, LLNL report UCRL-JC-135621, Lawrence Livermore National Laboratory, Livermore, CA (1999).
5. Trombino, C. N., Elastic properties of sand-peat moss mixtures from ultrasonic measurements, LLNL report UCRL-ID-131770, Lawrence Livermore National Laboratory, Livermore, CA (1998).
6. Berge, P. A., J. G. Berryman, B. P. Bonner, J. J. Roberts, and D. Wildenschild, Comparing geophysical measurements to theoretical estimates for soil mixtures at low pressures, LLNL report UCRL-JC-132893, *Proceedings of the Symposium on the Application of Geophysics to Engineering and Environmental Problems*, ed. M. H. Powers, L. Cramer, and R. S. Bell, March 14-18, 1999, Oakland, CA, Environmental and Engineering Geophysical Society, Wheat Ridge, CO (1999) pp. 465-472.
7. Bachrach, R., J. Dvorkin, and A. Nur, High-resolution shallow-seismic experiments in sand, Part II: Velocities in shallow unconsolidated sand, *Geophysics*, **63**, 1233-1240 (1998).
8. Marion, D., A. Nur, H. Yin, and D. Han, Compressional velocity and porosity in sand-clay mixtures, *Geophysics*, **57**, 554-563 (1992).

Mat. Res. Soc.Symp. Proc. Vol. 627 © 2000 Materials Research Society

Rheology of a granular column

Evelyne Kolb, Guillaume Ovarlez, Pascal Sausse, Eric Clément
Laboratoire des Milieux Désordonnés et Hétérogènes
Université Pierre et Marie Curie, Boîte 86
4, Place Jussieu, F-75252 Paris

ABSTRACT

We study the rheology of a granular material in a confined geometry. The grains are stacked in a vertical cylinder and pushed at different driving velocities. The resistance force encountered by the bottom piston is monitored while the piston is pushing the granular column upwards. Above a critical velocity, the motion is characterized by a steady sliding and by a force level increasing rather slowly with the pushing velocity. For driving velocities under this threshold, the system undergoes a dynamic instability and then, a stick-slip motion occurs. The amplitude of the slipping events, and thus, the elastic energy release, increase strongly when the velocity decreases. The critical velocity depends on the stiffness of the driving system and on the height of the granular column. This transition can be shifted towards higher velocity values by increasing the friction at the walls of the cylinder. It is also very sensitive to the state of compaction of the grains. Moreover, the mean energy release during a stick-slip motion seems to increase as a power-law when the pushing velocity is decreased. We also show that the distribution of energy release is strongly dependent on the level of disorder in the grains (polydispersity, friction, etc.). We argue that this complex phenomenology characterizes confined granular packing in connection with arching and aging phenomena.

INTRODUCTION

Granular materials have many unusual properties due to force chains and spatially inhomogeneous stress distributions [1]. The localization of stress paths is expected to lead to complex rheological behavior and to jamming effects [2]. The response of a confined granular material to an external applied force have been investigated by different groups both in compression and/or shear experiments [3][4][5][6], showing that a stick-slip motion can occur under certain shear rates. In the same way systematic studies of the force network by use of birefringent granular disks have revealed the dramatic effect on the mechanical strength of a slight change of compacity of the 2 dimensional granular stacking [7].

In order to investigate systematically the velocity dependence of the shear force which gives insight into the different mechanisms of energy dissipation in granular matter confined by walls, we choose the simple geometry of the silo. The grains we use are dry, non cohesive and monodisperse beads of millimeter size, so that the predominant forces acting on beads are due to collisions or friction between grains or between grains and walls. The friction at boundaries and the presence of heterogeneous stress paths lead to "arching effect" and give rise to enhanced forces needed to push the grains and to a complex rheological behavior in connection with the dynamics of the vaults.

EXPERIMENT

Grains are piled into a vertical cylinder closed at the bottom by a movable piston. The geometry is the same as the one used for measuring the pressure at the bottom of a silo [8] and testing the validity of the Janssen's model. It is the generalization in 3 dimensions of our former experimental system where the beads were confined in a 2 dimensional rectangular cell [9]. A force probe of stiffness k = 40000 N/m is located under the piston and is led at a constant driving velocity between 5 nm/s and 100 µm/s by the translation stage of a stepping motor. The resistance force encountered by the piston while pushing the granular column upwards is monitored as a function of time by use of the force probe .

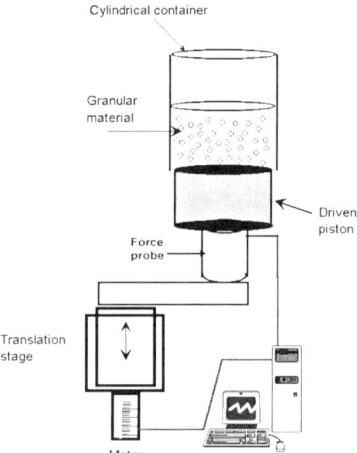

Figure 1 *Experimental display of the granular column pushed at the bottom by a piston mounted on an elastic displacement gauge (the force probe).*

Different couples of beads and walls have been investigated : frictional glass beads or steel beads for roll bearing device with a very low internal coefficient of friction combined with either walls of abraded Plexiglas or smooth duralumin. The diameter of the beads is 1.5 mm compared with 36 mm for the cylinder diameter.

RESULTS

Due to the redirection of stresses to the lateral walls and the effect of solid friction at the wall, the vertical force exerted by the piston is screened. Therefore the force needed to push and displace the granular column can be much larger than the simple weight of grains. In order to check this effect we have measured the mean resistance force as a function of the height of grains in the cylinder for a given velocity corresponding to a steady and continuous sliding of the beads. As expected the force does not grow linearly with the height of the column but exponentially. This behavior can be simply explained and fitted by a inverted Janssen law where the friction forces at the lateral walls are supposed to be fully mobilized in the downward direction.

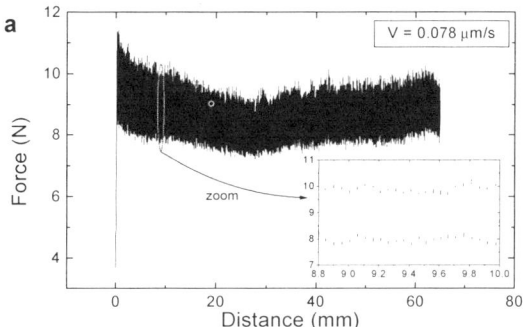

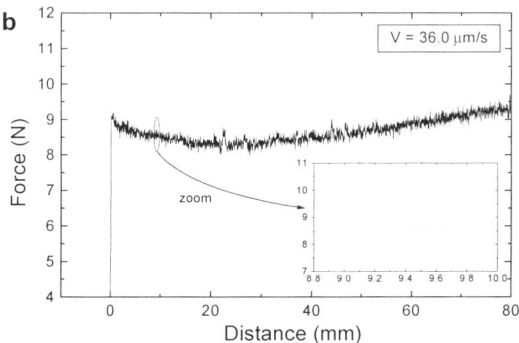

Figure 2 *Resistance force versus the displacement of the stepping motor for steel beads in an abraded plexiglass cylinder. The mass is m = 400 g. Fig. 2a: for a driving velocity*

V=0.078 μm/s. The inset clearly exhibits the stick-slip behavior. Fig. 2b: for a driving velocity of V=36.0 μm/s. The inset shows the continuous sliding regime

For a given height of beads we have extensively studied how the force signal depends on the driving velocity. Above a critical velocity, the motion is characterized by a steady sliding (fig.2b) and by a force level increasing rather slowly with the velocity. For driving velocities under this threshold, the system undergoes a dynamic instability and then a stick-slip motion occurs (fig2a).

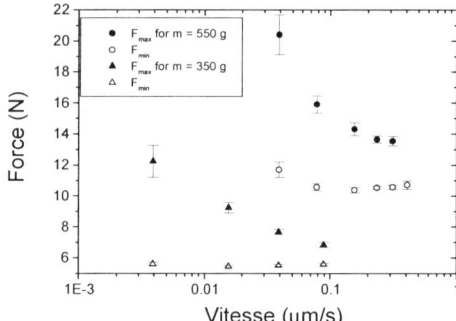

Figure 3 *Forces as a function of velocity in the stick-slip regime. Each full (respectively open) symbol represents the mean maximum F_{max} (resp. minimum F_{min}) force obtained just before (resp. after) a slipping event for the corresponding velocity. The difference between F_{max} and F_{min} at the same velocity is the mean force release during the slip. The circles correspond to a mass of m = 550 g of steel beads in an abraded plexiglass cylinder and the up triangles correspond to a mass of m = 350 g for the same beads and walls. The critical velocity above which the continuous sliding occurs is $V_c = 0.12$ μm/s for m = 350 g and $V_c = 0.35$ μm/s for m = 550 g.*

The amplitude of the slipping events (fig.3), and thus, the elastic energy release increases strongly as the velocity reaches values as small as 510^{-3} μm/s (i.e. of the order of the typical velocities of tectonic plates (cm/year). This energy release during a stick-slip motion seems to increase as a power-law as the pushing velocity is decreased (fig.4). The onset of instability below a given driving velocity is reminiscent of solid-on-solid friction experiments and this regime was already observed in experiments done with beads confined in a 2 dimensional cell [9]. By taking into account an aging effect of the contacts at the side walls and the redirection of force to these walls, we derive a relation between the velocity and the amplitude of the slipping events and explain the rather large increase of energy release with decreasing velocity in comparison to solid-solid friction.

The critical velocity depends on the stiffness of the driving system and on the height of the granular column. This transition can be shifted towards higher velocity values by increasing the friction at the walls of the cylinder, i.e. by using a abraded Plexiglas cylinder instead of a duralumin one. The critical velocity is also very sensitive to the state of compaction of the grains : in the case of glass beads where dilatancy effects are much more important than for steel beads, the transition can occur during one experiment at a given velocity simply because the compacity of the stacking evolves during the pushing .

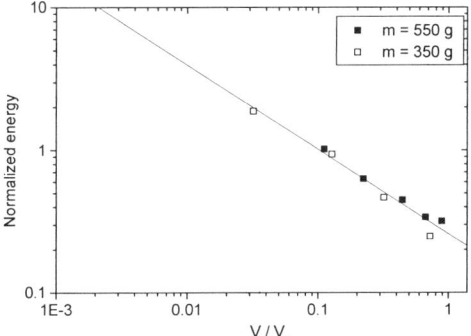

Figure 4 *Mean elastic energy release as a function of velocity in the stick-slip regime. The energy is normalized by the mean minimum elastic energy of the probe compression, while the velocity is normalized by the critical velocity for the corresponding mass of beads. The line is the power-law fit, that gives a slope of -0.6.*

CONCLUSIONS

We study the dynamical behavior of granular assemblies confined in a cylinder and pushed vertically from the bottom. The resistance to the pushing motion shows different regimes which are manifestations of the internal force transmission properties inside the column as well as the friction interactions with the wall. At the highest driving velocities we observe a continuous sliding regime with a resistance value increasing slowly with the driving velocity and exponentially with the column height. At low velocity, depending on the value of the force probe elasticity, we evidence a transition to a stick-slip instability which is very well characterized for the most regular metallic spheres. The existence of a dynamic transition from a stick-slip regime at low velocity to a continuous sliding one for velocity above a critical threshold is generic of all the systems beads/walls investigated. The energy release in the stick-slip regime increases with decreasing velocity. This effect can be related to contact aging at the walls simply by applying laws of solid-solid friction. However the amplitude of this effect is much more amplified in a confined granular material because of the redirection of the applied force to the lateral walls. Details on the dynamical behavior and the aging model will be presented elsewhere[10].

ACKNOWLEDGMENTS

We acknowledge the financial support of the CNRS-PICS grant n°563.

REFERENCES

[1] D. Mueth, H. Jaeger, S. Nagel, Phys. Rev. E 57, 3164 (1998)

[2] M. Cates, J. Wittmer, J.P. Bouchaud, P. Claudin, Phys. Rev. Lett. 81, 1841 (1998)

[3] S. Nasuno, A. Kudrolli, J.P. Gollub, Phys. Rev. Lett. 79, 949 (1997)

[4] V.K.Horwarth, I.M.Janosi. and P.J.Vella Phys. Rev.E **54**, 2005 (1996).

[5] R. Albert, M. Pfeifer, A. Barabasi, P. Schiffer, Phys. Rev. Lett. 82, 205 (1999)

[6] M. Lubert, A. de Ryck, *preprint*

[7] D. Howell, R.P. Behringer, C. Veje, Phys. Rev. Lett. 82, 5241 (1999)

[8] L.Vanel, E. Clément, Eur.Phys. J B 11, 525-533 (1999)

[9] E. Kolb, T. Mazozi, E. Clément, J. Duran, Eur.Phys. J B 8, 483-491 (1999).

[10] G.Ovarlez, E. Kolb, E. Clément, *in preparation.*

Mat. Res. Soc. Symp. Proc. Vol. 627 © 2000 Materials Research Society

Unsteady Heat Conduction in Granular Materials

Watson L. Vargas and Joseph J. McCarthy
Department of Chemical and Petroleum Engineering
University of Pittsburgh
Pittsburgh, PA 15261, U.S.A.

ABSTRACT

Heat transfer in granular materials impacts a variety of industrial applications, such as calcination, drying kilns, packed bed and multiphase reactors, etc. and may yield insight into the thermal response of some porous materials (in combustion synthesis or sintering, for example). In a dense bed of granular material, conduction occurs almost exclusively through the particle-particle contacts over a wide range of conditions. We have developed a novel Thermal Particle Dynamics (TPD) Simulation technique which incorporates both contact mechanics and contact conductance theories in order to model the dynamics of flow and heat conduction through granular materials. This model is uniquely suited to studying the effects of microstructure and flow on the dynamics of heat conduction in particulate materials. In this paper, we present experimental as well as numerical results of transient heat conduction through a bed of cylinders.

INTRODUCTION

A solid fundamental understanding of transport in particulate systems is not yet complete. However, the clear potential to benefit a wide variety of industries -pharmaceuticals, metallurgy, ceramics, polymers, and agriculture, to name a few - has recently spurred researchers in the physics and engineering communities to redouble their efforts (the recent work in this area has been the subject of several review papers [1-3]). The vast majority of this recent work has been focused on the flow of cohesionless particulates, however, and the transport of heat in granular systems has remained largely unstudied (notable exceptions include Patton, Sabersky and Brennen [4], Sullivan and Sabersky [5], and recent work by Hunt [6], Wassgren [7], and Massoudi and co-workers [8,9]).

To a significant extent the study of heat transfer in granular materials has been hampered by the lack of a universally accepted set of governing equations for granular flow. This difficulty has typically spurred previous studies to rely on effective medium approximations or simplifying assumptions (e.g., the particle bed is well-mixed). In contrast, soft-particle Particle Dynamics Simulations (henceforth simply termed Particle Dynamics) [10,11] - which calculate the flow of a granular material from contact mechanics considerations - inherently treat particles as discrete entities. This technique has been successfully applied to a wide variety of granular flows [11-14] and comprises the foundation of the new computational technique used in this study.

THEORY

Heat transfer in granular materials affects a variety of industries. Some specific applications where this type of phenomena is important include packed bed reactors, multi-phase reactors, calcining/drying kilns, storage of bulk reactive materials, sintering of powdered metals, and firing of green ceramics. Due to the diversity of the unit operations that are in some way affected by heat transfer via particles, the problem of heat transfer through individual particle contacts (contact conductance) is well established. The key idea in this paper is that contact conductance theory may be used as the building blocks of a new kind of discrete particle simulation (Thermal Particle Dynamics or TPD).

Contact Conductance

Contact conductance refers to the ability of two touching materials to transmit heat across their mutual interface. While much of the early work in this area was devoted to the study of conductivity in micro-electronic devices [15], there has been considerable research directed towards applications as varied as composite materials [16], cryogenic super-insulators [17], and nuclear reactors [18].

The most basic problem in contact conductance is that of heat transport between two smooth particles in elastic contact - where it is assumed that the radius of curvature of the spheres is much larger than that of the contact spot. This problem is well understood, and approximate analytical solutions have been proposed independently by Yovanovich [15], Holm [19], and Batchelor and O'Brien [16]. A slightly more accurate, completely rigorous, numerical solution is attributed to Chan and Tien [17]. All of these models predict that the conductance, $H = hA$ (where h is a heat transfer coefficient and A is the contact area), through a smooth, elastic contact varies with imposed normal force as $H \propto F_n^{1/3}$.

Thermal Particle Dynamics (TPD)

Particle Dynamics, a discrete method of simulation, captures the macroscopic behavior of a particulate system via calculation of the trajectories of each of the individual particles within the mass; the time evolution of these trajectories then determines the global flow of the granular material. The particle trajectories are obtained via explicit solution of Newton's equations of motion for every particle [10]. The forces on the particles - aside from gravity - typically are determined from contact mechanics considerations [20]. In their simplest form, these relations include normal, Hertzian, repulsion and some approximation of tangential friction (typically due to Mindlin).

The key feature of a Particle Dynamics Simulation is that many simultaneous two-body interactions may be used to model a many-body system [10]. This idea works because the time-step is chosen to be sufficiently small such that, during one time-step, a disturbance (in this case a displacement-induced stress on a particle) does not propagate further than the particle's immediate neighbors (i.e., disturbances are quasi-steady on the time-scale of the time-step). Generally, this criterion is met by choosing the time-step to be smaller than R/λ, where R is the particle radius and λ represents the relevant disturbance wave speed (for example, dilational, distortional or Rayleigh waves [21]).

In much the same way that contact mechanics for a two-body interaction is well understood [20] - allowing PD simulations to accurate reflect particle mechanical properties - contact conductance models are also well established (see above). It is appealing, therefore, to make a direct analogy with a Particle Dynamics Simulation's use of contact mechanics in the context of heat transfer.

As a condition of using this method a second restriction is imposed in addition to the analogous quasi-steady assumption of PD (i.e., that the particle's *temperature* is quasi-steady on the time-scale of the time-step). In order to assure that a particle's temperature does not vary significantly from one contact point to another, we must assume that the resistance to heat transfer *inside* the particle is significantly smaller than the resistance *between* particles – $Bi = HR/k \ll 1$. It is important to note that this relation is satisfied as long as the contact radius is small relative to the particle radius – a condition required by most contact mechanics relations. In order to satisfy the quasi-steady criterion, the time-step is chosen to be small enough that the change in a particle's temperature during a time-step is small relative to the difference in temperature between it and its neighboring particles, i.e.

$$\frac{dT_i}{(T_j - T_i)} = \frac{Hdt}{\rho_i c_i V_i} \ll 1, \tag{1}$$

where dT_i is the change in the temperature of particle i during the time-step, T_i (T_j) is the temperature of particle i (j), dt is the time step, and $\rho_i c_i V_i$ is the "thermal capacity" of particle i. In this paper we use the simplest model for contact conductance, which assumes that the particles are smooth, elastic, and in perfect thermal contact. In this case, an approximate expression for the contact conductance (where resistance is due to heat flow constriction only) is given by

$$H = hA = 2ka, \tag{2}$$

where k is the thermal conductivity of the particle and a is the radius of the contact spot (obtained from the applicable contacts mechanics theory – in this case for Hertzian elastic particles). Plugging this relation into our validity criterion (Equation 1) gives

$$\frac{2kadt}{\rho_i c_i V_i} \ll 1, \tag{3}$$

which can be satisfied by choosing a sufficiently small time-step, dt.

If this condition is met, then by obtaining H from contact conductance theories, one may implement a Thermal Particle Dynamics (TPD) Simulation by numerically solving for each particle's temperature at each time-step using

$$\frac{dT_i}{dt} = \frac{Q_i}{\rho_i c_i V_i}, \tag{4}$$

where

$$Q_i \approx \sum_j Q_{ij} = \sum_j H_{ij}(T_j - T_i). \tag{5}$$

EXPERIMENTAL DETAILS

As a test of this model we examine an initially isothermal, psuedo-two-dimensional, rectangular bed of particles under an uni-axial load (see Figure 1). The bed is aligned horizontally in order to minimize natural convection ($Ra < 10^{-2}$). Three side walls are insulated (as

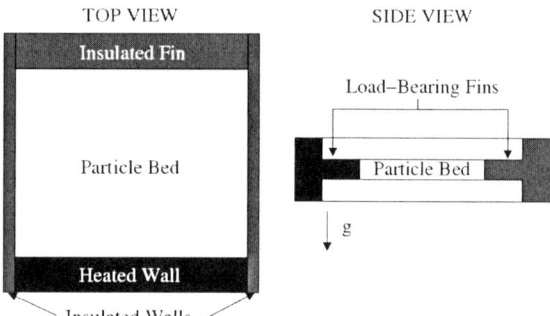

Figure 1. *Experimental schematic. A monolayer of cylindrical particles is positioned between two plates and a uni-axial load is imposed. Three walls are insulated and one is heated.*

well as the top and bottom plates - in the experiment), while the fourth may be heated. The bed begins at an initial temperature of roughly 20C and then one wall is heated to 50C. The width of the device - both computationally and experimentally - is roughly 200 particle diameters and the height is 100 particle diameters. The particles in the experiment are 304 stainless steel, roughly 6 mm long and 3 mm in diameter. The apparatus is attached to a vacuum pump so that experiments may be carried out at low interstitial gas pressure (roughly 28kPa). Temperature profiles within the bed are obtained using liquid-crystal thermography [22]. The parameters in the simulation are taken directly from the literature and consist solely of the particle material properties (i.e., *there are no freely adjustable parameters within the model*).

RESULTS

For a homogeneous system - or one for which a suitable effective medium assumption (EMA) may be made - this experiment/simulation represents an essentially one-dimensional heat transfer problem. In a (roughly) hexagonally packed system (like the one used in our experiments), the local void fraction is approximately constant so that effective properties may naively be expected to be essentially independent of space. In this case, the temperature front should propagate uniformly through the bed. Figure 2 shows a comparison of snapshots (essentially contour plots, color-coded by temperature) for a TPD simulation and an experiment after 30 minutes of heating. One can see for both the TPD simulation and the experiment that the temperature front is, in fact, *non-uniform* across the width of the container. This suggests that a qualitative examination of the packing or simple measurements of the void fraction are insufficient to describe the conduction through the bed. Instead, it becomes necessary to examine the microstructure of the bed in detail.

In granular materials stress "chains" [23] - localized structures which carry a disproportionate fraction of the total imposed load - have been shown to occur in all but the most perfect of crystalline particle packings and often span many particle. Because of this, and the dependence of contact conductance on imposed load, stress inhomogeneities can play an important role in granular heat conduction. In this case, even a seemingly uniform bed displays a

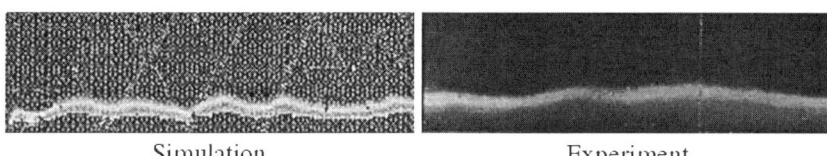

Simulation	Experiment

Figure 2. *Thermal maps of a two-dimensional particle bed. The temperature front in a transiently heated granular bed does not propagate uniformly, as may be expected using an effective medium approximation. This figure shows snapshots of a TPD simulation and an experiment (using liquid crystal thermography). Grey-level is a rough indication of temperature where black refers to both hot and cold.*

non-uniform temperature front. In fact, the temperature front can be seen to be jagged, with the vertical position of the front at a particular point along the container's width oscillating as stress chains converge and diverge along the bed height.

CONCLUSIONS

At present, this simulation technique (TPD) includes only the simplest contact conductance model (for smooth, elastic contact [16]) although incorporating more complex models represents a relatively modest extension. Also, the model currently assumes the impact of both the interstitial fluid and thermal radiation are negligible (i.e., the particles are in a vacuum and are below some temperature threshold, T_r, above which radiation plays a significant role). According to Batchelor [16], the first assumption is valid as long as

$$\frac{ka}{k_f R} \gg 1, \tag{6}$$

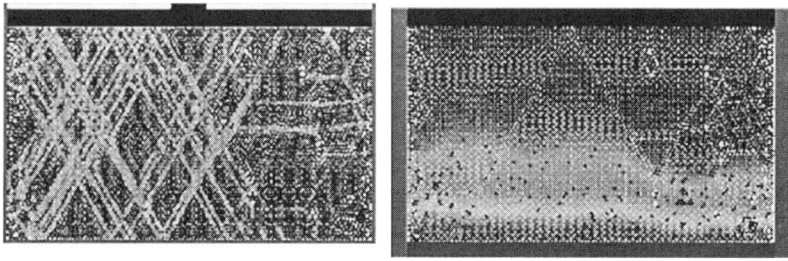

Figure 3. *The effect of stress chains. On the left is the stress distribution for a simulation coded by grey-level (light corresponds to high stress); on the right is a snapshot of the temperature distribution coded by grey-level (dark is cold). Although the void fraction and even contact distribution are qualitatively similar for the right and left hand sides of this bed the temperature front propogates much faster in the region of dense stress chains (left).*

where k_f denotes the fluid (interstitial medium) conductivity. This is identically true when $k_f \to 0$

(in a vacuum), and approximately true for high conductivity ratios (k/k_f). Even in its simplest form, however, TPD is capable of capturing details of particle-level temperature profiles which have not been previously reported, without requiring adjustable parameters. We find that stress chains, in even moderately imperfect lattices, may cause dramatic changes in the way that heat is transported by conduction. These results may have implications in materials processing of powders where heating or cooling rates affect the resultant microstructure. One advantage of this technique is that it is extensible in a variety of ways: incorporation of numerous contact mechanics/conductance models to account for particle roughness, plastic deformation, etc.; inclusion of interstitial fluid effects and gas-solid convection. Doing so will allow similarly detailed study of more complicated fluid/particle systems.

ACKNOWLEDGMENTS

The authors would like to acknowledge the support of the Central Research Development Fund of the University of Pittsburgh for partial support of this work.

REFERENCES

1.   J. Bridgwater, *Powder Technol* **15**, 215—231 (1976).
2.   H. M. Jaeger, S. R. Nagel, and R. P. Behringer, *Rev. Mod. Phys*, **68**, 1259—1273 (1996).
3.   P. G. de Gennes, *Rev. Mod. Phys.*, **71**, S374--S382 (1999).
4.   J. S. Patton, R. H. Sabersky, and C. E. Brennen, *Int. J. Heat Mass Transfer*, **29**, 1263—1269 (1986).
5.   W. N. Sullivan and R. H. Sabersky, *Int. J. Heat Mass Transfer*, **18**, 97—107 (1975).
6.   M. L. Hunt, *Int. J. Heat Mass Transfer*, **40**, 3059—3068 (1997).
7.   C. R. Wassgren, D. E. Beasley, and R. N. DeWachter, *1998 International Mechanical Engineering Congress and Exposition (IMECE) Conference Proceedings*, (1998).
8.   M. Massoudi and T. X. Phuoc, *Int. J. Non-Lin. Mech.*, **34**, 347—359 (1999).
9.   R. Gudhe, K. R. Rajagopal, and M. Massoudi, *Acta Mech.*, **103**, 63—78 (1994).
10. P. A. Cundall and O. D. L. Strack, *Geotechnique*, **29**, 47—65 (1979).
11. O. Walton, *Int. J. Engng. Sci.*, **22**, 1097--1107, (1984).
12. J. J. McCarthy and J. M. Ottino, *Powder Technol*, **97**, 91—99 (1998).
13. J. J. McCarthy, D. V. Khakhar, and J. M. Ottino, *Powder Technol*, **109**, 72—82 (2000).
14. C. R. Wassgren, C. E. Brennen, and M. L. Hunt, *J. Appl. Mech.*, **63**, 712—719 (1996).
15. M. M. Yovanovich, *J. Spacecraft Rockets*, **4**, 119—125 (1967).
16. G. K. Batchelor and R. W. O'Brien, *Phys. Rev. Lett.*, **355**, 313-333 (1977).
17. C. K. Chan and C. L. Tien, , *J. of Heat Transfer*, **42**, 302—308 1973).
18. Y. M. Lee, A. Haji-Sheikh, L. S. Fletcher and G. P. Peterson, *J. of Heat Transfer*, **116**, 17—27 (1994).
19. R. Holm, *Electrical Contacts: Theory and Application*, (Springer-Verlag, 1967).
20. K. L. Johnson, *Contact Mechanics*, (Cambridge University Press, 1987).
21. C. Thornton and C. W. Randall, in *Micromechanics of Granular Materials*, ed. M. Satake and J. T. Jenkins, 75—89(Elsevier Science Publishers, 1988).
22. D. Dabiri and M. Gharib, *Exper. Fluids*, **11**, 77—86 (1991).
23. C. Thornton and D. J. Barnes, *Acta Mech.*, **64**, 45—61 (1986).

Mat. Res. Soc. Symp. Proc. Vol 627 © 2000 Materials Research Society

Impulse acoustics based ejection of ferrofluid grains from a ferrofluid: the blueprint of a concept for a nozzle-free inkjet printer

Felicia S. Manciu, Marian Manciu and Surajit Sen
Department of Physics, State University of New York at Buffalo, Buffalo, NY 14260-1500, USA

Abstract

We present numerical simulations to demonstrate that it may be possible to eject ferrofluid grains from a ferrofluid using non-linear acoustic impulses. The study considers a container with some dilute ferrofluid that is placed in a strong, vertical, homogeneous magnetic field. The field induces the formation of magnetic dipoles into vertical chains that approximately span the region between the base and the surface of the container. We use particle dynamical simulations to show that an impulse generated at the base of any chain, will typically travel as a weakly dispersive bundle of energy. When the impulse magnitudes are appropriate (typically ~60 m/s or more) the ferrofluid grain nearest to the surface of the liquid may be ejected by the impulse. Since all ferrofluid grains possess a coating of the liquid host, the ejected grain can be used as an ink-drop, with typical diameter of 15 or so nanometers. The velocities of the ejecting grains can be controlled and hence the method, if experimentally feasible, may have wide ranging applications. One of these applications is likely to be in designing special-purpose nozzle-free inkjet printers of unprecedented resolution.

I. Introduction

Ferrofluids [1] are stable colloidal systems that are composed of solid, magnetic, predominantly single-domain grains in a non-magnetic solvent such as water or oil. A large variety of available ferrofluids have grain diameters that are less than ~20 nanometers. In the present study, we shall assume, without any loss of generality in the underlying physics, that the grains are monodisperse and spherical [2].

One such ferrofluid, γ-Fe_2O_3 , possesses grains that are approximately 8.5 nanometers in diameter, predominantly single-domain [3] and contain very large magnetic moments of some 2×10^4 μ_B. When such a ferrofluid is subjected to a homogeneous magnetic field of modest strength (say ~ 200 Gauss), the grains tend to align in vertical chains (Fig. 1). Since many of the host liquids can be colored, an ejected ferrofluid grain can be thought of as a "nanodrop" of ink.

In this study, we consider the following question: can one generate an impulse at the

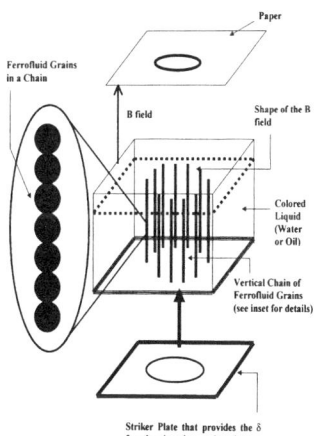

Figure 1: Schematic diagram of a ferrofluid based nanodrop production machine.

bottom of a chain of ferrofluid grains in a ferrofluid and arrange for the grain located nearest to the liquid-air interface to eject out of the interface? Since the grain carries a thin layer of the liquid, and since the liquid can presumably be colored, if the proposed idea works, then one can expect the impulse to generate a nanodrop of ink [4]. We shall discuss issues related to possible industrial applications later in this work. In what follows, we show that if the chains are rigid enough (which depends upon approximate monodispersity and the strength of the external magnetic field), appropriately generated impulses of reasonable magnitudes are expected to eject nanodrops of ink from a ferrofluid. We outline the consequences of single versus multiple pulsing to efficiently eject ferrofluid grains [5] for plausible experimental realizations. We also discuss the consequences associated with varying grain sizes.

II. The Model System

The interactions between magnetic dipoles in the ferrofluid include the following forces: magnetic dipolar forces, gravitational forces, Hertzian forces and surface forces [4-5]. We discuss these forces below.

The dipolar interaction forces between two parallel magnetic grains at distance r from one another is given by

$$F_m = 3\mu_0\mu^2/2\pi r^4, \qquad (1)$$

where $\mu_0 = 1.26 \times 10^{-6}$ H/m, $\mu = 2 \times 10^4 \mu_B$, $\mu_B = 9.27 \times 10^{-24}$ J/T. We assume the mass m of our monodisperse ferrofluid grains to be 2.72×10^{-21} kg and the radius to be 10^{-8} m. These magnitudes imply that the magnetic dipolar force at the chain edges is $\sim 10^{-12}$ N and inside the chain is $\sim 10^{-13}$ N.

The masses are rather small. Hence the gravitational force on each grain,

$$F_g = mg = 2.67 \times 10^{-20} \text{ N}, \qquad (2)$$

and can hence be neglected.

When two spherical grains of radius r are pressed against one another, they repel according to Hertz law [6], which yields a force

$$F_h = (5/2)a\delta^{3/2}, \qquad (3)$$

where

$$a = (2/5D)(r/2)^{0.5}, \qquad (4)$$
$$D = (3/2)(1-\sigma^2)/Y. \qquad (5)$$

We use $Y = 1.0 \times 10^{-11}$ $Nm^{-3/2}$, $\sigma = 0.3$, $r = 1.0 \times 10^{-8}$ m to find that for $\delta \sim 1$ Angstrom, F_h is $\sim 10^{-8}$ N [4-5].

The surface force associated with the surface tension of the liquid, e.g., water, is

$$F_s = 2\pi r\gamma, \qquad (6)$$

where $\gamma = 7.3 \times 10^{-2}$ N/m is the coefficient of surface tension of water. For the ferrofluid particle sizes of interest, the surface force is $\sim 10^{-9}$ N.

III. The Results

III.A. Solitary Waves

The particle dynamical simulations reported below have been carried out using the Gear algorithm [7] to solve the particle-dynamical equations,

$$m \, d^2r_i/dt^2 = k[(d - \Delta_0 - r_i + r_{i-1})^{3/2} - (d - \Delta_0 - r_{i+1} + r_i)^{3/2}]$$
$$+ 6\mu_0^2[\, 1/(r_{i+1} - r_i)^4 - 1/(r_i - r_{i-1})^4\,], \qquad (7)$$

where k = 5a/2 (see Eq. (4)) and the quantity Δ_0 gives the distance of closest approach between the grains in the absence of the pulse and is a parameter that describes the "loading" of the chain. In our studies, we generate an impulse by imparting an initial velocity v_0 to the grain at the bottom of the chain. All other grains are kept at rest at t = 0. The equilibrium positions of the grains are obtained by balancing the dipolar and Hertzian forces and these forces yield a column of grains with constant loading. An impulse generated at the edge produces a compression that is more significant than the loading. As shown by Nesterenko [8-10], Coste et al., Sen and coworkers and others, such an impulse leads to the generation of a *solitary wave* that propagates through the chain [8-12]. When the ferrofluid grains are spherical and monodisperse, these solitary waves propagate at velocity, c, where c depends upon the amplitude of the displacement associated with the impulse as $A^{1/4}$.

The solitary wave is about 5 grain diameters wide with most of its energy concentrated at the central grain [8-12]. Eventually, as the solitary wave reaches the surface, if the central grain of the solitary wave possesses enough kinetic energy to overcome the work done $\gamma \pi r^2$ to vacate the liquid surface by an area πr^2, where r is the grain radius and $\gamma \approx 7 \times 10^{-2}$ N/m. is the surface tension of water, the central grain may be able to escape the liquid [4-5]. Once ejected, one can presumably construct the appropriate magnetic field environment to direct the grain toward a surface of interest.

III.B. The Role of Restitution

It may be noted that granular contacts typically involve energy loss due to restitution. We define, w = $F_{unloading}/F_{loading}$ < 1, where $F_{unloading}$ and $F_{loading}$ refer to force between two grains in contact during the processes of loading and unloading associated with passage of an impulse through the contact. Thus, w < 1, if restitution is present. We find that the net effect of restitution is to attenuate the amplitude of the traveling impulse rather than to broaden the width of the impulse and make it dispersive. This is because the dispersion is controlled by the index of the Hertz force law and by the

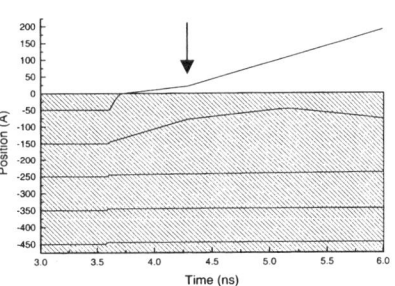

Figure 2: Position of 5 topmost grains with surface grain escaping, as functions of time, v_0=67.0 m/s. Arrow marks the time of escape.

presence of loading (Δ_0), if any. Thus, when $\Delta_0 \rightarrow 0$, in spite of attenuation, the signal suffers no dispersion. To overcome the predicted restitution in a system and eject the surface grain from the ferrofluid, one may generate the impulse with appropriate magnitude. The key limitation of this approach may be that some grains, which may not be strong enough, may crack and disintegrate, thereby affecting the solitary wave dynamics as it would propagate through the chain. One possible way to resolve such a difficulty may be to consider monodisperse, spherical grains of larger diameter. As we shall see later, one can generate solitary waves with smaller impact velocities when the grain sizes are larger. These questions will be best addressed via systematic experimental studies with the appropriate ferrofluid in a confined cylinder with a c-axis magnetic field used to align the chains [13].

III. C. Single and Multiple Grain Ejection

We solve Eq. (7) numerically using the Gear algorithm [7]. The initial conditions we use in our calculations are as follows: $dr_1/dt |_{t=0} = v_0$ m/s, $dr_i/dt |_{t=0} = 0$ for $i>1$ where $i = 1$ defines the grain that is nearest to the striker plate in Fig. 1, i.e., the bottom grain. As mentioned before, the equilibrium positions of the grains are obtained by balancing the dipolar and Hertzian forces and these forces yield a column of grains with a small but constant loading. A perturbation to an edge, which produces compression much bigger than initial loading (say ~ 1 Angstrom), results in an approximately solitary wave that propagates from the bottom of the chain to the surface grain at the water-air interface

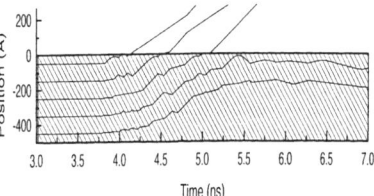

Figure 3: Position of 5 topmost grains as functions of time when 3 grains are ejected. We use v_0=50.0 m/s with 10 pulses generated at a frequency of 10 GHz at intervals of 0.1 ns.

For ferrofluid grains of diameter 100 Angstroms, we find that $v_0 \approx 67.0$ m/s or higher for the ejection of the ferrofluid grain that is nearest to the water-air interface [4]. The results showing the positions of various grains near the surface of the liquid-air interface is shown in Fig. 2. It is important to note that in the present approach, an impulse that is some 50 times larger is needed to successfully eject the second particle [4]. Thus, it is energetically highly inefficient to use single impulses to simultaneous eject multiple grains from the ferrofluid. For the purposes of designing an inkjet printer using ferrofluids, it may be desirable to eject a chosen number of grains from the liquid. As mentioned in Section 1, each grain contains a coating of the host liquid. If the water based host liquid is colored, then depositing several grains on some surface might help make a dot with sharper color and hence better visibility.

One possibility would be to send pulses after the column relaxes

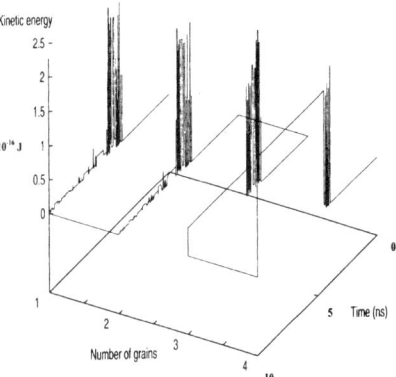

Figure 4: Complex time evolution of 4 grains nearest to the surface with grain no. 4 being closest to the interface for case shown in Figure 3.

following each topmost grain ejection. Since the relaxation time is of the order of 10^{-8}s [5], a successful pulsing should be in the MHz range. A numerical analysis of this process require a study, in which the dynamics of liquid itself must be taken into account along with that of impulse propagation in the chain of grains. This work is currently in progress.

Below we summarize a study [5] on the possibility of controllably ejecting multiple grains from the ferrofluid *before* the system relaxes (Figures 3 and 4). The physical idea underlying this approach is as follows. One must send impulses in the chain in such a way that a set of solitary waves can successively reach the end of the chain, which is near the liquid-air interface, to explore whether the combined effect of the clustered solitary waves can lead to the near simultaneous ejection of multiple grains from near the surface.

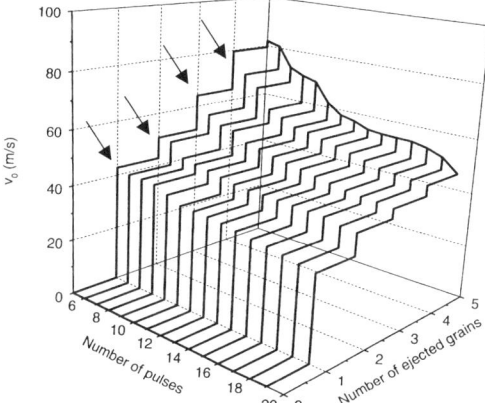

Figure 5: Minimum v_0 and number of pulses required to eject the first 4 grains.

In our calculations we consider three key parameters: (i) the initial velocity of the first grain due to the pulse, v_0, (ii) the number of pulses generated, and (iii) the frequency that characterizes how often the impulses are being initiated at the bottom grain of the chain of ferrofluid grains, v. Since the grain velocity is of the order of 10 m/s, the time for passing the liquid-air interface for a grain is about 10^{-9}s, which implies that in order to have multiple pulses interfering at surface, one should pulse the system with frequencies of the order 10 GHz.

Since the fine-tuning of the pulsing frequency or controlling the energy of each pulse might be quite challenging in practice, more important for application purposes is the dependence of number of ejected grains on the number of pulses. We repeat the study for v_0=37.5 m/s, 42.0 m/s and 50.0 m/s and 10 pulses at a frequency of 10 GHz (time window of 0.1 ns between pulses to find one, two and three grains are successively ejected, respectively (Figures 3 and 4). In our calculations, we find that the total energy is constant with accuracy better than 0.01%, thereby indicating that the study was reliable [5].

It is possible to construct a "phase diagram" with v_0, number of pulses needed for grain ejection and number of grains ejected along each axes. This is shown in Figure 5 for the case where the frequency v = 10 GHz.

Figure 4 presents data in which it is possible to see the microscopic dynamics of the grains near the surface when 10 pulses at 10 GHz are imparted to a chain. The results reveal that two of the grains nearest to the surface are about to leave the surface while the two farther away from the surface are unsuccessful in leaving the surface.

III.D. Scaling of v_0 with Grain Size

Let us consider spherical grains of some standard radius r_0 and let s be a scale factor to control grain radius $r = sr_0$. Then $m = s^3 m_0$, $F_m = s^2 F_{m0}$, $F_g = s^3 F_{g0}$ and $F_s = sF_{s0}$. The last scaling relation implies that $D_{esc} = sD_{esc0}$. In the absence of restitution, the intial kinetic energy imparted to a chain, $E_0 = mv^2/2$ must equal the minimum energy required for escape, $E_{esc} = \pi r^2 \gamma = s^3 m_0 v^2/2 = \pi s^2 r_0^2 \gamma = s^2 E_{esc0} = s^2 m_0 v_0^2$. Thus, $v^2 = v_0^2/s$ or $v = s^{-1/2} v_0$. The particle dynamics based results in Figure 5 confirm the validity of this scaling law, yielding an exponent of 0.497.

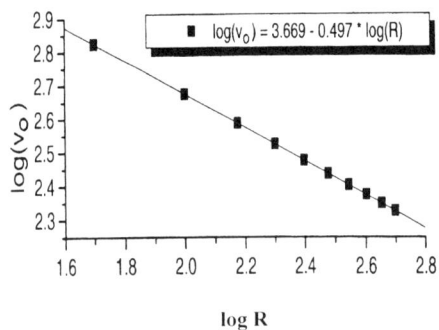

Figure 6: Scaling law showing that smaller v_0s are needed with increased grain size.

III.E. Comments on Polydispersity in Chains

One may worry that even the best available ferrofluid grains are seldom monodisperse and whether the predictions made in this work remain valid in the presence of polydispersity. It was shown by Spence [2], that the index of the Hertz law (Eq. (3)) depends upon the geometry of the grain-grain contact. Typically, this index lies in a range that runs roughly between 3/2 and 2. We have studied the propagation of solitary waves in chains of same r but with small (few percent) random variation in the index around 3/2 and 2. Our results reveal that the width of the solitary wave is weakly dependent upon the index in the above-mentioned range of values and hence the solitary wave is at best marginally affected by variations in grain contacts. The solitary wave is more significantly affected when the radius r varies significantly [14]. Thus, approximately spherical ferrofluid grains, which possess roughly the same radius, should be appropriate for our purposes.

IV. Summary and Conclusion

In this article we have presented calculations that support the hypothesis that it may be possible to use acoustic impulses to eject desired number of ferrofluid grains from dilute ferrofluids that are subjected to appropriate c-axis magnetic fields. Planned experiments are soon expected to test our hypothesis. If the approach is successful, it may allow one to make very small nanometer length scale imprints on surfaces. Perhaps, the ideas can also be used for processes such as controlled delivery of specialized drugs and many other applications.

This work has been partially supported by The Office of the Provost, SUNY at Buffalo, by Sandia National Laboratories through contract number DE-AC-04-85000 and by the National Science Foundation under grant number NSF-CMS-0070055. We are grateful to Professors V. Nesterenko, Daniel Pusiol, John Ho and Sara Majetich for helpful discussions.

References
[1] R. Rosensweig, Ferrohydrodynamics (Cambridge University Press, London, 1985).
[2] D.A. Spence, Proc. R. Soc. (Lond.) A **305**, 55 (1968).
[3] R.F. Ziolo, E.P. Giannelis, B.A. Weinstein, M.P. O'Horo, B.N. Ganguly, V. Mehrotra, M.W. Russell and D.R. Huffmann, Science **257**, 219 (1992).
[4] S. Sen, M. Manciu and F.S. Manciu, Appl. Phys. Lett. **75**, 1479 (1999).
[5] F.S. Manciu, M. Manciu and S. Sen, J. Magn. Magn. Mater. **220**, 285 (2000).
[6] H. Hertz, J. reine u. angew. Math. **92**, 156 (1881).
[7] M.P. Allen and D.J. Tildesley, Computer Simulation of Liquids (Clarendon, Oxford, 1989).
[8] V. Nesterenko, J. Appl. Mech. Tech. Phys. **5**, 733 (1983).
[9] A.N. Lazaridi and V. Nesterenko, J. Appl. Mech. Tech. Phys. **26**, 405 (1985).
[10] C. Coste, E. Falcon and S. Fauve, Phys. Rev. E **56**, 6104 (1997).
[11] S. Sen, M. Manciu and J.D. Wright, Phys. Rev. E **57**, 2386 (1998).
[12] S. Sen and M. Manciu, Physica A **268**, 644 (1999).
[13] D. Pusiol, private communication.
[14] M. Manciu and S. Sen, Physica D (submitted for publication).

Granular Flows II

Mat. Res. Soc. Symp. Proc. Vol 627 © 2000 Materials Research Society

Material Instability in Rapid Granular Shear Flow[1]

J.D.Goddard
Department of Mechanical and Aerospace Engineering
University of California, San Diego
La Jolla, CA 92093-0411

ABSTRACT

This is a survey of recent theoretical work on shear flow instabilities of dry granular media in the Bagnold or "grain-inertia" régime. Attention is devoted to steady homogeneous unbounded simple shear, with the goal of identifying *material* (constitutive) instabilities arising from the coupling of stress to granular concentration and temperature fields. Such instabilities, the dissipative analogs of thermodynamic phase transitions, are familiar in numerous branches of the mechanics of materials.

The current interest is motivated in part by the "dissipative clustering" found in various particle-dynamics ("DEM") simulations of granular systems. Since particle clustering may invalidate standard gas kinetic theory, it is pertinent to ask whether hydrodynamic models based on such theories may themselves exhibit clustering instability.

The present article is based largely on a recent review (Goddard and Alam 1999), which provides a unified linear-stability treatment for rapid granular flow, as well for slow flow of mobile particles immersed in viscous liquids. The analysis is based on a "short-memory" response of various fluxes to perturbations on steady uniform states, a feature characteristic of the most popular constitutive models for granular flow. In the absence of gravity, previous theoretical analyses reveal transverse "layering" and spanwise "corrugations" as possible forms of material instability (Alam and Nott 1998)

Based on current theoretical findings, further work is recommended, including the exploration of the effects of gravity and of stress relaxation, both of which are likely to be important in real granular flows.

INTRODUCTION

Particle clustering is found in various particle-dynamics simulations of dry granular media subject to rapid shearing (Hopkins and Louge, 1991). Other simulations (Goldhirsch and Zanetti, 1993; Tan and Goldhirsch, 1997) indicate that the density and extent of such clusters increase with collisional inelasticity, implicating microscale dissipation as the mechanism.

Temporal stress fluctuations have also been found in several experiments and computer simulations of shear flow (Savage, 1992a; Miller et al., 1996). Savage's simulation of a particulate mass sheared between rough walls exhibit régimes of quasi-turbulent stress

---

[1] Paper BB4.1, Materials Research Society, Spring Meeting, San Francisco, CA, April 24-28, 2000.

fluctuations with $1/f$ spectrum. Stress fluctuations have been observed experimentally by Behringer and Baxter (1991) in hopper flow and by Miller et al. in an annular shear cell.

Clustering and stress fluctuations may cast doubt on validity of continuum models, since smoothly varying fields are routinely invoked in the derivation of "hydrodynamic" equations from kinetic theory. Since the issue turns largely on the scale of fluctuations relative to experimental scales, there is a considerable motivation for theoretical investigation of the stability of various continuum models. (Savage, 1992b; Babíc 1993; Schmid and Kytömaa, 1994; Wang et al., 1996, 1997; Alam 1998; Alam and Nott, 1997, 1998; Nott et al., 1998). The dependence of instability on particle inelasticity found in these studies resembles that seen in particle-dynamics simulations and suggests that density fluctuations in the mean flow are precursors to clustering. In this respect, there are interesting analogies to density instabilities in otherwise uniform fluidized beds. (Goddard & Alam, 1999 - hereinafter denoted by G&A.).

As pointed out by G&A, it is important to know whether inhomogeniety and fluctuation represent ordinary dynamical instabilities of materially stable media (as with turbulence in Newtonian fluids), or whether they are a manifestation of material instability. Most existing work on material instability (Goddard 1996, G&A) deals with solids, while much less has been done on fluid-like materials. A common theme is the "soft-mode" instability of homogeneous states, of the type associated with thermostatic (i.e., equilibrium) phase transitions and the classical Hadamard instability. Such instability typically gives rise to short wavelength patterns resulting from change of type of the underlying field equations, e.g. loss of ellipticity in quasi-static equations with loss of hyperbolicity in the dynamics. Rapid granular shear flows are representative of a class of material media in which the diffusion of matter or energy are implicated in the stability phenomenon. The purpose of the present article is to review the recent work on instabilities in rapid granular flow, focusing mainly on the stability of unbounded homogeneous simple shear.

BRIEF SURVEY OF SHEAR-FLOW INSTABILITY

Most current constitutive models for stress, energy flux, dissipation rate and particle flux in rapid granular flow are inspired by the Newtonian limit of dense-gas kinetic theory. All these Newtonian-fluid models have similar structure, summarized by Alam and Nott (1998) and discussed by G&A, and appear to represent perturbations in a small inelasticity parameter $(1-r)$, where r is a representative coefficient of collisional restitution, of otherwise Hamiltonian systems.

The first systematic study of the stability of such models was considered by Savage (1992b) for unbounded simple shear flows, an analysis subsequently extended and elaborated on by Babic (1993) and Schmid & Kytömaa (1994). In these studies, a solution of the linearized disturbance equations was found in the form of Kelvin modes (Kelvin 1887). However, most of these analyses rely on approximating a time-dependent stability problem by a time-independent problem (method of `frozen coefficients'), which turns out to be invalid for the present problem, since the coefficients are found to vary on the time scale of the unstable growth rate. Despite the shortcomings, these studies, as amended by Wang et al. (1996) and Alam & Nott (1997), provide two important results: Streamwise perturbations (disturbance wave vectors) or "slugging modes" are always stable, and layering modes, with wave vector in the shear direction exhibit instability over a range of initial solids fraction. Moreover the layering modes unstable over finite range of transverse wave lengths. The instability is tied to the inelasticity ($1-r$) of particle collision, with growth rates vanishing as this parameter tends to zero.

Even when disturbances are asymptotically stable, the analysis of Schmid and Kytömaa (1994) reveals that they can grow to large amplitudes initially, owing to the strong non-normality of the stability operator. While such transient growth may play a role in subcritical flow transition by triggering finite-amplitude effects, this can only be confirmed by a proper non-linear stability analysis.

As a variant on the Newtonian-fluid model, Alam and Nott (1997) consider the influence of static friction, adopting the constitutive ("Bingham-plastic") model of Johnson and Jackson (1987).
They found that the inclusion of friction does not alter the stability of flow to streamwise disturbances but may enhance instability of layering modes. In particular, friction can stabilize or destabilize depending on the relative magnitudes of the restitution and Coulomb friction coefficients.

In addition to the 2D (planar) disturbances discussed above, Wang et al. (1996) consider the effect of spanwise disturbances or "corrugations", which subsequently are treated more completely by Alam and Nott (1998). Results from the latter, comprehensive stability analysis indicate that the finite range of unstable wave number found for both spanwise and transverse modes results from a competition between granular "heat" conduction and collisional dissipation (G&A). In that respect, the instability enjoys a kinship with "adiabatic" shear banding in solid plasticity.

Solid boundaries introduce complications beyond the domain of strict material instability (Wang et al., 1996, Alam & Nott 1998) and may have stabilizing or destabilizing effects, with a discrete spectrum of transverse wave numbers.

Based on the foregoing stability analyses, one can discern the possibility of a much more comprehensive analysis of linear material instability for more complex materials exhibiting history or relaxation effects, whose outlines are sketched here (Alam & Goddard, 2000)

TOWARD A GENERAL STABILITY ANALYSIS

As evident from the existing analyses (G&A), the transport equations (balance equations plus constitutive equations) are represented by PDEs of the general form

$$\dot{\psi} = f(\nabla, \psi)$$

(1)

where ψ is a vector array of the relevant field variables, including particle-phase density and temperature, and velocity vector, and the dot represents material (Lagrangian) time derivative. These suffice in the case of the standard Newtonian fluid, or other material without memory, and ψ is five-dimensional. In the case of more complex "viscoelastic" material response, ψ generally involves the stress rate, stress acceleration, etc., leading to a higher dimensional problem. The notation (1) indicates dependence on ψ and its spatial gradients $\nabla\psi$, $\nabla\nabla\psi$, ..., and the usual "non-local" models involve only those shown explicitly, i.e. only quadratic forms in ∇.

Material stability, as defined here, is concerned with the stability of (1) against arbitrary disturbances of a spatially uniform base state ψ^0, represented by any Gallilean-invariant solution of (1). Linear material stability refers, as usual, to the problem of infinitesimal disturbances ψ^1 and the linearized form of (1). As discussed in [1], this leads to the following time-dependent linear stability problem for the associated Fourier amplitude of ψ^1, say, $u(t, k^0)$:

$$\frac{d}{dt} u(t, k^0) = A(t, k^0) u(t, k^0)$$

(2)

where k^0 is an embedded "initial" wave vector, i.e. the image of a given wave vector k under the (inverse) affine deformation represented by the base flow at time t [1], and $A(t, k^0)$ is a square matrix derived from the derivatives ∂_ψ, $\partial_{\nabla\psi}$, $\partial_{\nabla\nabla\psi}$, of f evaluated at the base state (G&A). In the case of a simple shear this represents Kelvin's linear wave-vector stretching, and when f is quadratic in ∇ this gives

$$A = A_0 + A_1 t + A_2 t^2$$

(3)

where the matrices A_i depend on material parameters and k^0. As it turns out, the lower order terms in (3) may often determine asymptotic instability; e.g. the matrix A_2 may have a finite dimensional null space that is unstable because of the lower-order terms in (3) (cf. Alam, 1998, and Alam & Nott, 1998).

In the existing Newtonian models, the above stability matrices are 5-by-5, whereas the inclusion of relaxation effects in a Maxwell-type fluid model involves stress rate and leads to 11-by-11 matrices (with six components of stress joining the previous field variables). Even its direct effect on the evolution of perturbations is neglected, relaxation can still affect stability through the modification of the base flow stresses (Alam & Goddard, 2000). Inclusion of higher spatial gradients to represent weak non-locality would lead to a higher degree polynomial in t in (3) with the evident possibility of affecting stability. At any rate, it is hoped that this brief overview may lead to a systematic mathematical attack on more complex models.

SUMMARY

Key theoretical results are summarized above for the stability of rapid granular shear flows. According to existing models and analyses, such flows are generally stable against streamwise "slugging" but can be unstable against spanwise "corrugations" and/or transverse "layering". The unstable modes involve "cold" dense regions, which are re-thermalized by granular "heat flux", leading to a cut-off of the usual short-wave (Hadamard) instability.

Brief consideration is given to the effects of interparticle (Coulomb) friction and bounding walls on granular flows, but the potentially important effects of gravity have been neglected in most of the work reviewed here. It would be useful to study the possible effects of gravity on stability, although gravity, like bounding walls will generally rule out the idealized homogeneous base states considered above. One nevertheless may be able to distinguish material instabilities from more general dynamical instabilities associated with stratified fluids (G&A).

Simple kinetic-theory type estimates for relaxation time suggest that relaxation effects may cause important departures from Newtonian-fluid response and therefore may have a substantial effect on material stability. This is matter of ongoing theoretical investigation (Alam & Goddard, 2000).

ACKNOWLEDGEMENTS

Partial support from the U.S. National Aeronautics and Space Administration (Grant NAG3-1888), the U.S. Air Force Office of Scientific Research (Grant F49620-96-1-0246), and the National Science Foundation (Grant CTS-9510121) is gratefully acknowledged.

REFERENCES

Alam, M. 1998, *Stability of unbounded and bounded granular shear flows*. Ph.D. thesis. Indian Institute of Science, Bangalore, India.

Alam, M. & Goddard, J.D., 2000, work in progress and to appear.

Alam, M. & Nott, P. R. 1997, The influence of friction on the stability of unbounded granular shear flow. *J. Fluid Mech.* **343**, 267.

Alam, M. & Nott, P. R. 1998, Stability of plane Couette flow of a granular material. *J. Fluid Mech.* **377**, 99.

Babic, M. 1993 On the stability of rapid granular flows. *J. Fluid Mech.* **254**, 127.

Behringer, R. P. & Baxter, G. W. 1991, Pattern formation and complexity in granular flows. In *Granular Matter: An Interdisciplinary Approach* (ed. Anita Mehta), pp. 85-119. Springer-Verlag.

Goddard, J. D. 1995, *Material Instabilities*. Report No. 95-20. Institute for Mechanics and Materials, University of California, San Diego.

Goddard, J.D. & Alam, M. 1999, Shear-Flow and Material Instabilities in Particulate Suspensions and Granular Media. *Particulate Sci. and Tech.*, **17**, 69-96.

Goldhirsch, I. & Zanetti, G 1993 Clustering instability in dissipative gases. *Phys.Rev.Lett.* **70**, 1619.

Hopkins, M. A. & Louge, M. Y. 1991, Inelastic microstructure in rapid granular flows of smooth disks. *Phys. Fluids* **A 3**, 47.

Miller, T. M., O'Hern, C. & Behringer, R. P., 1996, Stress fluctuations for continuously sheared granular materials. *Phys. Rev. Lett.* **77**, 3110.

Nott, P. R., Alam, M., Agrawal, K., Jackson, R. & Sundaresan, S. 1998, The effects of boundaries on plane Couette flows of granular materials: a bifurcation analysis. *J. Fluid Mech.* **397**, 203-29

Savage, S. B.,1992a, Numerical simulations of Couette flow of granular materials: spatio-temporal coherence and 1/f noise. In *Physics of Granular Media* (eds. D. Bideau & J. Dodds), pp. 343-362. New York: Nova Science.

Savage, S. B. 1992b , Instability of unbounded uniform granular shear flow. *J. Fluid Mech.* **241**, 109.

Schmid, P. J. & Kytömaa, H. K. 1994, Transient and asymptotic stability of granular shear flow. *J. Fluid Mech.* **264**, 255.

Tan, M-L. & Goldhirsch, I. 1997, Intercluster interactions in rapid granular shear flows. *Phys. Fluids* A **9**, 856.

Thomson, W. 1887, Stability of fluid motion: rectilineal motion of viscous fluid between two parallel plates. *Phil. Mag.* **24**, 188.

Wang, C-H., Jackson, R. & Sundaresan, S. 1996, Stability of bounded rapid shear flows of a granular material. *J. Fluid Mech.* **308**, 31.

Wang, C-H., Jackson, R. & Sundaresan, S. 1997, Instabilities of fully developed rapid flow of a granular material in a channel. *J. Fluid Mech.* **341**, 161.

Mat. Res. Soc. Symp. Proc. Vol 627 © Materials Research Society

Analysis of shear bands in slow granular flows using a frictional Cosserat model

Prabhu R. Nott, K. Kesava Rao & L. Srinivasa Mohan

Indian Institute of Science, Bangalore 560012, INDIA

Abstract

The slow flow of granular materials is often marked by the existence of narrow shear layers, adjacent to large regions that suffer little or no deformation. This behaviour, in the regime where shear stress is generated primarily by the frictional interactions between grains, has so far eluded theoretical description. In this paper, we present a rigid-plastic frictional Cosserat model that captures thin shear layers by incorporating a microscopic length scale. We treat the granular medium as a Cosserat continuum, which allows the existence of localised couple stresses and, therefore, the possibility of an asymmetric stress tensor. In addition, the local rotation is an independent field variable and is not necessarily equal to the vorticity. The angular momentum balance, which is implicitly satisfied for a classical continuum, must now be solved in conjunction with the linear momentum balances. We extend the critical state model, used in soil plasticity, for a Cosserat continuum and obtain predictions for flow in plane and cylindrical Couette devices. The velocity profile predicted by our model is in qualitative agreement with available experimental data. In addition, our model can predict scaling laws for the shear layer thickness as a function of the Couette gap, which must be verified in future experiments. Most significantly, our model can determine the velocity field in viscometric flows, which classical plasticity-based model cannot.

1   Introduction

An important feature of granular flow in the slow flow regime, where shear forces arise primarily from frictional forces between grains, is the rate-independence of the stress: the stress remains unchanged if all elements of the rate of deformation tensor are scaled by the same factor. This feature is captured by continuum models traditionally used to describe slow granular flows, which are based on concepts in metal plasticity. However, these models fail to correctly describe viscometric flows, for which the direction of variation of the velocity is always orthogonal to the direction of flow. Experimental observations of viscometric flows, such as flow through vertical channels [8, 9] or shear in a cylindrical Couette cell [4] show large portions of the material remaining undeformed, and shear occurring only in thin layers [12, 9, 1]. In contrast, conventional plasticity-based models predict no shear layers [14, 2], but that the entire region moves as a block, with slip occurring at the walls.

The inability of frictional models to determine the velocity field in viscometric flows has been attributed to the absence of a material length scale in their constitutive equations [5, cited in [7]]. An attempt to correct this deficiency of the classical plasticity models was made recently by us [3], wherein the granular medium was modeled as a Cosserat continuum. This naturally brings in a material length scale in the constitutive equations. Cosserat plasticity models have been applied to problems in granular flow earlier [7, 6, 14, 13, 15], but the models in these studies are posed in terms of strain increments as they only addressed unsteady flows, and cannot be applied to analyze sustained flow.

The fundamental difference between a Cosserat continuum and a classical continuum is that the former involves two additional field variables, namely, the angular velocity, and the couple stress tensor. The couple stress is the couple per unit area exerted by the medium, which is assumed to be absent in a classical continuum. As a result, the mass and linear momentum balances in a Cosserat continuum must be supplemented by the angular momentum balance (which is satisfied identically in a classical continuum) relating the angular velocity, the Cauchy stress and the couple stress. Lastly, the angular velocity is not necessarily equal to half of the vorticity, as in a classical continuum. For steady, fully developed flow, spatial gradients of M cause σ to be asymmetric. These are effects that can perhaps be measured in the laboratory, and experiments in this direction are needed to settle this issue.

In this paper, we apply our frictional Cosserat model to a simple viscometric flow, namely plane Couette flow, with and without the presence of gravity. Our solutions illustrate the key features of our model, and appears to be in qualitative agreement with available experimental data.

2 Governing equations

As in a classical continuum, the balances for mass and linear momentum must be satisfied:

$$\frac{D\rho}{Dt} + \rho \boldsymbol{\nabla} \cdot \boldsymbol{u} = 0, \tag{1}$$

$$\rho \frac{D\boldsymbol{u}}{Dt} + \boldsymbol{\nabla} \cdot \boldsymbol{\sigma} - \boldsymbol{b} = 0, \tag{2}$$

where D/Dt represents the material derivative, $\boldsymbol{u}$ the velocity, ρ the bulk density of the granular medium, $\boldsymbol{\sigma}$ the Cauchy stress tensor (defined in the compressive sense) and $\boldsymbol{b}$ the body force. In addition to the above, the angular momentum balance must also be enforced for a Cosserat continuum. It takes the form

$$\frac{D(\rho\,\boldsymbol{\omega})}{Dt} + \boldsymbol{\nabla} \cdot \boldsymbol{M} + \boldsymbol{\varepsilon} : \boldsymbol{\sigma} - \rho\,\boldsymbol{\zeta} = 0, \tag{3}$$

where $\boldsymbol{\omega}$ is the intrinsic angular velocity, $\boldsymbol{M}$ is couple stress, $\boldsymbol{\varepsilon}$ is the alternating tensor and $\boldsymbol{\zeta}$ is the externally imposed body couple acting on the medium. In a classical continuum there is no independent angular velocity field, and the local rotation is equal to half the vorticity; with the further imposition of no surface or body couples, we arrive at the conclusion that the Cauchy stress is symmetric, and the angular momentum balance is identically satisfied. In a Cosserat continuum the intrinsic angular velocity may deviate from the vorticity and the stress in general is not symmetric.

To close the above set of equations, we require constitutive relations for $\boldsymbol{\sigma}$ and $\boldsymbol{M}$. As in the classical plasticity models, the constitutive relations comprise a yield condition and a flow rule. The former is, however, modified to account for the influence of the couple stress tensor. Similarly, the flow rule is modified to provide a relation for the angular velocity. We have adapted the model of [7] and [14], used for studying the development of shear bands in granular flow. While their models were posed in terms of strains and strain increments, we give the relations that are appropriate for sustained flow.

2.0.1 Yield condition

The yield condition for a classical continuum is a relation of the form $F(\boldsymbol{\sigma}, \nu) = 0$, where F is a scalar function of the stress tensor. It specifies the stress state of the granular medium undergoing plastic deformation. When the yield condition is not satisfied, the material is either rigid, or in a state of elastic deformation. The form of the yield condition we have chosen is a generalization of that for classical plasticity [3],

$$F \equiv \tau - Y = 0, \tag{4}$$

where

$$\tau \equiv \left(a_1 \sigma'_{ij} \sigma'_{ij} + a_2 \sigma'_{ij} \sigma'_{ji} + \frac{1}{(Ld_p)^2} M_{ij} M_{ij} \right)^{1/2}, \tag{5}$$

$\sigma'_{ij} = \sigma_{ij} - (1/3)\sigma_{kk}\delta_{ij}$ is the deviatoric stress. δ_{ij} is the Kronecker delta. a_1, a_2, and L are material constants, and d_p is the grain diameter. The parameter L defines a characteristic material length scale in terms of the grain diameter. [3] found that a value of 10 for L nicely fits experimental data for flow down vertical channels.

Following [7] we set $a_1 + a_2 = 1/2$, without loss of generality. We retain the choice of [3] of $A \equiv a_2/a_1 = 1/3$ and $L = 10$.

The yield limit Y depends on the mean stress $\sigma \equiv \sigma_{kk}/3$, and the solids fraction ν. We use the form for the yield limit proposed by [11],

$$Y = \sigma_c(\nu) \sin \phi \left(n\alpha - (n-1)\alpha^{(n/(n-1))} \right); \quad \alpha \equiv \frac{\sigma}{\sigma_c(\nu)}. \tag{6}$$

Here $\sigma_c(\nu)$ is the mean stress at the critical state, ϕ is a material constant called the angle of internal friction, and n is a material constant. This form for Y, in conjunction with the associated flow rule, ensures a dilation region on the yield locus, a critical state and a relatively steep compaction region, as is observed in dry granular materials.

A functional form for $\sigma_c(\nu)$ must, in general, be specified. However, in the problems that we have considered, we have assumed for the sake of simplicity that the solids fraction (or density) is constant. Hence, σ_c is determined as part of the solution, exactly like the pressure in an incompressible Newtonian fluid.

2.0.2 Flow rule

The flow rule relates the rate of deformation tensor to the stress. We extend the associated flow rule, used in plasticity theories for a classical continuum, to a Cosserat continuum by incorporating the angular velocity field and the couple stress (see [3] for a fuller description). In Cartesian tensor notation, the flow rule is

$$E_{ij} \equiv \frac{\partial v_i}{\partial x_j} + \varepsilon_{ijk}\omega_k = \dot{\lambda}\frac{\partial F}{\partial \sigma_{ji}}; \quad H_{ij} \equiv \frac{\partial \omega_i}{\partial x_j} = \dot{\lambda}\frac{\partial G}{\partial M_{ji}}, \tag{7}$$

Here D_{ij} is the rate of deformation tensor, $F(\boldsymbol{\sigma})$ is the yield function (6). When the couple stresses vanish and the angular velocity ω equals half of the vorticity, the above reduces to the associated flow rule for a classical continuum.

3   Application to plane Couette flow

We consider steady, fully developed flow between plane parallel plates of infinite extent, with shear generated by the motion of one plate relative to the other at constant speed. We choose a Cartesian reference frame such that flow is in the x direction and velocity gradient in the y direction (see figure 1). Gravity, if it exists, acts in the y direction. The mass balance (1) is trivially satisfied.

To begin with, we pose the governing equations in dimensionless form; all lengths are scaled with the Couette gap Δ, all velocities with the speed of the top boundary v_0, the angular velocity by v_0/Δ, the stress by the imposed normal traction N on the upper wall $(y = \Delta)$, and the couple stress by $d_p N$. The balance equations (2) and (3) then take the form

$$\frac{d\sigma_{yy}}{dy} = -\nu G, \tag{8}$$

$$\frac{d\sigma_{xy}}{dy} = 0, \tag{9}$$

$$\varepsilon\frac{dm}{dy} + \sigma_{xy} - \sigma_{yx} = 0, \tag{10}$$

where $G \equiv \rho_p \Delta g / N$ is the dimensionless gravitational force density and $\varepsilon \equiv d_p/\Delta$ is the ratio of microscopic to macroscopic length scales and $m \equiv M_{zy}$.

Since the diagonal components of E_{ij} are zero, it is easily seen [3] that the flow rule (7) implies that the normal stresses are equal, i.e.

$$\sigma_{xx} = \sigma_{yy} = \sigma_{zz} = \sigma_c(\nu). \tag{11}$$

The above result simplifies the yield condition (4)-(5) to

$$a_1(\sigma_{xy}^2 + \sigma_{yx}^2) + 2a_2\sigma_{xy}\sigma_{yx} + \frac{m^2}{L^2} = (\sigma_c \sin\phi)^2. \tag{12}$$

The remaining equations of the flow rule (7) are

$$E_{xy} = \frac{dv_x}{dy} + \omega_z = \frac{\dot{\lambda}}{\tau}(a_1\sigma_{yx} + a_2\sigma_{xy}), \tag{13}$$

$$E_{yx} = -\omega_z = \frac{\dot{\lambda}}{\tau}(a_1\sigma_{xy} + a_2\sigma_{yx}), \quad \text{and} \tag{14}$$

$$H_{zx} = \frac{d\omega_z}{dy} = \frac{\dot{\lambda}}{\tau}\frac{m}{(Ld_p)^2}. \tag{15}$$

BB4.3.4

On eliminating $\dot\lambda$ from the above we get

$$\frac{\mathrm{d}v_x}{\mathrm{d}y} = -\frac{(\mathrm{A}+1)(\sigma_{xy}+\sigma_{yx})\,\omega_z}{\sigma_{xy}+\mathrm{A}\sigma_{yx}}, \tag{16}$$

$$\frac{\mathrm{d}\omega_z}{\mathrm{d}y} = -\frac{\omega_z}{\varepsilon L^2}\frac{2(\mathrm{A}+1)\,m}{(\sigma_{xy}+\mathrm{A}\sigma_{yx})}. \tag{17}$$

For the sake of simplifying the analysis, we now make the assumption that ν is constant across the gap. Then, σ_c must be determined as part of the solution.

3.1   Boundary Conditions

When the material at a boundary is yielding, we use the usual friction boundary condition,

$$-\frac{\sigma_{yx}}{\sigma_{yy}} = \tan\delta, \tag{18}$$

which is the analog of the yield condition at the boundary. To determine the kinematic fields at the boundaries, we use the following condition that relates the angular velocity and the slip velocity at a boundary:

$$v_x - v_w = -K\omega_z \tag{19}$$

Here K is a dimensionless constant which reflects the roughness of the wall and v_w is the velocity of the wall. An explanation of this condition was provided by [3].

Conditions (18) and (19) suffice at a boundary when the material is yielding. When the stress state is such that the friction boundary condition is not satisfied, the material does not yield at the boundary. In this scenario, we apply the no-slip boundary condition at the boundary,

$$\boldsymbol{v} = \boldsymbol{v}_w. \tag{20}$$

The boundary condition for the stress we impose is (see [10] for an elaboration)

$$\sigma_{xy} = \sigma_{yx}. \tag{21}$$

3.2   Plane shear in the absence of gravity

Considering first the case where the properties of the upper and lower boundaries are identical, the symmetry of the problem allows us to solve for the fields in the half domain from the symmetry axis to the top wall. We then have the following conditions at the symmetry axis $y = 1/2$ (see [10] for details):

$$m = 0, \quad v = 1/2. \tag{22}$$

The normal stress on the upper wall is fixed by the external load N imposed on it, and hence

$$\sigma_{yy} = \sigma_c = 1 \quad \text{at} \quad y = 1. \tag{23}$$

At the wall ($y = 1$) we have the friction boundary condition (18) and the kinematic boundary condition (19).

The stress fields can be first obtained by solving the balance equations (8)-(10) in conjunction with the yield condition (12), the first of the symmetry conditions (22) and the friction boundary condition (18). The kinematic fields (velocity and angular velocity) can then be determined by solving the flow rule (17) with the second of the symmetry conditions and the kinematic boundary condition (19).

Figure 1 shows the profiles of all the field variables for various values of the wall angle of friction δ. For the case of a fully rough wall, $\tan \delta = \sin \phi$, the granular medium behaves as a classical continuum with zero couple stress, a symmetric stress tensor and the angular velocity equal to half the local vorticity. Further, the velocity varies linearly between the two walls, and the shear rate depends on the extent of slip at the walls. As the difference between $\tan \delta$ and $\sin \phi$ increases, the velocity profile deviates from the linear: the shear rate decreases with distance from the walls, resulting in the formation of a central core where there is little deformation.

When the angle of friction of one wall differs from that of the other, the solution is no longer symmetric about the mid-plane. If δ_L and δ_U are the angles of friction of the lower and upper walls, respectively, we suppose without loss of generality that $\delta_L > \delta_U$. In this case, the shear stress σ_{yx} is determined at the upper wall, where the friction boundary condition (18) applies. At the lower wall, the ratio of shear to normal stresses is less that $\tan \delta_L$, implying that the material does not yield there. We therefore impose conditions (20) and (21) at $y = 0$.

In this case, condition (21) implies that dm/dy vanishes at the lower wall. It is easy to verify that if m is finite, all higher derivatives of m also vanish at the lower wall, and therefore m remains constant across the Couette gap. The value of m is determined from the yield condition. The kinematic fields can be obtained by integrating (16) and (17), subject to the boundary conditions (19) at $y = 1/2$ and (20) at $y = -1/2$.

Figures 2 shows the profiles of the stress and kinematic fields for the case of asymmetric walls. For the case of a fully rough upper wall, the velocity profile remains linear, as in the case of similar walls. When $\delta_U < \delta_L$, the material shears near the upper wall,

4   Conclusions

The results of the preceding section nicely demonstrate the salient features of our frictional Cosserat model, foremost of which is that it is able to determine the velocity field for steady viscometric flows. As observed in experiments, deformation occurs in thin shear layers, while a large fraction of the material remains undeformed.

There are, however, significant shortcomings of the model that have to be addressed in future studies. One is that the model fails to predict dilation, or reduction in bulk density, within the region of high shear rate. There is clear qualitative evidence of dilation even in viscometric flows [8, 4], though quantitative measurements of the reduction in density is lacking. The second shortcoming of this model is that it does not predict normal stress differences, of which there is clear evidence in simulations, and also some experimental data.

However, we believe that these shortcomings can be overcome by further refinement of the model, and we are currently working in this direction.

References

[1] G. Gudehus and J. Tejchman. Some mechanisms of a granular mass in a silo-model tests and a numerical Cosserat approach. In O. Brüller, V. Mannel, and J. Najar, editors, *Advances in Continuum Mechanics*, pages 178–194. Springer, 1991.

[2] L. S. Mohan, P. R. Nott, and K. K. Rao. Fully developed flow of coarse granular materials through a vertical channel. *Chem. Engng Sci.*, 52:913–933, 1997.

[3] L. S. Mohan, P. R. Nott, and K. K. Rao. A frictional cosserat model for the flow of granular materials through a vertical channel. *Acta Mech.*, 138:75–96, 1999.

[4] D. M. Mueth, F. D. Georges, S. K. Greg, J. E. Peter, S. R. Nagel, and H. M. Jaeger. Signatures of granular microstructure in dense shear flows. *Cond-mat*, 00034333, 2000.

[5] H. B. Mühlhaus. Shear band analysis in granular materials by Cosserat theory. *Ing. Archiv.*, 56:389–399, 1986.

[6] H. B. Mühlhaus. Application of cosserat theory in numerical solution of limit load problems. *Ing. Arch.*, 59:124–137, 1989.

[7] H. B. Mühlhaus and I. Vardoulakis. The thickness of shear bands in granular materials. *Géotechnique*, 37:271–283, 1987.

[8] V. V. R. Natarajan, M. L. Hunt, and E. D. Taylor. Local measurements of velocity fluctuations and diffusion coefficients for a granular material flow. *J. Fluid Mech.*, 304:1–25, 1995.

[9] R. M. Nedderman and C. Laohakul. The thickness of shear zone of flowing granular materials. *Powder Technol.*, 25:91–100, 1980.

[10] P. R. Nott, K. K. Rao, and L. S. Mohan. *Document in preparation*, 2000.

[11] J. R. Prakash and K. K. Rao. Steady compressible flow of granular materials through a wedge-shaped hopper: the smooth wall radial gravity problem. *Chem. Engng Sci.*, 43:479–494, 1988.

[12] K. H. Roscoe. The influence of strains in soil mechanics. 10th Rankine Lecture. *Géotechnique*, 20:129–170, 1970.

[13] J. Tejchman and G. Gudehus. Silo-music and silo-quake experiments and a numerical Cosserat approach. *Powder Technol.*, 76:201–212, 1993.

[14] J. Tejchman and W. Wu. Numerical study of patterning of shear bands in a Cosserat continuum. *Acta Mech.*, 99:61–74, 1993.

[15] J. Tejchman and W. Wu. Numerical study on sand and steel interfaces. *Mech. Res. Comm.*, 21(2):109–119, 1994.

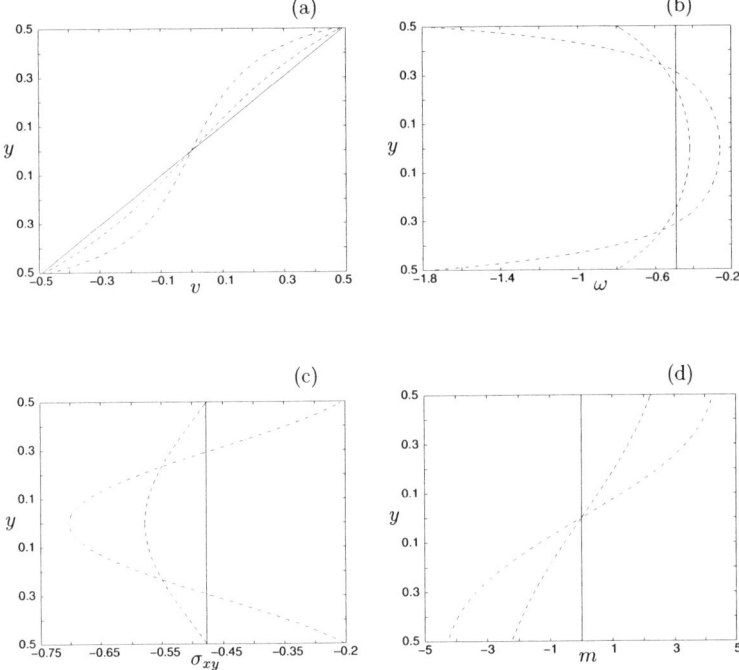

Figure 1: Profiles of the kinematic and stress fields for shear between identical walls in the absence of gravity. The solid line is for $\delta = 25.5°$, the dashed line for $\delta = 20°$, and the dot-dash line for $\delta = 25.5°$. Other parameters are: $\phi = 28.5°$, $A = 1/3$, $L = 10$, $K = 0.5$ and $\varepsilon = 0.39$. The panels (a)-(d) show the velocity, angular velocity, the shear stress σ_{xy} and the couple stress, respectively. The shear stress σ_{yx} is constant, equal to $\tan \delta$. Note that the velocity is linear and Cosserat effects vanish when the wall is "fully rough" ($\delta = 25.5°$).

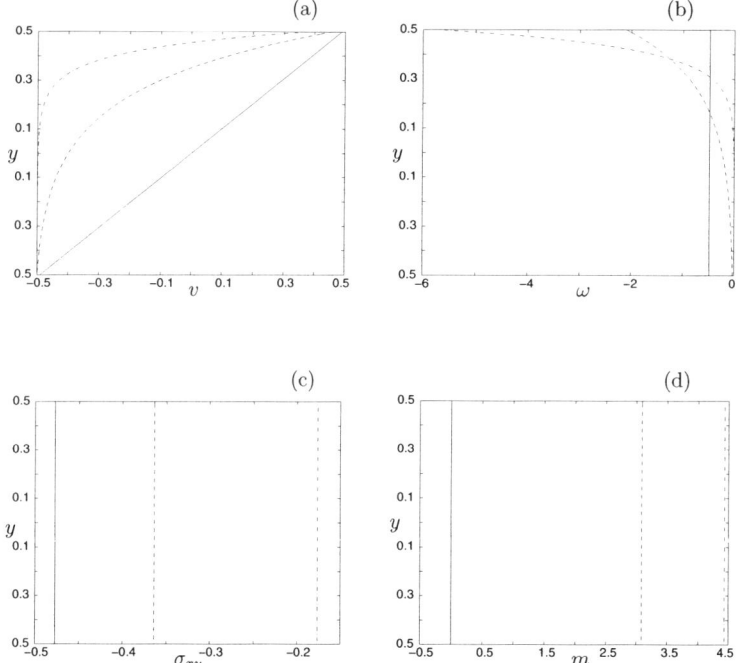

Figure 2: Profiles of the kinematic and stress fields for shear between dissimilar walls (the lower wall is rougher) in the absence of gravity. Parameters and lines are as in figure 1. In all cases, the couple stress is constant and the Cauchy stress is symmetric.

Mat. Res. Soc. Symp. Proc. Vol. 627 © 2000 Materials Research Society

Granular Flows in the Elastic Limit

Charles S. Campbell
Department of Mechanical Engineering
University of Southern California
Los Angeles, CA 90089-1453

ABSTRACT

This paper describes recent computer simulation studies into the rheological behavior of granular materials in the regime that lies between the quasistatic and rapid-flow regime. This investigation was prompted by studies of landslides, hopper flows and the "phase change" (i.e. the change between solid-like and fluid-like behavior) all of which indicated that the shear-to-normal stress ratio (the effective friction coefficient for the material) increased with shear rate. The results presented herein do demonstrate that the stress ratio varies with a dimensionless parameter created by scaling the shear rate with the stiffness of the interparticle contacts. In dense regimes, the stresses themselves scale with the stiffness indicating that they are generated by the elastic response of particle networks. Such speculation is supported by studies that show that the normal stresses are strongly dependent on the interparticle friction coefficient which affects the ability of internal elastic particle structures to support load and by the time variation of stress, which shows a spiky behavior as the structures form and break. However, analyses also indicate that these observations cannot explain the hopper, landslide and phase-change as these systems operate in regimes different from those in which the effect was observed. Finally, the effects of non-linear contacts are investigated and an appropriate scaling that takes the non-linearity into account is proposed.

INTRODUCTION

The history of granular flow modeling began with *quasistatic flow* theory (see the review in [15]), which, as the name suggests deals with slowly moving flows and generates a model by incorporating a Coulomb failure criterion into plasticity models. Implicit in these models is that the flowing material is always at the point of imminent yield with a ratio of maximum shear to normal stress equal to the internal angle of friction of the material. In the last 20 years or so, there has been a great deal of development at the other end of the spectrum, in the slightly misnamed field of *rapid granular flows*, which would more accurately be called rapidly-shearing granular flows (see the review in [5].) These theories took the point of view that in rapidly shearing systems, the particles acted like molecules in the kinetic theory of gases and could be analyzed using formalisms devised for kinetic theory. In such systems, transport is controlled by a "granular temperature" which is a measure of the kinetic energy contained in the random motion of the individual particles. Early work showed these ideas can account for measurements of heat, [22] mass [7] and momentum transfer [4,10].

However, it eventually became apparent that very few granular flows, especially those under Earth's gravity, were actually rapid granular flows. The first indication of this came from studies of the phase change between fluid-like and solid-like behavior of granular materials [6,10,22] which indicated that this transition could not be described in terms of rapid granular flow ideas but was instead demonstrated a quasistatic yield behavior. However, later studies on hoppers [19], which are clearly not rapid granular flows, indicated that the general accepted assumption of quasistatic yield at the internal angle of friction of the material could not explain the stress state within the hopper. Finally, large scale computer simulations of landslides [9] indicated that the ratio of shear to normal stress on the base of the slide appeared to vary with the shear rate (i.e. the velocity gradient at the slide's base) even though both rapid flow and quasistatic flow theories suggest that this ratio should be independent of shear rate. Examinations of both the hopper and phase change data also support the notion that the stress ratio increases with the local shear rate. However, in none of these cases was the spatial resolution of the velocity profile adequate to make a quantitative determination of this shear rate required to evaluate this speculation.

This led to the current investigations which were originally designed to quantitatively determine the effect of shear rate on the stresses in dense granular flow and, in particular, on the shear to normal stress ratio. Like much of the above, this will be done with computer simulations of systems of 1000 spheres. (See the reviews of computer simulation techniques in [3,8]). The simulation will use the soft-particle technique in which the interaction between particles is modeled as a spring (which may be linear or non-linear) and dashpot which act in parallel in the direction along the particle centers for as long as they remain in contact. The spring serves to push the particle surfaces apart and the dashpot dissipates the collisional energy. The spring has an associated stiffness (which will be varied in the following data) while the dashpot coefficient is varied to keep a constant coefficient of restitution (the ratio of recoil to impact velocity for a binary collision in the center of mass). The Young's modulus of a bulk material consisting of many such particles are proportional to this stiffness [2]. In the direction tangential to the contact point the particles are connected with a frictional slider in parallel with another spring with an associated friction coefficient . (I.e. as the particle surfaces move relative to one another in the direction tangential to the contact point, the tangential spring will load until the tangential force reaches times the normal force, at which point the surfaces slip relative to one another against a force equal to times the normal force). The particles are confined in a control volume bounded in all directions by periodic boundaries. This means that as a particle passes through one periodic boundary it reenters from the opposite side with exactly the same position and velocity with which it left; this simulates a situation where the control volume and every particle within it are periodically repeated infinitely many times up and downstream, so that as a particle passes out the downstream boundary, it passes into the downstream periodic image and is replaced by a particle entering from the upstream periodic image. These will be rheological studies, and it is necessary to induce a uniform shear within the control volume. To do this, the periodic images above and below the control volume in the y-direction are set in motion with fixed velocity in the manner originally used by Lees and Edwards [17]. It was found that uniform shearing could be achieved in these systems up to a solids concentration of It was found that uniform shearing could be achieved in these systems up to a solids concentration of $v = 0.62$. Beyond that concentration, the shearing generally took on a shear-band type of deformation with only a small part of the material undergoing shear.

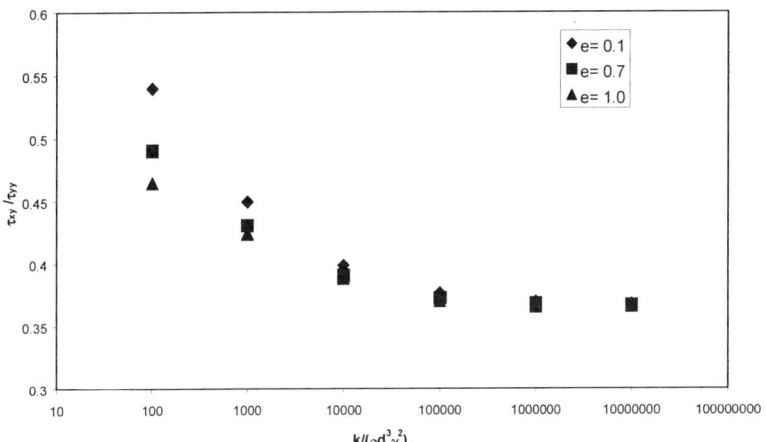

Figure 1: The ratio of shear to normal stress, τ_{xy}/τ_{yy}, as a function of the parameter, $k/(\rho d^3 \gamma^2)$. All of this data was taken from 1000 particle computer simulations at a constant solid concentration, $\nu = 0.6$ and particle surface friction, $\mu = 0.5$ for three different values of the coefficient of restitution, e. Note that τ_{xy}/τ_{yy} decreases with $k/(\rho d^3 \gamma^2)$ (i.e. increasing k, decreasing γ) eventually approaching a constant value indicative of quasistatic behavior.

RESULTS

Now the hopper, landslide and phase change simulations referred to above, all indicate that the apparent friction coefficient τ_{xy}/τ_{yy}, increases with the shear rate γ. At the outset, this presents a dimensional problem as γ has units of $(\text{time})^{-1}$, while the stress ratio τ_{xy}/τ_{yy} is dimensionless and can thus only be a function of dimensionless parameters. Now, the inverse shear rate represents the only time scale in rapid granular flow theory while no time scale appears in quasistatic theories. Thus some other parameter must be introduced into the analysis which contains units of time by which the shear rate can be scaled. The only possible candidates comes from the particle contact model and as the particle surface friction and the coefficient of restitution are dimensionless, the only remaining possibility is the contact stiffness, k This suggests a dimensionless parameter of the form:

$$\frac{k}{\rho d^3 \gamma^2}$$

where ρ is the particle density and d is the particle diameter. Note that this is similar to the parameter, B, studied by Babic, Shen and Shen [1]. Other parameters were explored, e.g., the shear scaled by the binary collision time, but there was no apparent advantage over the parameter proposed above.

The first test of whether this is an appropriate parameter is to examine its effect on the stress ratio . Figure 1 shows a plot of τ_{xy}/τ_{yy} vs. $\tau\, k/(\rho d^3 \gamma^2)$ at a constant solid concentration, $v = 0.6$, (i.e. 60% of a unit volume is solid material while the rest is void) particle surface friction coefficient $\mu = 0.5$, for three different coefficients of restitution, $\varepsilon = 0.1$, 0.7 and 1.0. The concentration $v = 0.6$ is used because all of the cases discussed above, phase-change, hoppers and landslides operate near the shearable limit, (about $v = 0.63$). Checks on this parameter by working with different diameters, shear rates and stiffnesses that yielded the same $k/(\rho d^3 \gamma^2)$, showed that the resultant stress ratios overlapped nearly exactly which indicates that no other parameter makes an appearance in this problem. Now, from the form of τ_{xy}/τ_{yy} it can be seen that increasing the stiffness k, moves one from left to right on the figure while increasing the shear rate, γ, moves one from right to left. One can see that for the lower values of $k/(\rho d^3 \gamma^2)$, the values of the stress ratio τ_{xy}/τ_{yy}, drop with increasing $k/(\rho d^3 \gamma^2)$, (i.e. with increasing k or with decreasing, γ). For large values of $k/(\rho d^3 \gamma^2)$, (i.e. large k, small γ) the stress ratio becomes constant which is indicative of quasistatic behavior. Note that the coefficient of restitution e, is only important at small $k/(\rho d^3 \gamma^2)$, which might be expected as this corresponds to conditions of large shear rate, γ, making the flows more rapid.

Note that while the effect of $k/(\rho d^3 \gamma^2)$ fits the basic understanding of quasistatic (small γ) and rapid-flow (large γ) behavior, the effect of the contact stiffness, k is somewhat contradictory. Large k implies a shorter binary contact time which approaches the instantaneous contact time assumption implicit in rapid-flow theories, yet increasing k (leading to large $k/(\rho d^3 \gamma^2)$) corresponds to more quasistatic behavior.

Figures 2 and 3 show the stresses scaled by the stiffness, $\tau_{xy}d/k$ and $\tau_{yy}d/k$ These data were generated for two different values of the stiffness although this is not apparent because the values overlay one another almost identically. Notice that both quantities decrease with $k/(\rho d^3 \gamma^2)$, becoming constant at large $k/(\rho d^3 \gamma^2)$, just as for the stress ratio, τ_{xy}/τ_{yy}. Note also that both quantities vary with the coefficient of restitution e, in a manner qualitatively similar to the stress ratio although the shear stress τ_{xy}, demonstrates a stronger effect of e than does τ_{yy}

The effect of the particle surface friction μ is shown in Figures 4 and 5. For a rapid granular flow, the surface friction largely affects only the energy dissipation (at least far from solid boundaries); so that the larger the surface friction, the larger the energy dissipation and, consequently, the smaller the granular temperature and all of the associated transport rates. But overall, this effect is weak. However, Figure 4 shows a relatively strong effect of the surface friction on the stress ratio τ_{xy}/τ_{yy}. As expected, τ_{xy}/τ_{yy} increases with μ. But notice that while there is a significant change in τ_{xy}/τ_{yy} by increasing μ from 0.1 to 0.5, there is a relatively minor change by going from $\mu = 0.5$ to $= 1.0$, that data for which nearly overlap in the "rapid" region (small $k/(\rho d^3 \gamma^2)$). (Note: curve fitting shows that all the lines will collapse together by plotting $(\tau_{xy}/\tau_{yy})\mu^{-0.085}$ as a function of $k/(\rho d^3 \gamma^2)$.)

Much more interesting is the effect on the normal stress $\tau_{yy}d/k$ plotted in Figure 5. In particular, in the quasistatic limit, (large $k/(\rho d^3 \gamma^2)$,) the low friction, $\mu = 0.1$ shows almost no generated normal stress (in actuality these stresses are about 40 times smaller than those for $\mu = 0.5$ or 1.0). This indicates the ultimate source of the stresses in this elastic limit may well be the

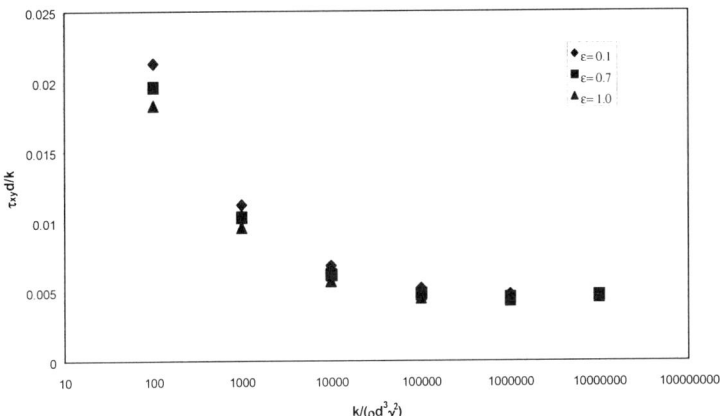

Figure 2: The shear stress scaled by the particle stiffness, $\tau_{xy}d/k$, for the data plotted in Figure 1. Note that the data scales with the stiffness indicating that the stresses are generated by the elasticity of the material.

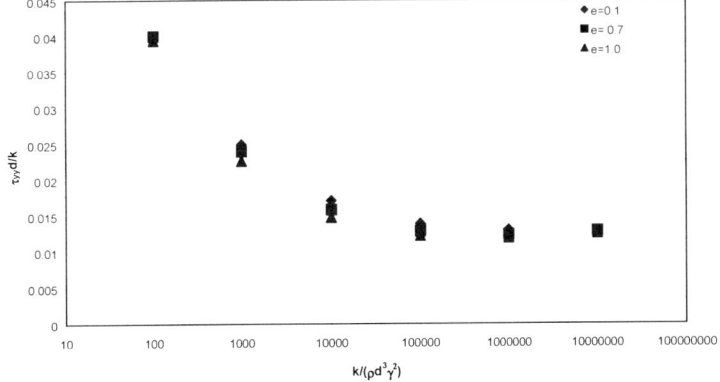

Figure 3: The normal stress scaled by the particle stiffness, $\tau_{yy}d/k$, for the data plotted in Figure 1. Note that, again, the data scales with the stiffness indicating that the stresses are generated by the elasticity of the material.

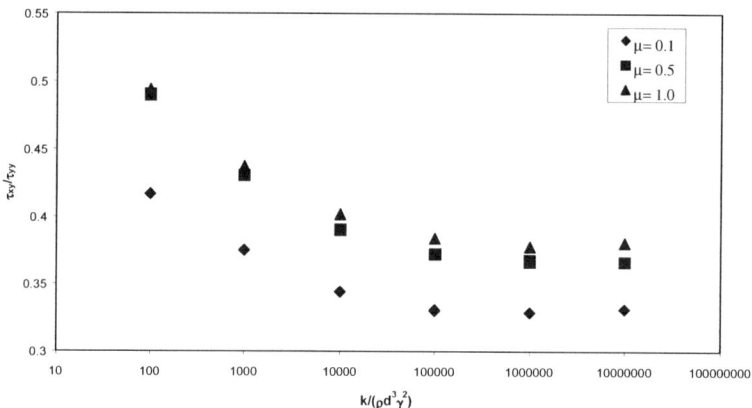

Figure 4: The effect of particle surface friction on the stress ratio τ_{xy}/τ_{yy}, all from 1000 sphere simulations at a concentration, $\nu= 0.6$. Note that while τ_{xy}/τ_{yy} varies strongly with μ, there is no direct relationship between the two.

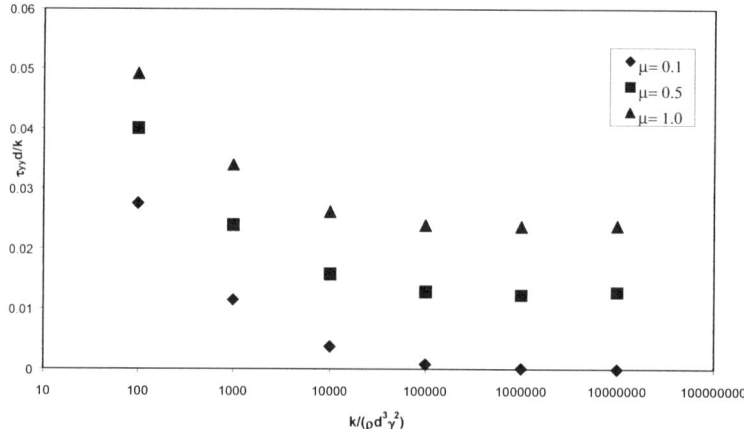

Figure 5: The effect of particle surface friction on the scaled normal stress $\tau_{yy}d/k$ for the data presented in Figure 4. Surprisingly, there is a very strong effect of μ on the normal stress. In fact the normal stress for $\mu= 0.1$ almost disappears in the quasistatic limit (large $k/(\rho d^3\gamma^2)$). This indicates that the stress is supported by elastic particle networks whose strength depends on their structural integrity which is strongly affected by μ.

structures that form within a static granular material (See for example [12,3]). The strength of the structures will be strongly affected by the interparticle friction. For the small friction case, these structures are weak and can only support very weak forces before the structure fails. Now the $\nu = 0.6$ concentration used for all of these examples, is smaller than the concentration of a random close pack of uniformly sized spherical particles. As a result, it is possible that particles need not be in intimate contact and thus may not be able to support an applied normal stress. This appears to be what is happening for the $\mu = 0.1$ cases shown in Figure 5. I.e. at large values of $k/(\rho d^3 \dot{\gamma}^2)$, only weak elastic structures form and little normal force can be supported. The effect shown here cannot be simply related to the friction coefficient, μ. Unlike Figure 4, it is not possible to collapse these curves together by plotting anything of the form $(\tau_{yy}d/k)f(\mu)$ as a function of $k/(\rho d^3 \dot{\gamma}^2)$; this can be easily seen as, while the stresses for $\mu = 0.1$ are about 40 times smaller than the $\mu = 1.0$ stresses at $k/(\rho d^3 \dot{\gamma}^2) = 10$, they are only about 1.8 times smaller at $k/(\rho d^3 \dot{\gamma}^2) = 100$.

This effect can be observed directly by examining the time traces of the normal stress $\tau_{yy}d/k$, shown in Figure 6. These were both taken out in the far quasistatic limit corresponding to $k/(\rho d^3 \dot{\gamma}^2) = 10^6$. Figure 6a shows a case corresponding to $\mu = 0.5$ while Figure 6b shows the corresponding case for $\mu = 0.1$; note that the vertical scale on Figure 6a is about 40 times larger than that for Figure 6b. Both show a spiked behavior that results from the formation and breakage of the internal elastic structures, but the structures not only support less force, they also persist for a much shorter time for $\mu = 0.1$ compared to $\mu = 0.5$. Those two observations are not unrelated. The shearing of the material results in compression of these elastic structures and the generation of the resultant stresses. Thus for $\mu = 0.5$, the structures are stronger and persist for longer periods of time before breaking. The longer life of the structures allows the shear to compress them to a greater extent, resulting in the larger stresses.

THE RELAVANCE OF THE PARAMETER: $k/(\rho d^3 \dot{\gamma}^2)$

The fact that the shear rate affects the stresses at all within these "elastic" flows, indicates an inertial effect. This may seem somewhat confusing because the elasticity of a material depends on deformation and not on deformation rate and from that point of view should be shear rate independent. However, from a more basic point of view, this effect should be anticipated. Simply put, the larger the shear rate, the larger the particle impact velocities, the larger the particle deformation and the larger the generated elastic forces. But there are still questions as to whether or not this effect is meaningful for granular flows of importance.

Note that Bathurst and Rothenburg [2] showed that for randomly packed spheres with linear contacts, the Young's modulus varies as:

$$E \propto \frac{nk}{d}$$

where n is the coordination number (the average number of contacts between particles) which in a packed bed can vary from a few, up to about 10. This means that:

$$\frac{nk}{\rho d} \propto \frac{E}{\rho} \propto (soundspeed)^2$$

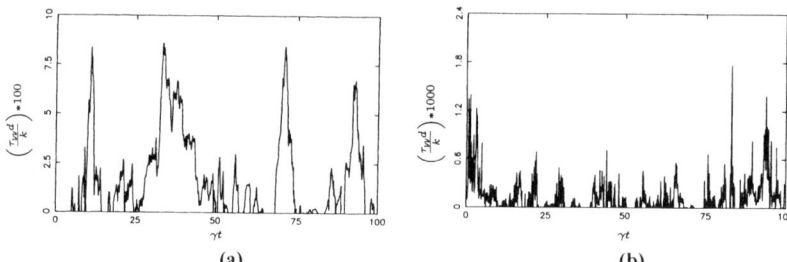

Figure 6: Two time traces of the scaled normal stress, $\tau_{yy}d/k$ plotted as functions of dimensionless time γt. (a) $\mu = 0.5$, (b) $\mu = 0.1$. Both cases are for $\nu = 0.6$ and $k/(\rho d^3 \gamma^2) = 10^6$
Putting them far out in the quasistatic regime. Note that the scale of (b) is magnified 40 times over that of (a).

But then, as γd is a shear velocity, the grouping:

$$\frac{k}{\rho d^3 \gamma^2} = \left(\frac{k}{\rho d} \right)\left(\frac{1}{\gamma d} \right)^2 \propto \left(\frac{1}{MachNumber} \right)^2$$

can be interpreted as an inverse square Mach number based on the shear velocity, γd. Now, examining Figure 1, one sees that the stress ratio can be observed to drop up until a value of $k/(\rho d^3 \gamma^2) = 10^4$ which implies a Mach number of the order of 10^{-2}.

The sound speed also gives us a means to calculate appropriate values for the stiffness k, for the sound speed in a loose sand is of the order of 100m/s [20]. To within an order of magnitude then:

$$\frac{k}{\rho d} \approx 10^4 \left(\frac{m}{s} \right)^2$$

This means that, to be in the range of stress ratio reduction, that $(d^2\gamma^2) \sim 1(m^2/s^2)$. Thus if one assumes particles of a size around 1mm, the shear rate would have to be close to $\gamma \sim 1000s^{-1}$, which is unlikely to ever be encountered in practice.

The only conclusion that can be drawn here, is that the transitional regime observed above, is most likely unimportant and that most flows that are to be encountered under Earth's gravity are quasistatic in nature. But what then of the phase-change, hopper and landslide results referred to above? By examining the range of parameters studied, only the landslide simulations fall into a flow regime where the stress ratio is changing with shear rate and thus those observations could be explained by these observations. But that is only an artifact of the simulation and is not a reflection of real landslide mechanics. This is because the landslide simulations were run at very small values of k, (a standard technique that allows larger time steps so is used to speed the execution of the simulations). Thus in reality, these results can explain none of the phenomena that instigated this investigation.

NON-LINEAR CONTACTS

So far we have only considered linear contact springs, i.e that the force generated on a contact is:

$$F = k\delta$$

Where δ is the overlap between the particles. Real materials do not behave this way. For example, if one were to assume that the contact between particles behaved in a Hertzian fashion, then, placed in this context, the force would vary as:

$$F = k_H \delta^{\frac{3}{2}}$$

Now many people would assume that materials behave in a Hertzian fashion, simply because Hertz theory is a solution to linear elastic theory. However, that does not appear to be born out by measurements. The sound speed in a granular material, depends on the elastic properties of the material and can be used as a probe of the contact dynamics. As pointed out by Goddard [14] the experimental evidence (e.g.[20,21]) indicate that the power of δ is actually larger than 3/2. Goddard [14] suggested that this was due to the fact that the particles actually interacted across sharp asperities and using known elastic solutions proposed a model that eventually transitioned to Hertzian behavior once the stress became large enough to press out the asperities. This type of behavior appears to be borne out by the direct load cell measurements [18]. One can also make a case that linear models are fine as it should always be possible to linearize a model around a base overlap δ_0 which is a function of the local stress levels, and indeed such ideas are usually used in sound propagation models.

There is no simple answer to these questions especially since the behavior of contacts undoubtedly varies from material to material, but it should be possible to have some general understanding of the behavior of non-linear contacts by investigating a generic contact with behavior:

$$F = k_0 \delta^n$$

Notice that k_0, k and k will all have different dimensions. Note that it is possible to assume that the generic non-linear contact model possesses a stiffness:

$$k = k_0 \delta^{n-1}$$

Note also that this stiffness is zero until there is some overlap and increases from then on.

Now what effect will this have on the rheology of the material? Well the first thing to notice is that since overlaps are generally small, the non-linear models produce a less stiff material than their linear counterparts, unless the constant k_0 is made proportionally larger as n is increased. Now if one imagines a shear flow with particles that are continually in contact, the flow will press particles together resulting in some sort of average displacement $\langle\delta\rangle$ that generates the stress response. If one assumes that the force is distributed over the area of a particle so then:

$$\tau = \frac{k_0 \langle\delta\rangle^n}{d^2}$$

Solving for $\langle\delta\rangle/d$ gives:

$$\frac{<\delta>}{d} = \left(\frac{\tau d^{2-n}}{k_0} \right)^{1/n}$$

Note that one expects that the degree of deformation $<\delta>/d$, that is induced by the shear, would be a function largely of the density. Note also that all the n's, i.e. all of the signs of nonlinearity appear on the right hand side. Thus, one might expect that:

$$\left(\frac{\tau d^{2-n}}{k_0} \right)^{1/n}$$

is a scaling for the stress that takes that takes the nonlinearity into account and would be an appropriate scaling for the stress. Note that the effects of n are quite interesting. Note that for $n=2$, the parameter, and hence the stresses are independent of particle diameter. For $n<2$ the stress, τ will, at constant values of this parameter, increase with particle diameter and for $n>2$, the stress will decrease with particle diameter.

Now the non-linearity will also affect the parameter, $k/(\rho d^3 \dot{\gamma}^2)$, where $k = k_0 \delta^{s-1}$. However, the overlap δ is a function of whatever generates the stresses, i.e. both by the deformation, and by inertial effects. Consequently, it is not possible dissociate the two effects in a nonlinear system so produce some characteristic value of k to use when defining $k/(\rho d^3 \dot{\gamma}^2)$. However, the conclusions drawn in the last section are still valid. That is, that it is unlikely that the inertial effects make an appearance in any of these flows. Consequently, it is possible to perform tests of the non-linear effects at large values of these parameter when the system is shear-rate independent.

A plot of the scaled yy stress component is shown in Figure 7. In this figure, at each density, the particle size was varied by a factor of 4 and the stiffness constant was varied by a factor of 10. All were run at very small shear rates to place them in the quasistatic regime. This scaling appears to work at the largest concentrations, but very quickly falls apart at concentrations below about $v=0.59$, where the data scatters with both the nonlinearity factor n and with particle size, d. This may be an indication that the material rapidly ceases to behave in a purely elastic manner for concentrations below $v=0.59$ and that rapid-flow effects are appearing.

CONCLUSIONS

This paper is the start of an investigation into the intermediate range of granular flow that lies between the rapid flow (which dominates at large shear rates and small solid concentrations) and the quasistatic limit which dominates at large concentrations and small shear rates. Previous studies of landslides, hopper flows, and of the "phase-change" between solid and fluid behavior indicate that the stress ratio τ_{xy}/τ_{yy} increased with the shear rate γ. Dimensional analysis provided an appropriate dimensionless parameter, $k/(\rho d^3 \dot{\gamma}^2)$ that scales the shear rate γ

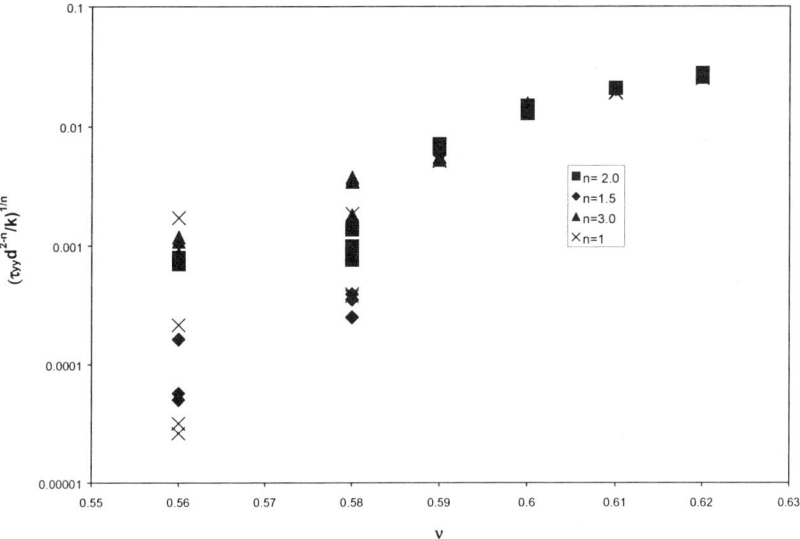

Figure 7: A check on the scaling for the nonlinear contact models. Note that the scaling holds for large concentrations of ν>0.59.

with the interparticle stiffness, k. The computer simulations shown in this paper were designed to generate constant shear rates and were thus designed to study this behavior in a rheological context.

The simulations demonstrated that the stress ratio τ_{xy}/τ_{yy} decreases with $k/(\rho d^3 \dot{\gamma}^2)$ and thus does indeed increase with the shear rate. In that region, the stresses scale with the interparticle stiffness in the form $\tau_{yy}d/k$, indicating that the forces are supported by networks of particles in intimate contact. The fact that the stresses vary with the shear rate at all appears to be a first order inertial effect, i.e. that the larger the shear rate, the larger the impact velocities, the larger the particle deformation and the larger the generated elastic forces.

However, further analysis indicates that while indeed this represents an important regime of granular flow, it does not explain the hopper or phase-change data, simply because these two cases operate in a regime of $k/(\rho d^3 \dot{\gamma}^2)$ where the stress ratio is a constant. Indirectly, it explains the landslide data, but only because approximations made for the landslide simulations put them into that regime, not because real landslides would be expected to behave that way.

Finally, an investigation was made of nonlinear contact models for which the interparticle force varies as $k_0 \delta^n$. These studies indicated that the non-linear effects could be taken into account by scaling the stresses as: $(\tau d^{2-n}/k_0)^{1/n}$.

ACKNOWLEDGMENT

This work was supported by the National Aeronautics and Space Administration under grant number NAG3-2358.

REFERENCES

1. M. Babic, M., H.H. Shen and H.T. Shen, J. Fluid Mech. **219** 81 (1990).
2. R. J. Bathurst, and L. Rothenburg, Journal of Applied Mechanics, **55**, 17 (1988).
3. C.S. Campbell, C.S. (1986) *Proc. 10th Nat. Cong. of App. Mech., Austin Texas.* June 1986, ASME, New York 327-38 (1986).
4. C.S. Campbell, J. Fluid Mechanics, **203** 449-473, (1989).
5. C.S. Campbell, Annual Review of Fluid Mechanics, **22** 57 (1990)
6. C.S. Campbell, C.S., The transition from fluid-like to solid-like behavior in granular flows, *Powders and Grains 93,* 289 A. A. Balkema, (1993).
7. C.S. Campbell, J. Fluid Mechanics, **348**, 85 (1997).
8. C.S. Campbell, *Powder Technology Handbook.* 2 ed., Marcell Dekker, New York 777 (1997).
9. C.S. Campbell, P. Cleary, and M.A.Hopkins, J. Geophysical Res., **100**, 8267 (1995).
10. C.S. Campbell and A. Gong, J. Fluid Mech. **164** 107-125 (1986).
11. C.S. Campbell and Y. Zhang, Y. *Advances in Micromechanics of Granular Materials - Proceedings of the Second U.S./Japan Seminar on the Micromechanics of Granular Materials,* Potsdam New York, August 5-9, 1991, (Ed. by H.H. Shen, M. Satake, M. Mehrabadi, C.S. Chang, & C. S. Campbell) Elsevier, Amsterdam, 261(1992).
12. P. A. Cundall and O.D.L. Strack, Geotechnique **29** 47 (1979).
13. A. Drescher and G. De Josselin de Jong, J. Mech. Phys. Solids, **20**, 337 (1972).
14. J.D. Goddard, Proc. Roy. Soc., **430** 105 (1990).
15. H. Hwang, and K. Hutter Cont. Mech. and Thermo **7** 357-384 (1995).
16. R. Jackson *Theory of Dispersed Multiphase Flow.* Academic Press, 291 (1983).
17. A. W. Lees and S. F. Edwards, J. Phys. C: Solid State Phys. **5** 1921 (1972).
18. M. Mullier, U. Tuzun, and O.R. Walton, Powder Technology **65**, 61 (1991).
19. A. V. Potapov and C.S. Campbell, Physics of Fluids A, **8** 2884 (1996).
20. F. E. Richart, R.D. Woods and J.R. Hall, 1970, *Vibrations of Soils and Foundations,* Ch. 6., Prentice Hall (1970).
21. F.E. Richart, *Dynamic Response and Wave propagation in Soils,* (ed. B. Prange) A. A. Balkema, (1978).
22. D.G. Wang and C.S. Campbell J. Fluid Mechanics, **244**, 527 (1992).
23. Y. Zhang, and C.S. Campbell, J. Fluid Mechanics, **237**, 541 (1992).

Mat. Res. Soc. Symp. Proc. Vol 627 © 2000 Materials Research Society

Liquid Filled Hourglasses:
A Study of Interstitial Fluid Effects on the Kinetic Behavior of Granular Materials

Benson K. Muite and Melany L. Hunt
Division of Engineering and Applied Science, California Institute of Technology,
Mail Code 104-44
Pasadena, CA 91125, U.S.A.

ABSTRACT

Experimental measurements of discharge rate and observations of flow pattern were made for spherical, monosized particles in a liquid-filled, hourglass shaped container. Different combinations of fluid properties, solid properties and vessel geometry exhibited a range of flow patterns that are described as lubricated, fluidized, oscillatory, or localized channeling discharges. A Bagnold number is used to present the discharge rate and flow regime results.

INTRODUCTION

For centuries, hourglasses (see Fig. 1) have been used as a simple method to measure the passage of time. The time-keeping principles are simple: the volume of material that has discharged from the upper bulb or hopper is linearly proportional to the time that has passed. Hence, the flow rate through the hopper is constant with time and independent of the amount of material above the orifice.

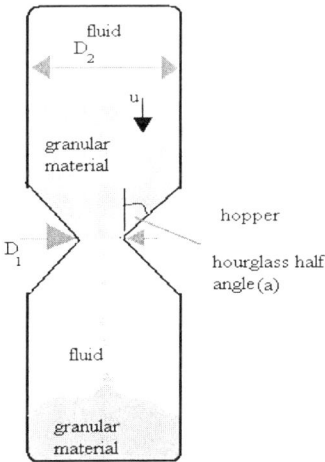

Figure 1 A Typical Hourglass

Most hourglasses are filled with sand and air. As the sand falls from the upper hopper, air moves upward through the permeable particle bed. Hence, the discharge rate in an hourglass may be weakly dependent on the upward flow of interstitial air. Comparable studies on hoppers, suggest that flow rates in an unventilated hopper are reduced when compared to ventilated flow if the particles are sufficiently small [1]. If the interstitial fluid is a liquid, the countercurrent liquid flow may have a pronounced effect on the discharge rate. Liquid-filled hourglasses may not be important for time keeping, but similar flows are found in industries that handle slurries or debris flows.

The current research uses a simple hourglass geometry to examine the effect of fluid properties on the discharge rate and discharge pattern of different interstitial fluids and granular materials. The important physical properties in determining hourglass discharge rate are gravity (g), particle density (ρ_p), fluid density (ρ), particle diameter (d), fluid dynamic viscosity (μ) and hourglass geometry. The surface roughness of the particles and the walls may also be important, especially for dry flows where frictional effects may dominate.

Research on ventilated hoppers using dry granular materials [2] has shown that the velocity of the material exiting (u_{Dl}) the hopper is proportional to:

$$u_{Dl} \propto [g(D_1 - kd)]^{1/2} \qquad (1)$$

where g is the gravitational acceleration, D_l is the orifice diameter, d the particle diameter and k the Beverloo correction. This correction factor is used to define an effective flow diameter that does not include the region near the wall [2]. In the current work k is assumed to be 1.3. This characteristic velocity is used to non-dimensionalize experimentally measured discharge rates. The velocity, u, in the upper hopper of diameter, D_2, is related to the velocity at the orifice by the ratio of the cross sectional areas. Hence, a characteristic discharge rate, U, at the fluid-solid interface is:

$$U = \frac{u}{u_{Dl}} \cdot \frac{(D_2 - kd)^2}{(D_1 - kd)^2} = \frac{u(D_2 - kd)^2}{g^{1/2}(D_1 - kd)^{5/2}} \qquad (2)$$

In granular flows with significant interstitial fluid effects, an appropriate dimensionless parameter is the Bagnold number, Ba [3, 4],

$$Ba = (\rho_p - \rho_f)d^2\gamma/\mu \qquad (3)$$

where γ is a shear rate. In shear flows, interstitial fluid effects are important for Bagnold numbers less than 500. For hopper flow, an extensional shear rate can be defined for a given geometry [3],

$$\gamma = (\sin a)g^{1/2}/[4(D_1 - kd)^{1/2}] \qquad (4)$$

where a is the hopper half angle (or the estimated angle of repose, of the stagnant granular material near the hopper in the 90° hourglass).

Material	Lead	Lead	Lead	Glass	Glass	Glass	Glass
Diameter (mm)	0.5	2.0	3.5	0.6	1.6	3.0	4.0
Density (g/cm^3)	11.0	11.0	11.0	2.5	2.5	2.5	2.5

Table I Particle Properties

Fluid	Air	Silicon Oil	Water-Glycerin mixtures
Viscosity (g/(cm.s))	0.0018	1 or 5 or 10	0.01 to 6
Density (g/cm^3)	0.0012	1	0.95 to 1.2

Table II Fluid Properties

EXPERIMENT

Three mass flow hourglasses were constructed for the experiment. Two were made from blown glass and had approximately 15° and 30° hopper half angles. The third was machined out of acrylic and had a 90° hopper half angle. All the hourglasses had a region with a constant cross sectional area (between 15 cm^2 and 20 cm^2) that was at least 20 cm in length. The hourglass neck was approximately 2 cm in diameter.

In the experiments, the hourglass was inverted by hand and the fluid/grain interface location recorded on video. This footage was digitized over a distance of 7 to 10 cm, which started approximately 5 cm below the top of the hourglass and ended 5 cm above the orifice. This eliminated end effects. A best-fit line for the variation of interface height and time was regressed on the position-time data. The slope of this line was the inverse of the average discharge velocity. Experiments were repeated at least seven times for each fluid, particle and hourglass combination after which an average velocity was found. For most flows, the standard deviation in the measurements was less than 7%. However, as described below, some experiments had larger variations in the discharge rate.

Physical properties of the monosized spherical particles and the interstitial fluids used in the experiments are found in Tables I and II. Glass beads were discharged in air, pure water, and water-glycerin mixtures, while lead shot was discharged in water and silicon oil. The density and viscosity of the water-glycerin mixtures were found using temperature and specific gravity measurements of these mixtures and tables.

RESULTS

The observed flow patterns are sketched in figure 2. The flows were classified as dry,

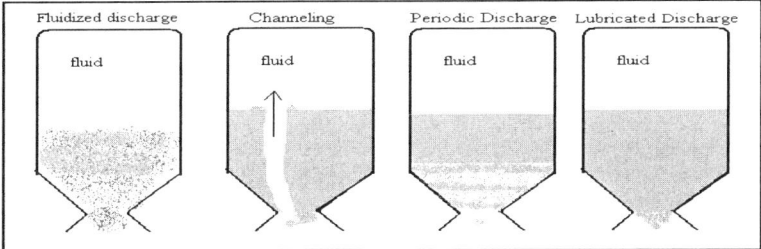

Figure 2 Discharge patterns for a fluid filled hourglass

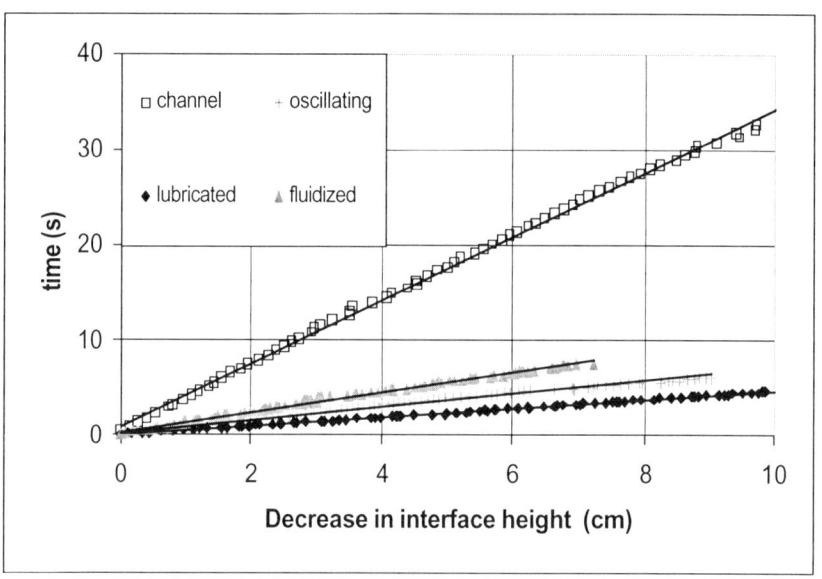

Figure 3 Interface Position-time graphs

lubricated, oscillatory, channeling or fluidized based on visual observations. Discharge velocities for flows in air corresponded with empirical relations found in the literature [2]. However, for the smallest glass particle size (0.6 mm), the flow rate was considerably smaller than predicted, suggesting that the discharge rate was reduced by the upward airflow. Lubricated discharge describes flows that appear dry-like, but where the interstitial liquid reduces friction by lubricating particle motion along the hopper walls. The fluid-solid interface receded linearly with time, as shown in Fig. 3, and the particles were in constant contact in this regime. For example, the flow of 3-mm glass spheres in water looked similar to discharge in air with the discharge velocity decreased by a factor of 8 due to the drag of the upward moving liquid. This drag can be estimated from Ergun's law [4], which relates the driving pressure gradient to a viscous (Darcy) term and an inertial term. Hence, discharge rates increased when interstitial fluid viscosity decreased or particle diameter increased. An analysis similar to Drucker et al. [1] based on Ergun's law satisfactorily predicts discharge rate in this regime.

Fluidized discharge occurred for more viscous fluids and for smaller particles. The upward fluid pressure was comparable to the weight of the particles. In most cases, fluidization was observed only at the constriction but for the smallest particles (such as the 0.6mm glass beads in water and glycerol), the entire bed was fluidized. Position-time data showed the greatest variation over short time scales for this regime (see figure 3), and resulted in the greatest standard deviations in average discharge rate. Several of the fluidized experiments had standard deviations greater than 17%. Oscillatory discharge occurred when liquid flow was large enough

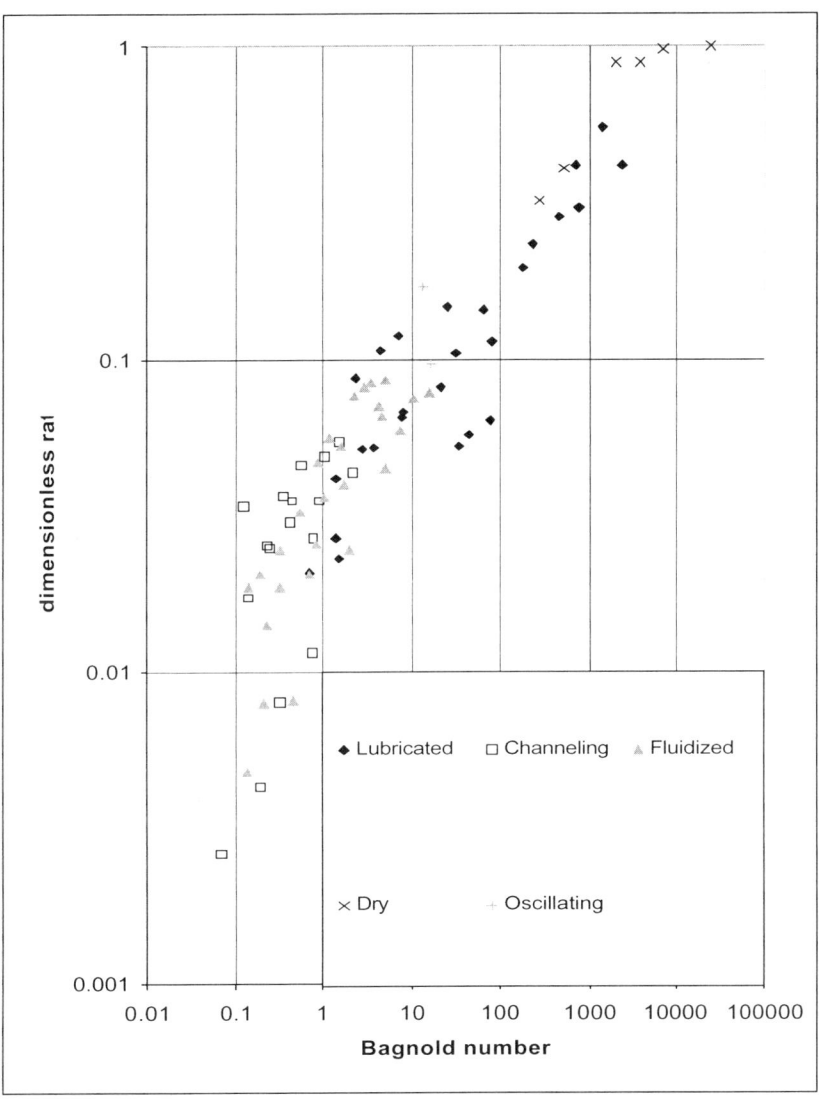

Figure 4 Graph of Dimensionless Discharge Rate against Bagnold Number illustrating the different regimes

to start fluidization at the orifice but not to maintain it. Hence, discharge shifted from a fluidized to a packed bed, creating density waves in the hopper. The discharge rate was constant over the distances recorded in all regimes; however, over short time scales, variation around the mean rate was observed as can be seen from figure 3.

A channel near the hourglass wall sometimes served as the conduit for upward fluid flow. The channel formed on the side of the hourglass that was facing upwards when the hourglass was inverted because of reduced solid fraction and hence larger permeability there both during and after inversion.

Figure 4 presents the dimensionless discharge rate, U, for all of the experiments as a function of the Bagnold number (Ba). The data are identified by the observed flow regimes. Dry flows correspond to $U \approx 1$, and are for the highest dimensionless numbers where viscous effects are negligible. Lubricated flows occur for dimensionless numbers between 1 and 1000. For some experiments with $Ba \leq 5$, the fluid created a channel or a fluidized bed. Waves were observed for some flows around $Ba \approx 5$. For $Ba < 0.02$, the discharge rate could not be measured consistently. The figure shows several points that are labeled as the lubricated regime with Bagnold numbers below 1. These points correspond to lead particles greater than 2 mm in size. Calculations of pressure at the orifice indicate that these particles should be fluidized; however, the large particle size and density may make particle-constriction interactions significant enough to prevent fluidization.

CONCLUSION

This paper presents preliminary results of a study that examines the effects of a viscous fluid on the discharge of granular material in an hourglass. Below a Bagnold number of approximately 1000, the discharge velocity decreases due to interstitial fluid effects. As the dimensionless number is decreased further, the flow exhibits different unsteady flow patterns including fluidization and channeling. Additional studies will focus on local measurements of pressure conditions at the orifice, and analytical descriptions of the discharge behavior.

ACKNOWLEDGEMENTS

Thanks go to Gustavo Joseph, John Van Deusen, Rodney Rojas, and Richard Gerhart and to Drs. R.F. Scott, C.E. Brennen, and C. Campbell for their help, encouragement and advice. The National Science Foundation and Schlumberger supported this research.

REFERENCES

1) J.R. Drucker, M.E. Drucker, R.M. Nedderman, The Discharge of Granular Materials from Unventilated Hoppers, *Powder Technology*, **42**, 3-14 (1985)
2) R.M. Nedderman, *Statics and Kinematics of Granular Materials*, (Cambridge University Press, Cambridge, 1992), pp 292-328
3) G. Zeininger, C.E. Brennen, Interstitial fluid effects in hopper flows of granular materials, *Proc. Cavitation and Multiphase Flow Forum*, Albuquerque, NM, (American Society of Mechanical Engineering, New York, 1985).
4) D. Gidaspow, *Multiphase Flow and Fluidization*, (Academic Press, San Diego, 1994), pp102-103, 304-305

Mat. Res. Soc. Symp. Proc. Vol. 627 © Materials Research Society

Effect of Polydispersity on Stresses in Granular Shear Flow

Meheboob Alam, Richard Clelland[1], and Christine M. Hrenya
Department of Chemical Engineering, University of Colorado
[1]Department of Mathematics, University of Colorado
Boulder, CO 80309

ABSTRACT

Event-driven molecular dynamics simulations are performed for the simple shear flow of smooth inelastic disks, with a focus on the effect of polydispersity on stresses. Simulations are conducted for both binary mixtures and polydisperse media with a Gaussian size distribution. For the binary mixture, the total solids volume fraction ($\phi = \phi_1 + \phi_2$), the solids fraction ratio (ϕ_1 / ϕ_2) the particle diameter ratio (d_1 / d_2) and the coefficient of restitution (e_p) are varied, and the simulation results are compared with an existing kinetic-theory model. The calculated stresses compare reasonably well with the model predictions for $d_1 / d_2 < 2$ in the nearly elastic limit, but the agreement deteriorates significantly at larger size ratios. Furthermore, the assumption of equipartition of granular energy between the two particle sizes appears to have a very limited range of validity in terms of both particle-size ratio and inelasticity. For a granular mixture characterized by a Gaussian size distribution, simulations are carried out in which the standard deviation in particle diameter, the coefficient of restitution and the total solids fraction are varied. The calculated stresses are in good agreement with predictions obtained from an existing kinetic-theory model for an "equivalent" monodisperse mixture.

INTRODUCTION

Over the past two decades, computer simulation studies and kinetic-theory models have been widely used to probe the rheological behavior of granular materials in the rapid flow regime with moderate success (see, for a review, Campbell, 1990). Most of the simulation studies to date are confined to monodisperse media, whereas a real granular system is typically characterized by some degrees of polydispersity in size and/or density. In powder processing, polydispersity often leads to unwanted size segregation in an otherwise homogeneous mixture. From both a practical and fundamental viewpoint, there is a strong need to better understand the effects of polydispersity on the rheology of such systems. This can be achieved by conducting "ideal" experiments using molecular dynamics (MD) simulations of simple flow configurations (e.g. simple shear flow, Poiseuille flow etc.). The results of such simulations can then be used to probe the range of validity of the currently available constitutive models, and thereby eventually lead to improved theories.

Several kinetic-theory models for rapid granular flows of binary mixtures have been proposed by extending earlier studies on monodisperse mixtures (Jenkins and Mancini, 1987; Jenkins and Mancini, 1989; Arnarson and Willits, 1998; Willits and Arnarson, 1999). Jenkins and Mancini (1987) assumed a Maxwellian velocity distribution to derive the constitutive relations for rapid flows of smooth, nearly elastic disks and spheres. An extension to this model for spheres was later proposed by Jenkins and Mancini (1989) using the revised Enskog theory, in which the assumption of a Maxwellian velocity distribution was lifted. The resulting constitutive model was recently corrected by Arnarson and Willits (1998), and then later extended to disks (Willits and Arnarson, 1999). The predictions for the shear viscosity coefficient obtained using the constitutive model of Willits and Arnarson showed excellent agreement with the Monte Carlo simulation data for a system of perfectly elastic hard disks with a diameter ratio of 1.25 over a range of densities. A similar comparison with the model of Jenkins and Mancini (1987) demonstrated poor agreement, thereby indicating that the correction due to a non-Maxwellian velocity distribution can be very significant.

Several computer simulation studies have been performed to unfold the rheology of a binary granular mixture. Ladd and Walton (1989) performed hard-sphere simulations for a simple shear flow of binary mixtures of smooth inelastic spheres. They reported the corresponding shear stress as a function of particle size ratio for a relatively dense system at an overall solids fraction (ϕ) of 0.5 and a coefficient of restitution (e_p) equal to 0.95. Their simulation data were later compared to the kinetic model predictions of Jenkins and Mancini (1989) with reasonable agreement. More recently, Louge and Jenkins (1997) carried out hard-sphere simulations for binary-sized rough spheres in an effort to isolate the mechanism behind the size-segregation phenomenon and also to guide the design of a microgravity shear cell. They considered a flow configuration between two oppositely moving walls characterized by different levels of roughness, resulting in a nonuniform shear field. The results of this study showed the tendency of large particles to migrate to regions of low granular energy. Another simulation study of relevance considers soft, rough disks (Karion and Hunt, 1999) for a similar flow field in the dense limit ($\phi = 0.75$). The comparison of their simulation data for stresses with the model predictions of Jenkins and Mancini (1987) for simple shear flow showed that the theoretical model significantly underpredicted the calculated stress levels. This study also revealed that the equipartition of energy, which is a major assumption in kinetic-theory models, does not hold even at small diameter ratios.

Although the previous computer simulation studies have shed light on the effect of a binary size distribution on certain aspects of rapid granular flows, a comprehensive assessment of the predictive ability of the existing kinetic-theory models has not been performed. Hence, one of the objectives of the current work is to perform a systematic comparison between the constitutive model developed by Willits and Arnarson (1999) for binary mixtures of smooth inelastic disks and the results obtained from discrete particle simulations using a hard-disk model. In addition, simulations are also performed for granular mixtures characterized by a Gaussian distribution of particle sizes. In both cases, the flow configuration is that of a simple shear, and the effects of particle size distribution on the normal and shear stresses are examined.

COMPUTATIONAL MODEL

The simulated system consists of smooth inelastic hard disks of constant density (ρ_p) in a period box of size $H \times H$ which is shearing in the x-direction. The state of simple shear flow (SSF) is attained using Lees-Edwards boundary condition (Lees and Edwards, 1972), which accounts for both the periodicity in the y-direction and the momentum transfer imparted by shearing. The top and bottom image boxes are driven in the opposite directions with a relative velocity of U to maintain a uniform shear rate ($\gamma = U / H$) in the y-direction. Essentially, the SSF represents an *extended* doubly-periodic system where the periodicity in the transverse direction is in the local Lagrangian frame. The box size H is set equal to one after non-dimensionalization and sets the length scale for the system; the top plate moves with a dimensionless velocity of 0.5 and the bottom plate with –0.5.

The disks are initially placed in a nearly square lattice with small random displacements in both directions. The initial velocity field is composed of the uniform shear component and a Gaussian random component. Given the initial positions and velocities of the particles, an event-driven algorithm is used to simulate instantaneous binary collisions by exactly solving a quadratic equation for collision time. The link-cell algorithm (Allen and Tildesley, 1989) is used for the efficient search of future collision partners. Once the minimum collision time is determined, all particle positions are advanced following the well-known kinematic relation. At this point, only two particles are in contact and the collision dynamics is implemented by updating their velocities and simultaneously calculating the collisional impulse. The collisional component of the stress is calculated by accumulating collisional impulse over time, and the kinetic component from the change in particle velocities. The details are omitted for the sake of brevity and the reader is referred to Hopkins and Louge (1991). The average fluctuating kinetic energy, referred to as granular energy (E), is also calculated to ascertain the steady-state of the system. The simulation is allowed to proceed until a constant granular energy is achieved (in the statistical sense). Generally, it takes about 500 to 1000 collisions per particle to reach a statistical steady-state, and thereafter the stresses are calculated by running the simulation for another 5000 collisions per particle. Note that the stresses are averaged over the central periodic box, and are nondimensionalized by $\rho_p d^2 \dot{\gamma}^2$; for a binary mixture, d is the diameter of the large particle and for the polydisperse case, d is the mean-square particle diameter.

There are five control parameters for the binary mixture: the total solids fraction (ϕ), the solids fraction ratio (ϕ_1 / ϕ_2) the diameter ratio (d_1 / d_2), the diameter of large particle (d_1), and the coefficient of restitution (e_p). For a given ϕ, the particle diameters can be controlled by controlling the number of particles (N_1 and N_2). For all the simulations on binary mixtures, the diameter of the larger particle is fixed at 0.05046 which corresponds to a box size (H / d_1) of 19.817; all other control parameters are varied. For the polydisperse media with a Gaussian size-distribution, there are four control parameters: the total solids fraction (ϕ), the mean square particle diameter (d_{ms}), the standard deviation in diameter (σ_d) and the coefficient of restitution (e_p). For this case, d_{ms} is kept constant at 0.01595 which corresponds to a box size of 62.695 and the simulations are performed by varying the remaining three control parameters.

RESULTS AND DISCUSSION

Before considering the results for binary mixtures, the stress calculations for monosize granular media are presented. Figure 1 shows the variation of dimensionless stresses with solids fraction for two different values of e_p. The circles and triangles denote the two components of the normal stress (Σ_{xx} and Σ_{yy}), while the squares denote shear stress (Σ_{xy}). In each subplot, the calculated stresses are compared with the kinetic model predictions of Jenkins and Richman (1985). The simulated results show excellent matching with theory for $e_p = 0.95$. As the value of e_p is decreased to 0.8, however, the agreement deteriorates and a significant normal-stress difference is observed. Note that the constitutive model of Jenkins and Richman is valid only in the nearly elastic limit ($e_p \sim 1$) and the presence of normal stress difference can be explained considering a higher-order model (Jenkins and Richman, 1988).

For the case of a binary mixture, the variation of dimensionless stresses with the particle diameter ratio (d_1 / d_2), is shown in figure 2 for two solids fraction ratios ($\phi_1 / \phi_2 = 1,4$) and two different coefficients of restitution ($e_p = 0.95, 0.8$); the total solids fraction is set to 0.1. In each subplot, the corresponding predictions of the constitutive model of Willits and Arnarson

(1999) are also plotted for comparison; the solid and dashed lines represent pressure and shear stress, respectively. For all cases, there is good qualitative agreement between the simulation results and model predictions. It is seen in figure 2a that the shear stress data matches quite well with model predictions up to a diameter ratio of 3, but the model overpredicts normal stresses beyond $d_1 / d_2 \geq 2$. Changing the solids fraction ratio by a factor of 4 results in a better matching with model predictions as seen in figure 2b. The results for the more dissipative case ($e_p = 0.8$) are similar to that of the nearly elastic case as seen in figures 2c and 2d, but the agreement with model predictions is inferior over the whole range of size ratios due to reasons outlined in the preceding paragraph.

Figure 3 displays the ratio of the granular energy between the two species with diameter ratio at a total solids fraction of 0.1 for two different values of ϕ_1 / ϕ_2. It is observed from figure 3a that the granular energy is not equipartitioned at large size ratios irrespective of the value of e_p. Even at $d_1 / d_2 = 2$, there is a considerable non-equipartition of energy for $e_p = 0.8$, and it continues to become more prevalent with decreasing e_p. Increasing the solids fraction ratio from 1 to 4 does not affect this overall trend, as observed in figure 3b. (The results for two other solids fractions ($\phi = 0.3, 0.5$) are similar.) Thus, it appears that the assumption of equipartition of energy has a very limited range of validity in terms of both the size ratio and inelasticity.

Lastly, the variation of stresses for a polydisperse mixture with Gaussian size distribution is shown in figure 4 at a total solids fraction of 0.1 for two different values of e_p. The mean square particle diameter (d_{ms}) is set to 0.01595. The circles and squares represent pressure ($= \frac{1}{2} (\Sigma_{xx}$ and $\Sigma_{yy})$) and shear stress, respectively. The solid and dashed lines correspond to the kinetic model predictions of Jenkins and Richman (1985) for an equivalent monosize mixture with a particle diameter equal to d_{ms}. The calculated stresses decrease only by a few percent even at a 30% standard deviation of particle size, and kinetic-theory predictions for an equivalent monodisperse mixture provide excellent agreement.

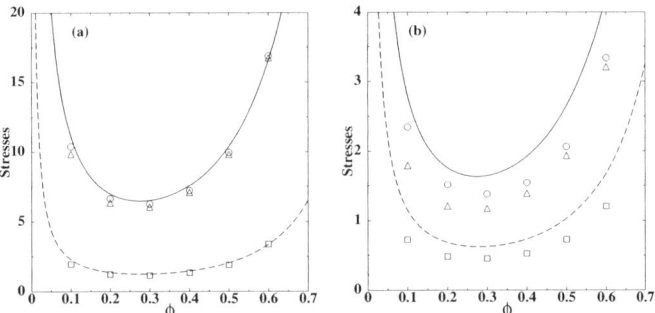

Figure 1. *Variation of dimensionless stresses with solids fraction for a monosize granular mixture in simple shear flow: H / d = 19.817; (a) e_p = 0.95, (b) e_p = 0.80; (o, Σ_{xx}; Δ, Σ_{yy}; $\square$, Σ_{xy}). The solid and dashed lines represent the corresponding kinetic model predictions of Jenkins and Richman (1985).*

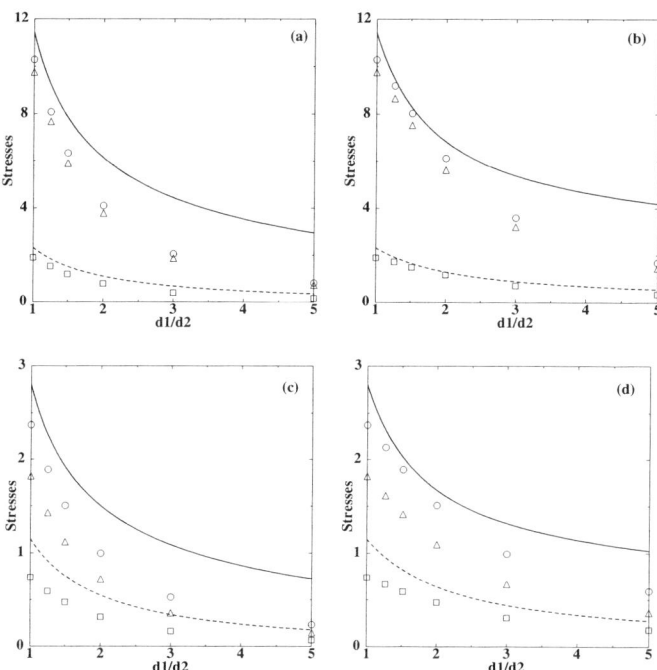

Figure 2. *Comparison of calculated stresses for a binary granular mixture with kinetic model predictions: $\phi = 0.1$; (a) $e_p = 0.95$, $\phi_1 / \phi_2 = 1.0$; (b) $e_p = 0.95$, $\phi_1 / \phi_2 = 4.0$; (c) $e_p = 0.8$, $\phi_1 / \phi_2 = 1.0$; (d) $e_p = 0.8$, $\phi_1 / \phi_2 = 4.0$. Symbols are the same as in Figure 1.*

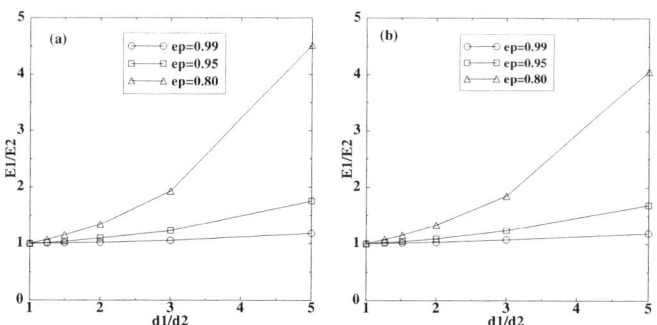

Figure 3. *Non-equipartition of granular energy in simple shear flow: $\phi = 0.1$; (a) $\phi_1 / \phi_2 = 1.0$, (b) $\phi_1 / \phi_2 = 4.0$.*

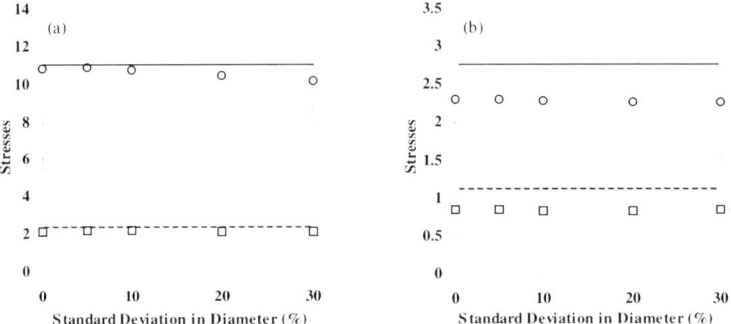

Figure 4. *Effect of Gaussian particle size distribution on stresses: $\phi = 0.1$, $d_{ms} / H = 0.01595$; (a) $e_p = 0.95$, (b) $e_p = 0.8$; (o, Pressure; $\square$, Σ_{xy}). The solid and dashed lines correspond to model predictions of Jenkins and Richman (1985) for an equivalent monodisperse mixture.*

SUMMARY

The present work represents a systematic MD simulation study to probe the rheological behavior of a two-dimensional rapid shear flow of binary and polydisperse granular mixtures. The calculated stresses for the binary mixture were compared with a recent kinetic-theory model (Willits and Arnarson, 1999) with reasonable matching in the appropriate parameter range. The stresses for the mixture with a Gaussian size distribution showed excellent agreement with an equivalent monodisperse mixture having the same mean-square particle diameter.

It has been shown that an equipartition of granular energy is not exhibited at $d_1 / d_2 \geq 2$ even in the nearly elastic limit. The assumption of *molecular chaos* is also not likely to hold at large size ratios due to the possibility of repeated collisions of smaller particles with large ones, signifying the presence of strong positional and velocity correlations. A proper constitutive model should incorporate such effects in a rigorous manner to be useful as a predictive tool.

REFERENCES

Allen, M. P. and D. J. Tildesley, *Computer Simulation of Liquids*, Oxford Univ. Press (1989).
Arnarson, B. O. and J. T. Willits, *Phys. Fluids*, **10**, 1324 (1998).
Campbell, C. S., *Ann. Rev. Fluid Mech.*, **22**, 57 (1990).
Hopkins, M. and M. Louge, *Phys. Fluids A*, **3**, 47 (1991).
Jenkins, J. T. and F. Mancini, *J. Applied Mech.*, **54**, 27 (1987).
Jenkins, J. T. and F. Mancini, *Phys. Fluids A*, **1**, 2050 (1989).
Jenkins, J. T. and M. W. Richman, *Phys. Fluids*, **28**, 3485 (1985).
Jenkins, J. T. and M. W. Richman, *J. Fluid Mech.*, **192**, 313 (1988).
Karion, A. and M. L. Hunt, *Powder Tech.*, submitted (1999).
Ladd, J. C. and O. R. Walton, *Proc. Joint DOE/NSF Workshop of Fluid-Solids Transport*, Pleasanton, CA, 1 (1989).
Lees, A. W. and S. F. Edwards, *J. Phys. C.*, **5**, 1921 (1972).
Louge, M. Y. and J. T. Jenkins, *Mechanics of Deformation and Flow of Particulate Materials*, ASCE, New York, (1997).
Willits, J. T. and B. O. Arnarson, *Phys. Fluids*, **11**, 3116 (1999).

Mat. Res. Soc. Symp. Proc. Vol 627 © 2000 Materials Research Society

A Nonequilibrium Approach for Self-Diffusion in Unbounded Rapid Granular Flows

Payman Jalali[1], Piroz Zamankhan[1] and William Polashenski, Jr.[2]
[1]Department of Energy Technology, Lappeenranta University of Technology, Lappeenranta, Finland
[2]Lomic, Inc., PA 16803, U.S.A

ABSTRACT

A nonequilibrium simulation scheme is introduced to investigate the transverse diffusive motion in unbounded shear flows of smooth, monodisperse, inelastic spherical particles. A certain labeling algorithm is used in this scheme to extract a one-way particle mass flux which results a concentration gradient for the labeled particles. The self-diffusion coefficient can then be obtained from Fick's law. Using this scheme, one may find that the self-diffusion phenomenon across any layer inside the granular shear flow is analogous to the classic diffusion problem across a membrane. Under steady conditions, the current simulation results revealed that the particle diffusivity can be described by a linear law. This finding justifies the assumption of a linear law relationship in the kinetic theory type derivation of an expression for self-diffusivity. Moreover, it is shown that the results of self-diffusion coefficient obtained from the computer simulations are in agreement with the predictions of kinetic theory formulations in the range of solid volume fractions less than 0.5.

INTRODUCTION

Study of mass transfer in powders and granular materials is of interest in many industrial applications or research fields. In this context, a precise study of diffusion can improve our understanding of mass transport and mixing process. The self-diffusion phenomenon in the flows of granular materials has been studied experimentally, theoretically or by computer simulations. In the experimental efforts [1-2] the self-diffusion coefficient in different types of flows were found for monosized spherical particles. Meanwhile, computer simulation techniques have also been developed to study of granular flows. The self-diffusion coefficient was extracted from the results of computer simulations [3-5] using the velocity autocorrelation function (VAF) or mean-square displacement formula (MSDF). However, MSDF and VAF are completely equivalent, so that one can be derived from the other [6]. These formulas are based on a linear law, but it is not clear whether the linear or non-linear terms are governing the diffusion process, especially, the linearity of self-diffusion in bounded flows may be questionable. Rapid granular flows may be modeled based on the kinetic theory to obtain a continuum description in which the kinetic theory formulations of gases is modified to consider the energy dissipated in the collisions between particles [7]. The dimensionless granular temperature, which is a measure of the fluctuating energy of particles, could be expressed as a function of coefficient of restitution and solid volume fraction and consequently the self-diffusion coefficient is evaluated using the value of granular temperature. In the present paper, a method is introduced to analyze the self-diffusion process in the rapid granular flows, which are simulated using monodisperse, inelastic spherical particles. In this approach a nonequilibrium method has been used from which the self-diffusion coefficient can be extracted [8] by a certain labeling rule. A one-way flow of labeled particles may be obtained in a slice of bulk flow at equilibrium. When the steady state is achieved, the gradient of the concentration of marked particles can be obtained and the self-diffusion coefficient may be calculated using the proportionality of mass flux and concentration gradient. The results showed that the predictions of self-diffusion coefficient by kinetic theory formula are in a good agreement with the results of nonequilibrium simulation.

COMPUTER SIMULATION

A system of monodisperse, smooth, inelastic spherical grains is simulated inside a shear cell of dimensions $1 \times 1 \times H$. The number of particles in the cell changes from 4296 to 9679. The location of the box could be imagined to be at the middle of a homogeneous shear flow in which horizontal layers have relative motions and a layer consists of rectangular cells. Each cell can exchange some particles with the neighboring cells according to the proper boundary conditions. The mid-cell shown in figure 1(a) represents the simulation box. In this box, an average streamwise relative velocity of $2U_0$ is given between the top and bottom faces. Two types of Periodic Boundary Conditions (PBC) have been implemented [9]. For the four vertical faces of the box, the usual type of PBC is used in which any particle leaving one face is substituted by another one coming at the same position on the opposite face with the same velocity vector. For example, particle b in figure 1(a) is replaced by particle b′ by an identical velocity vector. The rule of particle exchange between the cells across the horizontal faces is different from the vertical ones. A modified PBC has been proposed by Lees and Edwards [5,9] in which not only does the longitudinal position of the incoming particle depend on the relative displacement of moving cells in the upper and lower layers, but also its streamwise velocity component has to be adjusted in accordance with the velocity of the front face. In figure 1(a), particle a is exchanged by a′ between the horizontal faces. This type of PBC is used to set up and maintain a steady velocity profile, which is approximately linear.

The hard sphere potential function has been used in the current simulation. Since all the grains fall with a zero relative acceleration in the presence of gravity, the effects of gravitational force may be neglected in the simulation of unbounded shear flows. The simulation is started from a randomly distributed initial configuration. The event-driven algorithm has been used in simulations in which successive binary collisions are determined and after a collision, the velocity components of the pair in contact will be changed according to the laws of collision dynamics while the rest of particles will continue their individual motions [9]. The formulation of collisions is described by Lun and Bent [10] in detail, so it is not discussed further here. The coefficient of restitution is assumed to be a function of normal impact velocity of colliding particles [11].

SELF-DIFFUSION COEFFICIENT

In a system of identical noncohesive particles, the random walk mechanism may govern the percolation of particles. If the particles in a certain region of the system are labeled by a different color than the others at any origin of time, the particles will diffuse into neighboring regions by the random walk mechanism. In the macroscale, the distribution of labeled particles along the transverse direction may be described by Fick's law. The self-diffusion coefficient, D, can be obtained using the numerical values of mass flux and concentration gradient. The nonequilibrium method presented by Holian [8] may be applied in granular flows for calculating self-diffusion. Consider an arbitrary rectangular slice cut with thickness l inside the simulation box somewhere

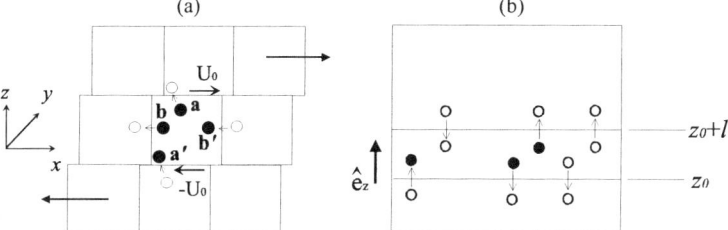

Figure 1. *Schematic of simulation box and labeling rule. (a) Different layers are shown in relative motion and each layer consists of periodic cells. The middle box is the simulated shear cell in which two types of PBC are shown. (b) Demonstration of labeling rule. Black circles represent the labeled particles and white circles for unlabeled ones.*

along the z-axis as sketched in figure 1(b). The main flow is at steady state and the system is in equilibrium. In such a condition, a nonequilibrium subsystem can be created by assuming the aforementioned slice as a control volume, which initially does not contain any particles. A certain direction is selected to label the particles entering to the control volume. Therefore, an incoming flow of particles is expected across the appropriate face of slice. The labeling and unlabeling rule of particles is schematically shown in figure 1(b). It is obvious that the marked particles, which are also in an upward mean motion, can be found just inside the layer. After some time, the diffusion process throughout the control volume will reach equilibrium and the mean diffusive flux of the labeled particles will remain almost constant. The labeling rule can be performed with a certain probability, p, in which particles are labeled with different colors rather than black and white.

The influx and efflux of particles are denoted by $n_i(z_0,t)$ and $n_o(z_0+l,t)$, respectively, in a slice from z_0 to z_0+l. The number of tagged particles in the layer at any time t, can be calculated by,

$$N(t) = \int_0^t n_i(z_0,t')\,dt' - \int_0^t n_o(z_0+l,t')\,dt' \qquad (1)$$

$N(t)$ is initially zero, but it increases with time and reaches to a steady value beyond a transient time. Therefore, $N(t)$ is an important parameter which characterizes the steady state of diffusion process in the slice. However, another quantity, $n(t)=(N(t)/N_0)(H/l)$, may be defined which shows the ratio of the number of labeled particles in the given layer to the total number of particles in the box. The concentration profile of diffusive labeled particles is linear which confirms that the effect of nonlinear factors such as the super-Burnett terms [8] are unimportant. Consequently, the diffusive flux may be simply related to the gradient of concentration via Fick's law.

If the concentration is replaced by solid volume fraction φ, and the vertical mean velocity of labeled particles is denoted by $\bar{v}_z$, the instantaneous volumetric flux of labeled particles per unit area flowing at z could be calculated simply by

$$j_z(z,t) = \varphi(z,t).\bar{v}_z(z,t) \tag{2}$$

The fluctuating nature of various macroscopic properties within the sequential time steps, requires to express them as a time average, which is denoted by $\langle ... \rangle$. By calculating the time average of the volumetric flux in the labeling layer and the time average of the concentration gradient in the layer, the diffusion coefficient corresponding to the labeling layer can be determined as

$$\overline{D} = -\overline{\langle j_z \rangle} / \overline{\langle \nabla \varphi_z \rangle} \tag{3}$$

Where the overbar shows the spatial average over the main layer of thickness l.

The kinetic theory analysis of rapid granular flows is taken from the ordinary nonequilibrium kinetic theory for dense gases, with a modification of considering the inelasticity of the particles in the corresponding theory for granular flows. Inelasticity dissipates the kinetic energy during every collision, and thus governs the granular temperature and diffusivity in granular flows [7]. The granular temperature is determined from the balance of the rate of energy dissipation due to the inelastic collisions to the rate of energy production. In steady, homogeneous rectilinear shearing flow of grains with a constant shear rate, the value of dimensionless granular temperature can be obtained as [7],

$$T^{*} = \frac{T}{\sigma^2 \gamma^2} = \frac{\dfrac{\mu}{\omega}}{9(1-e)} \tag{4}$$

Where μ is the shear viscosity and ω represents the bulk viscosity. Also, γ represent the shear rate and the coefficient of restitution and the diameter of a particle are denoted by e and σ, respectively. The ratio of shear viscosity to the bulk viscosity may be found as follows [7],

$$\mu / \omega = \frac{3}{5} + \frac{5\pi \left[1 + \dfrac{2}{5}\varphi g_0(3e-1)(e+1)\right]\left[1 + \dfrac{4}{5}\varphi g_0(1+e)\right]}{32\varphi^2 g_0^2 (1+e)^2 (3-e)} \tag{5}$$

Here, g_0 is the radial distribution function at contact, which is a function of solid volume fraction. The expression given by Goldshtein et al. [12] has been used to calculate the contact value of the equilibrium radial distribution function,

$$g_0(\varphi) = 1/(1 - (\varphi / \varphi_m)^{4\varphi_m/3}) \tag{6}$$

Where φ_m represents the maximum shearable solid fraction for hard spheres, which may be taken as 0.641. The dimensionless self-diffusion coefficient can be expressed as a function of dimensionless granular temperature, solid volume fraction and coefficient of restitution [3],

$$D^* = \frac{D}{\sigma^2 \gamma} = \frac{\sqrt{\pi T^*}}{8(1+e)\varphi g_0} \tag{7}$$

RESULTS AND DISCUSSION

In this part the results of the nonequilibrium approach for a sample case are presented, from which the self-diffusion coefficient is extracted. Moreover, the predictions of kinetic theory formula are compared with the nonequilibrium simulation results.

In the steady state of a sample case of the present simulations, the time-averaged profiles of flow properties were found to be constant. The mean solid volume fraction in this case is $\varphi = 0.185$, with $e = 0.93$ and $\gamma = 2$ s^{-1}. The size of simulation box is $1\times1\times1$ and the number of particles inside the box is 9679. The simulation box is divided into a certain number of bins along the z-axis and the mean values of each variable are calculated inside the bins.

The results of the nonequilibrium approach for the sample case are presented in figure 2, for some different labeling probabilities. The labeling layer is 0.2 times of the height of simulation box located in the middle. The parameter n changes with dimensionless time, defined as $t^* = (\gamma/2)t$, as shown in figure 2(a). The concentration profiles of labeled particles is shown in figure 2(b). The profile of concentration is linear which implies that the linear-response theory, which is summarized in equation 3, properly describes the mass transport phenomenon in the system [8]. The slope of this line gives the steady value of concentration gradient of the labeled particles. The time average of vertical velocity component of labeled particles is plotted for different bins of the layer in figure 2(c). This time-average velocity of labeled particles can be fitted by a hyperbolic function in the form of:

$$v_z = \frac{a_1}{z - z_0} + a_2 \tag{8}$$

Multiplying this velocity profile by the concentration profile will give the mean profile of diffusive mass flux. Now, the mean value of diffusion coefficient across the layer can be obtained using the values of concentration gradient and mass flux in equation 3. A mean value for the self-diffusion coefficient may be calculated by aver aging on different layers. The mean value of D^* for the middle layer is obtained as 0.750 and the mean value over 5 different labeling layers is D^*=0.718. Using equation 7 one can obtain $D^* = 0.758$, which predicts the self-diffusion coefficient to within 1 percent of that obtained for the middle layer and 5 percent of that obtained from the averaging over some layers.

The values of D^* for different solid volume fractions are illustrated in figure 2(d), which shows a good agreement with the kinetic theory. The given values for the simulations are calculated by averaging on different layers and different probabilities. Using the different probabilities causes to get several series of the results of nonequilibrium approach from one simulation. [8]

The aforementioned procedure may be used to calculate the self-diffusion coefficient in the bounded granular shear flows. The concentration and vertical velocity profiles corresponding to the labeled particles and the entire particles in a layer adjacent to the wall are shown in figure 3. Note that the moving wall is located in $z=0.5$ and the labeling layer is limited between $z=0.3$ to $z=0.46$. The average solid volume fraction in the simulation box is 0.35 with 9679 particles and the apparent shear rate is $2\ s^{-1}$. Although the entire profile of solid volume fraction is not constant through the layer, the linear concentration profile of labeled particles is observed beyond a certain distance from the wall as well as a hyperbolic velocity profile at the same region. However, the linearity of concentration or hyperbolic form of velocity collapses near to the wall. This may imply that the self-diffusion phenomenon may not be simply described by equation (3) in the Couette flows within a certain distance from the wall. However, the validity of this fact should also be confirmed by the experimental observations.

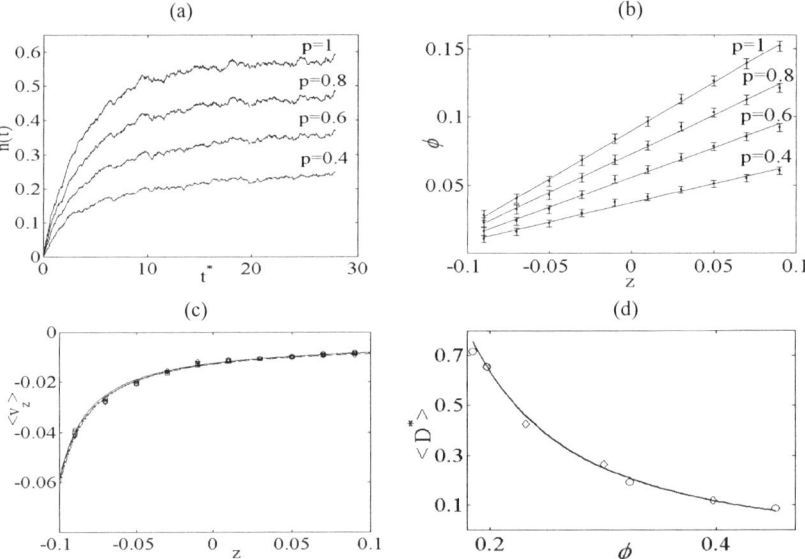

Figure 2. (a) Variation of parameter n(t) with the dimensionless time t^* for different probabilities. Note that $H/l=5$ and $N_0=9679$. (b) The concentration profile of labeled particles with different probabilities. (c) The vertical velocity profile of labeled particles with different probabilities. All the profiles coincided to each other regardless of the value of probability parameter. (d) The calculated average values of dimensionless diffusivity in the given layer from different probabilities. The solid line shows the prediction by kinetic theory. Diamonds and circles represent the simulations with 4296 and 9679 particles, respectively.

CONCLUSIONS

A nonequilibrium method has been developed to calculate the self-diffusion coefficient in shear flows of granules. In this scheme, the self-diffusion phenomenon across any layer inside the granular shear flow has been treated analogous to the classic diffusion problem across a membrane. The linearity of concentration profile implies that a linear law governs the self-

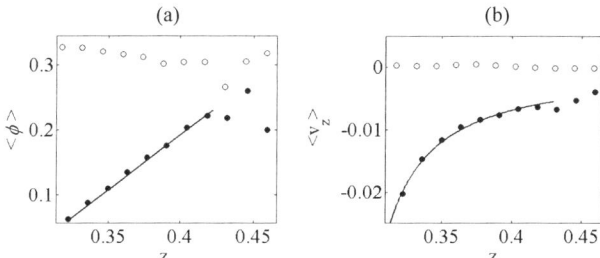

Figure 3. (a) Solid volume fraction profile in a layer adjacent to the wall, which is located at z=0.5. White and black circles represent for the concentration of entire and labeled particles, respectively. (b) Vertical velocity profile in the given layer close to the wall. White and black circles represent for the velocity of entire and labeled particles, respectively.

diffusion phenomenon in unbounded granular shear flows. However, the assumption of a linear law for the self-diffusion phenomenon may fail in the bounded flows close to the wall.

REFERENCES

1.   V.V.R. Natarajan, M.L. Hunt, and E.D. Taylor, J. Fluid Mech., **304**, 1-25 (1995).
2.   N. Menon, and D.J. Durian, Science, **275**, 1920-1922 (1997).
3.   S.B. Savage, and R. Dai, Mechanics of Materials, **16**, 225-238 (1993).
4.   C.S. Campbell, J. Fluid Mech., **348**, 85-102 (1997).
5.   A.W. Lees, and S.F. Edwards, J. Phys., **C5**, 1921-1929 (1972).
6.   J.P. Boon, and S. Yip, Molecular Hydrodynamics, p. 48. New York: Dover Publications, Inc., 1980.
7.   J.T. Jenkins, and M.W. Richman, Arch. Rat. Mech. Anal., **87**, 355-377 (1985).
8.   J.J. Erpenbeck, and W.W. Wood, "Molecular Dynamics Techniques for Hard-Core Systems" in *Statistical Mechanics, Part B: Time-dependent Processes,* Chapter 1. B. J. Berne, Ed., New York: Plenum 1977.
9.   M.P. Allen, and D.J. Tildesley, Computer Simulation of Liquids, p. 24, 246. Oxford: Clarendon Press, 1997.
10.   C.K.K. Lun, and A.A. Bent, J. Fluid Mech., **258**, 335-353 (1994).

11.   T. Schwager, and T. Pöschel, Phys. Rev. E, **57**, 650-654 (1998).
12.   A. Goldshtein, M. Shapiro, and C. Gutfinger, J. Fluid Mech. **316,** 29-51 (1996).

Mat. Res. Soc. Symp. Proc. Vol 627 © 2000 Materials Research Society

Particle Fluctuation Velocity in Gas Fluidized Beds - Fundamental Models Compared to Recent Experimental Data

G. D. Cody,
Visiting Professor Mechanical and Aerospace Engineering, Rutgers University
Mail Address: 30 Bainbridge St, Princeton NJ 08540, E-mail; gdcodypva@att.net

ABSTRACT

The first measurements of the mean squared fluctuation velocity, or granular temperature, of monodispersed glass spheres in gas fluidized beds were recently obtained by two independent techniques: Power Spectral analysis of wall vibrational energy excited by random particle impact or Acoustic Shot Noise (ASN), and Diffusing Wave Spectroscopy (DWS) of reflected laser light multiply scattered by random particle motion. We explore the relevance of this data to the initial stability of the uniform fluidized state and to recent fundamental models for the magnitude, gas flow, and particle diameter dependence of the steady state granular temperature.

INTRODUCTION

In 1948 Wilhelm and Kwauk[1] defined two modes of fluidization in gas and liquid fluidized beds: "Particulate" which exhibits, at and above initial fluidization, a single uniform particle phase and "Aggregative" which exhibits at fluidization gas rich bubbles moving through a dense particle phase. In 1973 Geldart [2] in a series of comprehensive experiments over a wide range of particle diameters and densities for relatively monodispersed particles defined four fluidization regimes for gas fluidized beds of which the Geldart A fluidization regime defined as "fluidizes before bubbling" corresponds to "Particulate" fluidization and Geldart B defined as "bubbles at fluidization" corresponds to "Aggregative" fluidization.

The stability of the uniform fluidized state was first addressed by Jackson and Anderson in the early '60s through spatially averaged equations of motions for the gas and particles and their interactions[3] and an analysis of the initial one dimensional stability of the uniform particulate phase against first order perturbations in particle concentration. A critical parameter for the stability of the particulate phase against the formation of the aggregative phase is the particle pressure, P_S within the uniform fluidized state, which, if sufficiently large, can stabilize the particulate phase. This important observation remained a conjecture for more than 30 years due to the absence of experimental data on, or fundamental models for, P_S.

In this paper we discuss two recent experiments that directly measure the particle mean squared fluctuation velocity $<V^2>$, or granular temperature, $T^*=(1/3)<V^2>$, at and near the wall of gas fluidized beds consisting of monodispersed glass spheres as a function of sphere diameter D and gas superficial velocity U_S. We also present data on the initial bed expansion, $\alpha(D)$ $=\delta \ln BH(U_S)/\delta U_S$ where $BH(U_S)$ is the bed height, obtained in the same experiments and use that data as an independent experimental measure of initial bed stability. From $T^*(D,U_S)$, the granular pressure, $P^*(D,U_S)$ of the uniform particulate phase is easily obtained by dense kinetic theory[4] and we will use it to determine the initial stability of the uniform fluidized state as a function of particle diameter and concentration[5] for the two experiments and compare these results with the initial bed expansion data. Finally we compare the data for $T^*(D,U_S)$ obtained in these experiments with the predictions of recent fundamental models for the steady state granular temperature driven by random impact and shear forces.

EXPERIMENTAL

We define $V_n(D,U_s)^2 = (1/3)<V(D,U_s)^2> = T^*(D,U_s)$ where V_n is the fluctuation velocity toward the wall in the ASN data, and one component of $<V^2>$ in the DWS data. In **Figs. 1 and 2** we exhibit the results of two recent measurements of $V_n(D,U)$ as a function of $U=(U_s/U_{mf})$ where U_{mf} is the value of U_s at fluidization and is presented in **Fig. 4**.

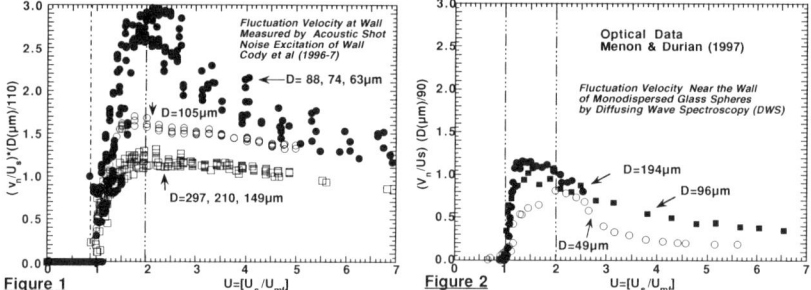

Figure 1   **Figure 2**

The data of **Fig. 1** were taken by Power Spectral analysis of the vibrational wall energy of the wall of a gas fluidized bed utilizing the theory of Acoustic Shot Noise (ASN)[6, 7, 8]. The data of **Fig. 2** were obtained from the analysis of the interference of reflected laser light multiply scattered by the random motion of the spheres utilizing Diffusing Wave Spectroscopy.(DWS)[9]. The trend of all the data is similar: a sharp rise at $V_n(D, U=1)$ reflecting the transition from the fixed to the fluidized bed, reaches a well defined value near $U=2$ and then falls with increasing U_s.

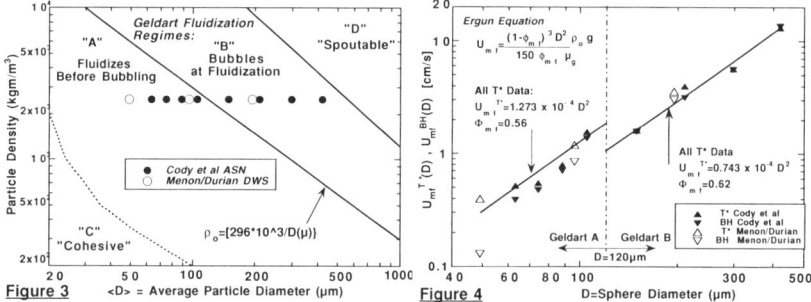

Figure 3   **Figure 4**

The ASN data of **Fig. 1** into three groups at $U=(U_s/U_{mf})=2$, in each of which $[V_n(D,U=2)/(2U_{mf})][D/D_o]=1$ where D_o is a length scale: $D_o =120\mu m$ for $D=149-297\mu m$; $D_o =190\mu m$ for $D=105\mu m$; and $D_o =300\mu m$ for $D=63, 74, 88\mu m$. Given the difference in techniques there is more than satisfactory agreement between the two sets of data for $D=194\mu m$. The *ASN* *data predicts* $[V_n(194\mu m, U=2)/(2U_{mf})]=0.6$ about 20% higher *than the DWS experimental data* with $[V_n(194\mu m, U=2)/(2U_{mf})]=0.5$. The agreement is considerably poorer for spheres with $D=49\mu m$. The *ASN data predicts* $[V_n(49\mu m, U=2)/(2U_{mf})]=6$ about a factor of 4 higher *than the DWS experimental data* with $[V_n(49\mu m, U=2)/(2U_{mf})]\approx1.5$! Since both experiments use monodispersed glass spheres obtained from the same manufacturer, and since neither should be sensitive to changes in diameter by a factor of 2, the difference of a factor of 16 in $T^*(D,U=2)$ demands a fundamental explanation.

Fig. 3 places the spheres in the context of the Geldart fluidization regimes and we note that the data of **Figs. 1, 2** can be summed up in the observation that the Geldart B spheres of **Fig. 1** behave like the Geldart B and Geldart A spheres of **Fig. 2**! In **Fig. 4** we exhibit the Minimum Fluidization Velocity, U_{mf}, obtained in the two experiments. We may define $U_{mf}^{T^*}$ as the value of U_s at the onset of the increase in the granular temperature T^* and U_{mf}^{BH} as the onset of bed expansion. Since the minimum fluidization velocity is physically defined by equating the viscous pressure drop across a vertical region of the fixed bed to the weight of particles in that region, the quantity, U_{mf}, by either criteria, should be described by the semi-empirical Ergun equation[2], exhibited in **Fig. 4**. We note from **Fig. 4** that the data for $U_{mf}^{T^*}$ from both experiments are in good agreement with each other and with an estimated solids concentration, in Geldart B, of $\Phi_{mf}=0.62$, and, in Geldart A, of $\Phi_{mf}=0.56$. We also note good agreement between $U_{mf}^{T^*}$ and U_{mf}^{BH} for all the data except $D=49\mu m$ in the DWS experiments where $(U_{mf}^{T^*}/U_{mf}^{BH})\approx3$!

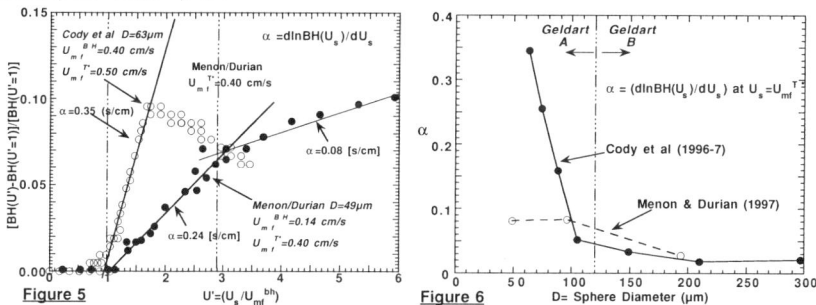

In **Fig. 5** we show bed expansion as a function of $U'=[U_s/U_{mf}^{BH}]$ for $D=63\mu m$ in the ASN experiments and for $D=49\mu m$ in the DWS experiments. The $63\mu m$ data can be interpreted as uniform bed expansion in the particulate phase where there is no alternative path for gas flow. The fall in bed height for $U_s >2U_{mf}^{T^*}$ is characteristic of all the Geldart A spheres of the ASN experiment and can be ascribed to the onset of significant bubbling and subsequent reduction of gas flow expanding the particulate phase. The bed expansion for the DWS monodispersed glass spheres ($D=49\mu m$) in **Fig. 5** is complex. The initial bed expansion, defined by the quaintly $\alpha = [\delta \ln BH(U_s)/\delta U_s]$ *for the fluidized state* $(U_s>U_{mf}^{T^*})$ is $\alpha=0.08$, about a factor of 4 less than the initial slope for $D=63\mu m$, $\alpha=0.35$ in the ASN experiment. However the initial bed expansion *before the fluidized state* $(U_{mf}^{BH} <U_s<U_{mf}^{T^*})$, is $\alpha=0.24[s/cm]$ about 70% of the initial slope

for D=63μm, α=0.35 in the ASN experiment. Presumably *even before the on-set of fluidization* the bed has been expanded by the granular pressure exerted on adjacent spheres. for D=49 μm in the DWS experiments .

In **Fig. 6** we summarize the initial bed expansion data from the two experiments and note the significant increase in bed expansion, α, across the Geldart A/B boundary for the monodispersed glass spheres of Cody et al, that is absent for the fluidized state data of Menon and Durian. An increase in initial bed expansion in the Geldart A regime is expected since bubbles do not offer an alternative path for the gas flow. The absence of such an increase for the monodispersed glass spheres in the DWS experiments is consistent with the absence of a jump in T* across the Geldart A/B boundary exhibited in **Fig. 2**.

THEORY

As noted earlier the granular temperature T*(D,U) is a critical quantity for the determination of the initial stability of the particulate phase of the fluidized bed. Until the mid '90's there were no fundamental calculations or measurements of this quantity for gas fluidized beds. In the discussion that follows we will focus on the well defined diameter dependence of $V_n(D, U=2)$. The Jackson stability criteria considers the one dimensional, first order Navier Stokes equations for the gas and particle motion with physical model for the coupling between the two phases. Following Wallis[10] we define the Jackson first order stability criteria[5] in terms of two velocities, a "dynamic wave velocity" $V_w(D,U)$, and a "continuity wave velocity" $U_e(D,U)$. The uniform particulate phase is only stable with respect to first order, one dimensional perturbations in particle concentrations when $V_w(D,U)>U_e(D,U)$, for $V_w(D,U)<U_e(D,U)$ shock waves are produced from the smallest perturbations in particle concentration.

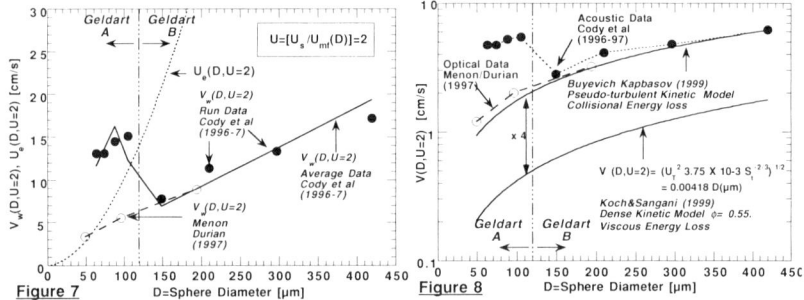

Figure 7   D=Sphere Diameter [μm]

Figure 8   D=Sphere Diameter [μm]

We can define both quantities in terms of two fundamental properties of the uniform particulate phase[5] its bulk modulus E*(D,U) and its concentration $\Phi= [\rho/\rho_0]$, the ratio of the mass density of the particulate phase ρ, to the density of the particle, ρ_0,

$$V_w (D,U)^2 = (E^*/\rho_0) \qquad (1)$$

where $E^*=(\delta P^*(D,U)/\delta\Phi)$ and $P^*(D,U)$ is the particle pressure given in the dense kinetic theory by, $P^*(D,U) = \rho G(\phi)T^*(D,U)$ and from the Carnahan Starling hard sphere statistical model[11] $G(\Phi) = [(1+\Phi+\Phi^2-\Phi^3)/(1-\Phi^3)]$. The continuity wave velocity $U_e(D,U)$ is defined by

$$U_e(D,U) = -\Phi(\delta U_S/\delta\Phi) \qquad (2)$$

which can be closed with the empirical Richardson Zaki equation[2] $U_s = U_t (1-\Phi)^n$ where the Stokes velocity $U_t = (D^2\rho_o g/18\mu_g)$, and g is the gravitational constant and μ_g is the viscosity of the fluidizing gas. The numerical constant n, from the Φ data deduced from the Ergun equation of **Fig. 4** is given by n=4.7 for D>149 μm and n=4.85 for D<105μm. We obtain finally for the condition of stability

$$V_w(D,U_s) = T^*(D,U_s)^{1/2} (\delta\Phi G(\Phi)/\delta\Phi) > U_e(D,U_s) = U_t [n(1-\Phi)^{n-1}] \qquad (3)$$

The quantities $V_w(D,U_s=2U_{mf})$ and $U_e(D,U_s=2U_{mf})$ are plotted in **Fig. 7** for the monodispersed glass spheres of Cody et al and Menon and Durian utilizing the data of **Fig. 4** for U_{mf}. The experimental data for $T^*(D,U=2)$ of the ASN Geldart A monodispersed glass spheres makes the uniform particulate phase stable at U=2 and the Geldart B glass spheres unstable. For the DWS experiments both the Geldart A and Geldart B spheres should exhibit an unstable particulate phase.

If one considers instability to be source of the Geldart B description, "bubbling at fluidization", the data displayed in **Fig. 7** account for the striking difference in bed expansion between the two sets of data displayed in **Fig. 5 and 6**. As noted earlier one explanation of the bed expansion observed prior to fluidization for the D=49μm glass sphere in the DWS experiments is that the granular pressure exerted on the light particles, while sufficient to expand the *granular state* of the "yet to be fluidized bed", is not sufficient to produce Geldart A behavior in the fluidized state.

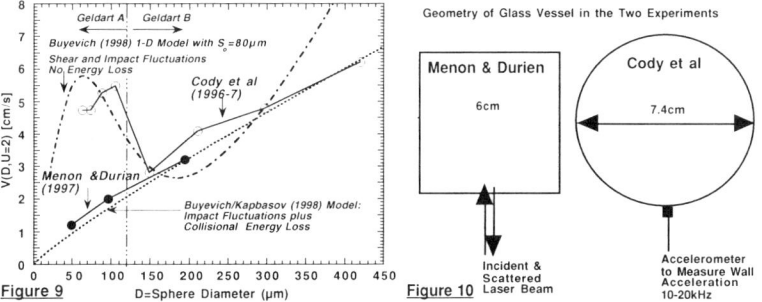

Figure 9   Figure 10

In **Fig. 8** we compare both sets of experimental data with a fundamental statistical model due to Buyevich and Kapbasov[12, 13, 14, 15, 16]. The model applies the power spectral formulation of steady state turbulence to the simpler problem of calculating the steady state mean squared fluctuation velocity of particles in the dense phase of a fluidized bed. It imposes isotropy on the particle fluctuation velocity, assumes collisional energy loss, and utilizes dense kinetic theory for the collision frequency and particle pressure. **Fig. 8** exhibits the excellent one parameter fit to the Geldart B ASN data of Cody et al, and all the DWS data of Menon and Durian. The fit is defined by an coefficient of restitution of e= 0.99995 ≈ 1 as expected for hard glass sphere collisions at low velocities[17]. The significance of the magnitude of "e" is not energy loss, but rather its contribution to collisional correlation's that have a significant effect on T^* as found by Bizon[18] in simulations of vibrating granular material. In **Fig. 8** we also show the results of a fundamental kinetic calculation due to Koch and Sangani[19, 20]. It is satisfying that the linear dependence of $V_n(D,U=2)$ on D is the same for both models. The factor of 4 difference in the magnitude of V_n between the two first principles calculations may be due to the inclusion of an additional sources of fluctuation energy in the Buyevich Kapbasov model -the random vertical motion of concentration fluctuations driven by gravity. Further research is required to separate this contribution compared from other differences in the two models. As expected both theories predict that the uniform fluidized state dominated by particle impact collisions is unstable in both the Geldart A and B regimes.

Since random collisional forces are not sufficient to account for the apparent bifurcation in $T^*(D,U=2)$ in the ASN experiments for D≤88μm that is responsible for ths stabile particulate phase predicted by **Fig. 7** other contributions to $T^*(D,U=2)$ are required. To account for the bifurcation in $V_n(D,U=2)$ Buyevich developed a two parameter pseudo turbulent model[16, 21] which includes random collisional **and random shear forces** driving the steady state granular temperature of the monodispersed glass spheres. As shown in **Fig. 9** the shear fluctuations produce a significant increase in T^* for small particles in the range of d≈60-100μm and could be the source of the bifurcation in T^* for the Geldart A monodispersed glass spheres in the ASN data. The magnitude of the shear fluctuations depend on the flow circulation with the fluidized bed which is strongly geometry dependent. The significant difference in bed geometry's between the two experiments indicated in **Fig. 10** and the consequent difference if flow circulation, could thus account for the failure of the Geldart A monodispersed glass spheres of Menon and Durian to exhibit Geldart A behavior in either the granular temperature (**Fig. 2** compared to **Fig, 1**) and initial bed expansion (**Figs. 5 and 6**) despite the remarkable similarities in Geldart B behavior exhibited in **Figs. 8 and 9**.

CONCLUSIONS

The new theoretical and experimental results strongly suggest, but do not prove, that the initial fluidized state of monodispersed glass spheres in the Geldart A regime, may be primarily determined by shear fluctuation forces on these small particles resulting from fluctuations in the circulation flow that establishes itself within the bed and whose magnitude is highly sensitive to both bed geometry and dimensions. This result, if validated by further experimental and theoretical research, will impact the theoretical and experimental investigation of the gas fluidized state for all particles whose diameter and density place them in the Geldart A regime, from the catalyst particles of petrochemical and refining interest to the monodispersed glass spheres examined in this paper

ACKNOWLEDGMENTS

In appreciative memory of the three year collaboration with colleague, co-author and friend - Yuri Buyevich (1937-1998)

REFERENCES

1. R. K. Wilhelm, M. Kwauk, *Chem. Eng. Progress*, **44**, (1948) 201-218.
2. D. Geldart, Ed., *Single particles, fixed, and quiescent beds and characterization of fluidized powders,* John Wiley, New York, 1986, pp. 11-51.
3. R. Jackson, in *Fluidization*, J. F. Davidson, R. Clift, D. Harrison, Eds., Academic Press, New York, 1985, pp. 47-72.
4. D. Gidaspow, *Multiphase Flow and Fluidization - Continuum and Kinetic Theory Descriptions*, Academic Press, San Diego, 1994, p. 239-354.
5. M. Nicolas, J.-M. Chomaz, E. Guazzelli, *Phys. Fluids*, **6**, (1994) 3936-3944.
6. G. D. Cody, D. J. Goldfarb, G. V. Storch Jr., A. N. Norris, *Powder Technology*, **87**, (1996) 211-232.
7. G. D. Cody, D. J. Goldfarb, in *Dynamics in Small Confining Systems-III*, M. Drake, J. Klafter, R. Kopelman, Eds., vol. 464, Materials Research Society, Pittsburgh, Pa, 1997, pp. 325-338.
8. G. D. Cody, D. J. Goldfarb, in *Fluid.-IX*, L.-S. Fan, T. M. Knowlton, Eds., Eng. Found., New York, 1998, pp. 53-60.
9. N. Menon, D. J. Durian, *Phys. Rev. Lett.*, **79**, (1997) 3407-3410.
10. G. B. Wallis, *One-Dimensional Two-Phase Flow*, McGraw-Hill, New York, 1969, p. 122-242.
11. N. F. Carnahan, K. E. Starling, *J. Chem. Phys.*, **51**, (1969) 635-637.
12. Y. A. Buyevich, S. K. Kapbasov, *Chem. Engng. Sci.*, **49**, (1994) 1229-1243.
13. Y. A. Buyevich, *Chem. Engng. Sci.*, **52**, (1997) 123-140.
14. Y. A. Buyevich, S. K. Kapbasov, *Int. J. Fluid Mech. Res.*, **26**, (1999) 72-97.
15. Y. A. Buyevich, Ind. Eng. Chem. Res., **38**, (1999) 731-743.
16. G. D. Cody, S. K. Kapbasov, Y. A. Buyevich, in *AIChE 1999 Symposium Series: Advanced Technologies for Fluid-Particle Systems*, H. Arastoopour, Ed., vol. 95 #321, AIChE, New York, 1999, pp. 7-12.
17. K. L. Johnson, *Contact Mechanics*, Cambridge University Press, Cambridge, 1986, p. 353.

18. C. A. Bizon, *Simulations of Wave Patterns in Oscillated Granular Media*, Thesis, Department of Physics, University of Texas at Austin (1998).

19. D. L. Koch, *Phys. Fluids A*, **2**, (1990) 1711-1723.

20. D. L. Koch, A. S. Sangani, *J. Fluid Mech.*, **400**, (1999) 229-263

21. Y. A. Buyevich, G. D. Cody, Particle fluctuations in homogeneous fluidized beds, Paper 207 World Congress on Particle Technology-3, Brighton, UK (1998).

Vibrated and Rotated
Granular Media

Mat. Res. Soc. Symp. Proc. Vol. 627 © 2000 Materials Research Society

Size Segregation in Granular Beds Subject to Discrete and Continuous Vertical Oscillations

Dimuth N. Fernando and Carl R. Wassgren
School of Mechanical Engineering
Purdue University
West Lafayette, IN 47907-1288, U.S.A.

ABSTRACT

Size segregation of particulates is of concern in a number of industries that handle materials such as chemicals, pharmaceuticals, fertilizers, and food products. Of particular interest in this paper is segregation resulting from externally applied vibration. In industrial applications this vibration may either be applied intentionally in devices such as vibrating conveyors or "live wall" hoppers, or unintentionally during material handling and transport. This paper investigates size segregation in granular beds subject to discrete "taps" and continuous, sinusoidal vertical vibration. The results from discrete element computer simulations indicate that the rise rate of a single impurity increases monotonically with amplitude for discrete vibrations but for continuous vibrations the rise rate increases, reaches a maximum value, then decreases as the oscillation amplitude increases.

INTRODUCTION

Size segregation of particulates is of concern in a number of industries that handle materials such as chemicals, pharmaceuticals, fertilizers, and food products. As a result, researchers have investigated size segregation for a number of geometries including rotating drums [1], rotating blenders [2], along inclined chutes [3], and in hoppers [4]. Of particular interest in this paper is size segregation occurring in granular beds subject to external vibration. In industrial applications this vibration may either be applied intentionally in devices such as vibrating conveyors or "live wall" hoppers, or unintentionally during material handling and transport.

Rosato *et al.* [5] utilized 2D Monte Carlo simulations to investigate the rise rate of a single, large impurity in an otherwise monodisperse bed of particles subject to discrete, vertical "taps." Their studies indicate that the position of the impurity increases linearly with the number of vibration taps until reaching the free surface of the bed. The rise rate, defined as the average increase in the impurity's vertical position over a single oscillation cycle, increases monotonically with the amplitude of the taps. Consequently, the rise rate is independent of the position within the bed. The impurity rises to the free surface because the small particles are more likely to fill the voids formed underneath the large particle when the particle bed is in flight and thus the large particle is forced toward the free surface.

The results of Rosato *et al.* stand in contrast to the recent experimental investigations by Hsiau *et al.* [6]. Their experiments investigated size segregation of binary mixtures subject to continuous, sinusoidal vertical vibration. They found that the degree of segregation of the material first increases as the amplitude of the vibrations increases, reaches a maximum value, and then decreases as the amplitude continues to increase. Thus, the experiments of Hsiau *et al.* indicate that there is a maximum segregation rate for a particular value of the vibration amplitude whereas Rosato *et al.*'s simulations indicate that the rate of segregation increases monotonically.

In order to better understand these previous observations, a 2D, soft-particle, discrete element computer simulation [7] was developed to investigate size segregation in vertically oscillated granular beds.

METHOD OF INVESTIGATION

The simulated experiment consists of a 2D, rectangular box containing a single, large impurity in a bed of smaller particles. The container has a width of 20 small particle diameters with periodic, lateral boundaries. Periodic boundaries are used in order to avoid sidewall convection effects on segregation [8]. The container is subject to two types of vertical, sinusoidal oscillations in order to investigate how the vibration profile affects the rise rate of the impurity. The first type of vibration consists of a single oscillation cycle, or "tap," after which the particle bed is allowed to come to rest before tapping the container again. The second type of vibration consists of continuous sinusoidal oscillations. For the simulations presented here, the dimensionless amplitude of the oscillations, a/d_s where a is the oscillation amplitude and d_s is the mean diameter of a small particle, is varied while the dimensionless oscillation frequency remains constant at $\omega(d_s/g)^{1/2}=1.3$, where ω is the oscillation radian frequency and g is the acceleration due to gravity.

The granular bed consists of frictionless, circular particles with a coefficient of restitution of 0.5. A low restitution coefficient was chosen in order to reduce the time required for particles to come to rest after an oscillation cycle. The small particles have a uniform distribution of diameters within 10% of the mean diameter. Large to small particle diameter ratios of 2/1 and 3/1 were investigated. The density of the large particle is adjusted so that the mass of the large particle is equal to the mass of a small particle. This is done in order to avoid segregation effects due to differences in particle inertia. One thousand small particles are used in the simulations giving an approximate bed depth of $50d_s$.

Measurements of the vertical position of the large particle were made as a function of the number of oscillation cycles. In order to make well-defined measurements, the bed is allowed to come to rest before the impurity position is measured. For the discrete oscillation case, the position measurement is made after the bed comes to rest between taps. For the continuous vibration case, the bed is allowed to come to rest after a particular number of oscillation cycles before the position measurement is made. The simulation then begins anew using the same initial conditions and continues until the total number of oscillation cycles is increased by one.

RESULTS AND DISCUSSION

The first set of simulations examined the rise rate of the impurity as a function of the number of oscillation cycles. While in most of the simulations the height of impurity varies nearly linearly with the number of oscillation cycles, it was not uncommon to observe the rise rate of the large particle decreasing significantly when its vertical position was near the middle of the bed. Figure 1 shows a plot of the impurity vertical position as a function of oscillation cycles using discrete taps for two different initial conditions. For the first initial condition, the impurity rises to the free surface at a nearly constant rate. For the second initial condition the impurity rises at a nearly constant rate except near the middle portion of the bed where the rise rate is nearly zero. Observations of the particle bed indicate that when the decrease in the rise rate occurs, the small particles surrounding the impurity form into hexagonal packing (HP) structures. Since this packing arrangement is difficult to shear, the voids that form while the bed is in flight are not large enough to allow for relative particle movement to occur and the large particle remains trapped in the bed. After a sufficient number of vibrations the packing structure breaks and the large particle continues toward the free surface. This same behavior has been observed in the work of a number of other researchers [9-12]. Measurements of the bed void fraction while it is in flight indicate that the largest voids occur near the upper and lower free surfaces of the bed. Consequently, these regions have the least likelihood of having sustained HP structures. In contrast, the middle portion of the bed has a relatively low void fraction throughout the flight time of the bed and thus an HP structure will be more likely to persist.

The average rise rate of the large impurity as a function of the oscillation amplitude was also investigated. Although in many of the simulations the large particle reaches the free surface of the bed within a reasonable computation time, in several of the simulations the rise rate of the large particle is slow enough that running the simulation until the large particle reaches the surface is impractical. As a result, the average rise rate of the particle is defined as the vertical distance the large particle moves in 100 oscillation cycles divided by 100 oscillation cycles.

For discrete taps, the rise rate of the impurity increases monotonically with increasing oscillation amplitude for dimensionless amplitudes between $a/d_s=1.2$ and 3.1. This trend is shown in figure 2. The simulation results for the rise rate of the impurity are consistent with Rosato et al.'s [5] Monte Carlo simulation results.

For continuous vibration, however, the rise rate follows the trends observed by Hsiau et al. [6]. The rise rate of the impurity increases with increasing oscillation amplitude, reaches a maximum near $a/d_s\approx1.9$, and then decreases again. Hsiau et al. observed that the maximum segregation coefficient occurs near a dimensionless acceleration amplitude of $a/d_s\approx1.6$ regardless of the particle diameter or diameter ratio. For a particle diameter ratio of 2/1, the dimensionless amplitude at which the maximum occurs is $a/d_s=1.6$. Note that Hsiau et al. defined the segregation coefficient as the percentage of large particles contained in the upper half of the container after a fixed number of oscillation cycles. The maximum rise rate in the present work may occur at a higher oscillation amplitude than that found by Hsiau et al. since the present simulations are two-dimensional and a low coefficient of restitution is used. Both of these factors would require larger vibration amplitudes in order to create voids large enough for particles to interchange positions.

CONCLUSION

Two-dimensional DE simulations were used to investigate the rise rate of a single large impurity in a bed of smaller particles subject to both discrete and continuous sinusoidal, vertical oscillations. Although the rise rate of the impurity is generally constant, the occasional formation of hexagonal packing structures near the middle region of the bed can result in a large decrease in the impurity's rise rate. As a result, the impurity can remain "trapped" near the middle region of the bed. Measurements also indicate that for discrete oscillations, the average rise rate of the impurity increases monotonically with the acceleration amplitude; a finding that is consistent with the observations of Rosato et al. [5]. However, when continuous oscillations are used the rise rate increases as the oscillation amplitude increases, reaches a maximum value, and then decreases. This trend has also been observed in the experiments of Hsiau et al. [6]. Thus, the manner in which the vibration is applied to the bed can significantly affect the rate at which particles will segregate. Of particular concern is the applicability of extrapolating segregation rate trends from discrete or continuous sinusoidal oscillations to systems subject to continuous random vibrations.

ACKNOWLEDGEMENTS

The authors gratefully acknowledge the support of the National Science Foundation who supported this work under grant #9996025-CTS.

REFERENCES

1. K. M. Hill, A. Caprihan, and J. Kakalios, Axial segregation of granular media rotated in a drum mixer: Pattern evolution, *Phys. Rev. E*, **56**, 4386-4393 (1997).
2. M. Moakher, T. Shinbrot, and F. J. Muzzio, Experimentally validated computations of flows, mixing and segregation of non-cohesive grains in 3D tumbling blenders, *Powder Technology*, **109**, 58-71 (2000).
3. D. V. Khakhar, J. J. McCarthy, and J. M. Ottino, Mixing and segregation of granular materials in chute flows, *Chaos*, **9**, 594-610 (1999).
4. N. Standish, Studies of size segregation in filling and emptying a hopper, *Powder Technology*, **45**, 43-56 (1985);
5. A. Rosato, F. Prinz, K. J. Standburg, and R. Swendsen, Monte Carlo simulation of particulate matter segregation, *Powder Technology*, **49**, 59-69 (1986).
6. S. S. Hsiau and H. Y. Yu, Segregation phenomena in a shaker, *Powder Technology*, **93**, 83-88 (1997).
7. P. A. Cundall and O. D. L. Strack, A discrete numerical model for granular assemblies, *Geotechnique*, **29**, No. 1, 47-65 (1979).
8. J. B. Knight, H. M. Jaeger, and S. R. Nagel, Vibration-induced size segregation in granular media: The convection connection, *Phys. Rev. Lett*, **70**, No. 24, 3728-3731 (1993).
9. E. R. Nowak, M. Povinelli, H. M. Jaeger, and S. R. Nagel, Studies of granular compaction, *Powders & Grains 97*, Behringer, R. and Jenkins J. (eds.), 377-380 (1997).
10. S. J. Linz, Phenomenological modeling of the compaction dynamics of shaken granular systems, *Phys. Rev. E*, **54**, 2925-2930 (1986).
11. C. R. Wassgren, D. E. Beasley, and R. N. DeWachter, Heat transfer in vertically vibrated granular materials, *1998 International Mechanical engineering Congress & Exposition (IMECE) Conference Proceedings* (1998).
12. A. Rosato and D. Yacoub Microstructure evolution in compacted granular beds, *Powder Technology*, **109**, 255-261 (2000).

FIGURES

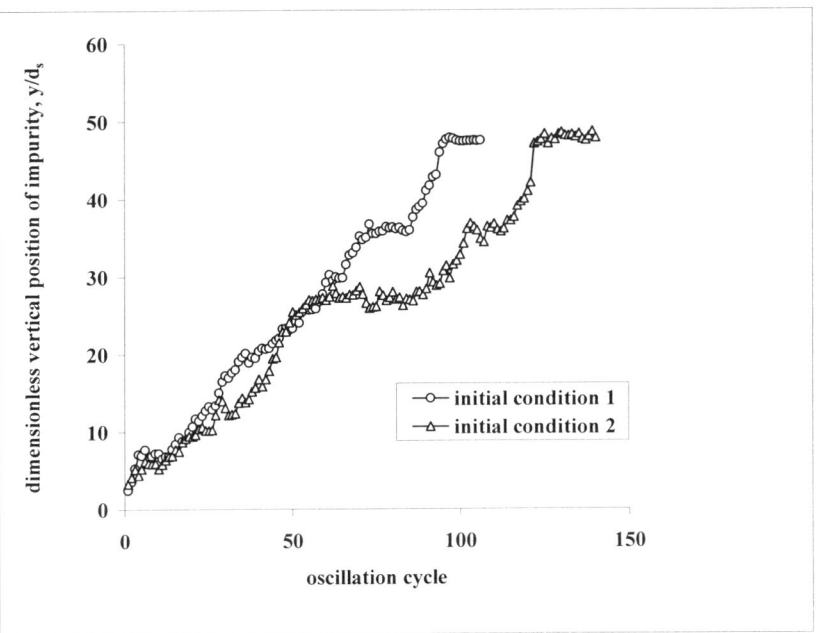

Figure 1. *The vertical position of the large impurity (diameter ratio of 3/1) as a function of the oscillation cycles for two different initial conditions. The dimensionless amplitude of vibration is $a/d_s = 1.9$ and the dimensionless vibration frequency is $\omega(d_s/g)^{1/2} = 1.3$. Discrete, sinusoidal taps are used in the simulations.*

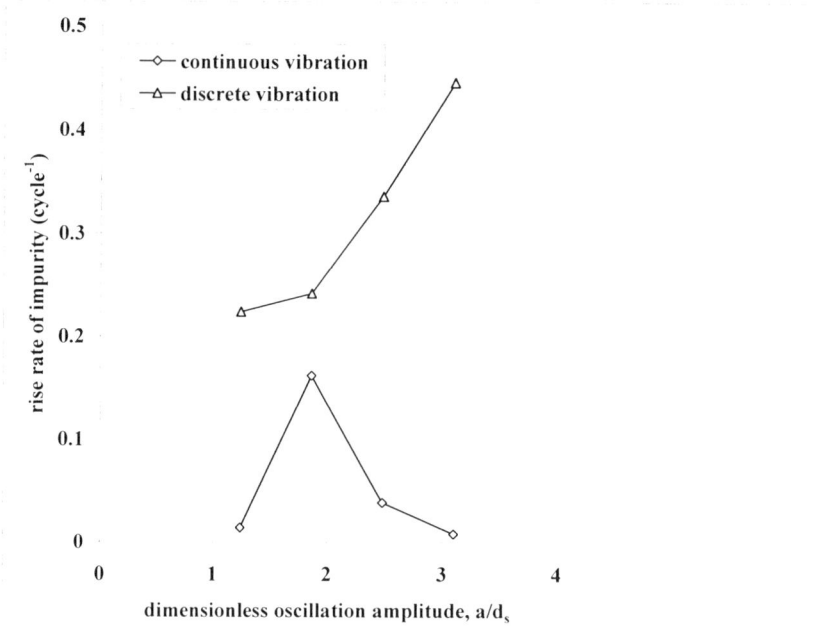

Figure 2. *The dimensionless average impurity rise rate over 100 oscillation cycles plotted against dimensionless oscillation amplitude for discrete and continuous vibration. The dimensionless vibration frequency for the oscillations is $\omega(d_s/g)^{1/2} = 1.3$. The large to small particle diameter ratio is 2 to 1.*

Mat. Res. Soc. Proc. Vol.627 2000 Materials Research Society

Friction and Flow in Granular Materials

R. P. Behringer, L. Kondic, G. Metcalfe, D. Schaeffer,
and Sarath G. K. Tennakoon

Department of Physics and Center for Nonlinear and Complex Systems
Duke University, Durham, North Carolina 27708-0305

We probe the transitions between solid-like and fluid-like granular states in the presence of shaking in the horizontal and vertical directions. These transitions are fundamental to other aspects of granular flow such as avalanche flow, in which there is a free surface. Key control parameters include accelerations in the vertical and horizontal directions, $\Gamma_i = A_i \omega 2_i/g$, for shaking of the form $s_i = A_i \cos(\omega_i t + \phi_i)$, $i = h, v$. Here, g is the acceleration of gravity. Also important is the relative phase between the two modes of shaking. We focus on low to moderate dimensionless accelerations, $0 < \Gamma_{v,h} < 1.6$. We consider first the case $\Gamma_v = 0$, i.e. pure horizontal shaking. In this case, there is a hysteretic transition between solid and fluid states, where the fluid state consists of a sloshing layer of material of height H plus additional transverse flow. The hysteresis is lifted in the presence of a modest amount of fluidization by gas flow, or if a slight overburden is provided. We also identify a time scale, τ, for the transition between the phases that diverges inversely as the distance $\varepsilon = (\Gamma_h - \Gamma_{hc})/\Gamma_{hc}$, from the appropriate transition points, i.e. as $\tau \propto \varepsilon^{-1}$. We identify a new convective mechanism, associated with horizontal shearing at the walls, as the mechanism that drives the transverse convective flow. For combined horizontal and vertical shaking, there exist a related set of novel dynamics and stability properties. These include the spontaneous formation of a static heap and a transition to flow, similar to the flow state under horizontal shaking, when the vertical acceleration $\Gamma_v < 1$. A simple friction model provides a good description of the steady states and a reasonably good description of the transition to flow. Horizontal and vertical shaking frequencies that differ by a small amount can lead to a novel switching state, as the relative phase, $\phi_h - \phi_v$, shifts over time.

1 Introduction

Although granular materials are extraordinarily common, a fundamental understanding of their dynamical behavior is still an open problem. Consequently, the dynamics of granular materials have attracted considerable recent interest. For reviews see ref.[1]. Granular materials can exhibit both fluid-like and solid-like properties depending on the circumstances: they resist shearing up to a point, but flow freely under strong enough shear or at low enough density. Here, we focus on the transitions back and forth between the solid-like and fluid-like states. The transition from solid to fluid is well studied in the context of Mohr-Coulomb models. But the reverse transition is not well characterized to our knowledge.

Much recent attention has been focused on the dynamics of vertically vibrated granular materials[1]. Although there have been some studies[2, 3, 4, 7, 6, 8], much less is known about the dynamics of granular materials subject to horizontal vibration. Regarding combined vertical and horizontal vibration, there is even less known[8], particularly when the phases between the two modes of shaking are varied. These flows are of interest because both horizontal and vertical vibration are commonly used in industries as an aid to mixing, segregating and transporting granular materials. Soil liquefaction during earthquakes is a common and destructive phenomena associated with horizontal shaking. Finally, the present experiments show a novel shear-induced

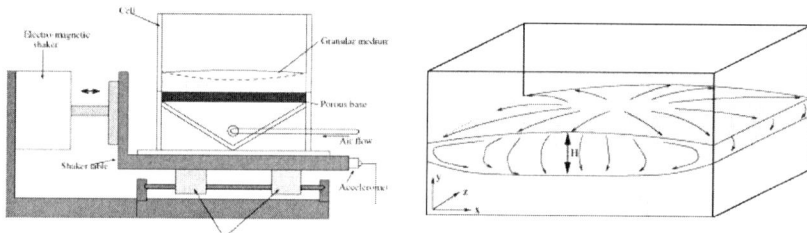

Figure 1: Left: Schematic of the apparatus. Right: Sketch of convection flow lines in the liquified layer induced by horizontal shaking (a) as seen from the top, and (b) as seen from the side. Grains rise in the middle of the cell, flow along the surface towards the side walls, and then sink at the wall boundaries.

flow which is the motor for convection transverse to the shaking direction; this mechanism is likely to be present in other shear flows as well.

2 Experimental Details

In these experiments, a container of granular material is subject to continuous horizontal and/or vertical displacements

$$s_i = A_i \cos(\omega_i t + \phi_i). \qquad (1)$$

For each component of shaking there is a relevant control parameter

$$\Gamma_i = A_i \omega_i^2 / g, \qquad (2)$$

where g is the acceleration of gravity, and $i = v$, h for vertical or horizontal. If $\omega_v = \omega_h$, the relative phase, $\phi_h - \phi_v$ is also important; otherwise, $\Delta\omega = \omega_v - \omega_h$ is important. Other parameters, such as $E = (A\omega)^2 / gd$, which are important in describing higher order phenomena in vertically shaken materials (e.g. traveling waves[10], coarsening[11]) are not necessary to describe the onset of flow in these experiments. Typically, the layer had a height $h/d \approx 40$; as long as $h/d > 20$, we found no dependence on h.

Our experimental setup is shown in Fig 1. The heart of the experiment is a rectangular Plexiglas cell with cross-sectional dimensions of 1.93 *cm* by 12.1 *cm*. The base of the cell is made of a porous medium through which gas can flow in order to fluidize the granular medium. This provides an independent control over the dilation of the material. The cell is mounted on a Plexiglas base of the same cross-sectional dimensions as the cell, which is in turn mounted on a movable table; the base acts as the gas distributor to the system. The table is mounted on four linear bearings sliding on horizontal shafts rigidly attached to the fixed bottom frame. An electromechanical actuator provides a sinusoidal drive at frequencies, ω, spanning of 3–15 Hz and at amplitudes A, spanning 0–20mm. A similar drive provides independent shaking in the vertical direction. In a typical run, we varied the vertical and horizontal amplitudes A_v and A_h and observed the evolution of the system while keeping ω_v and ω_h fixed.

We used several types of approximately monodisperse granular materials, including approximately monodisperse spherical glass beads, smooth nearly spherical Ottawa sand, and sieved rough sand. These particles had diameters ranging from 0.2$mm \leq d \leq$ 1.0mm. We prepared the system in one of two ways: either by pouring in the selected material, leveling the sample, and then compacting it at modest Γ, or by creating a heap at the angle of repose in the absence of shaking. The material did not form a perfectly ordered packing because the particles where neither highly spherical nor monodisperse.

We used several different ways to characterize the flows, including strobing, high-speed video (250 frames/s) and conventional video. Much of the data discussed here were obtained by observing the flow near the walls through the transparent Plexiglas container with a small surveillance CCD camera mounted on the shaker platform. To obtain the flow speeds at various locations and the overall flow fields, we seeded the granular material with darker tracer particles. (In this regard, considerable care must be taken, since coloring the grains can change the surface friction and lead to strong segregation.) Using video images, a framegrabber and computer software for particle tracking, we extracted information such as heap heights, shapes of the top surface, flow speeds, and mean flow fields.

3 Pure Horizontal Shaking

For simple horizontal shaking, we focus on the initial instability to fluidization, which results in a sloshing of the grains in the direction of the shaking, plus slower convective flow, including flow in the direction transverse to the shaking. In a typical run for pure horizontal shaking, we observed the evolution of the system as A_h was increased from zero while keeping ω_h fixed. Once flow was well initiated, we then decreased A_h. Fig 1 shows a sketch of the mean convective flow lines (i.e. discounting the sloshing flow) for Γ_h somewhat above onset (at Γ_{hu}) in a cell that is 30 grains wide and

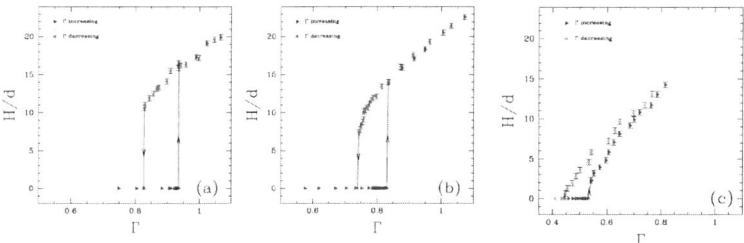

Figure 2: Bifurcation diagrams for the initial sloshing transition of horizontally agitated dry granular materials: (a) rough sand, (b) smooth Ottawa sand, and (c) glass beads. The transition points and size of the hysteresis loop tend to increase with increasing material roughness.

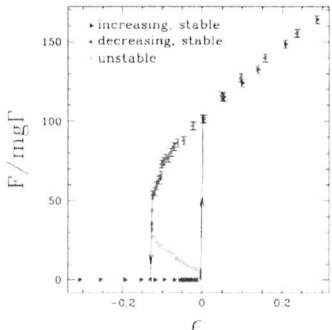 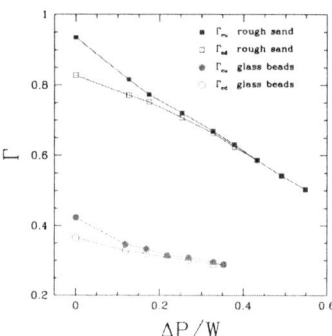

Figure 3: Left: Full bifurcation diagram, including the unstable (negative slope) branch, for Ottawa sand vs. Γ_h in reduced units. $F/mg\Gamma$ is a measure of the fluidized layer height. Right: Bifurcation points Γ_{cd} and Γ_{cu} versus dimensionless lifting pressure and air flow rate for rough sand and for glass beads.

50 grains deep. Grains rise up in the middle of the cell and flow along the surface towards the side walls and then sink at the wall boundaries. The top surface of the liquefied layer has a dome shape that is concave down, and the bottom surface of this layer is concave upwards near the wall boundaries. Thus, the thickness of the fluidized layer is largest in the middle of the cell and smallest at the end walls, as seen previously by Evesque[3]. In the case of thin cells, where the width of the cells are less than six grains wide, the convective flow transverse to the shaking direction is suppressed and the curvature of the top surface perpendicular to the motion direction vanishes.

A useful measure of the strength of the flow is the thickness, H, of the liquefied material in the middle of the cell, as seen at the long sidewall. We show typical hysteretic behavior for H as a function of Γ in Fig. 2. With increasing Γ_h, there is a well defined transition to finite amplitude (in particular $H \neq 0$) flow at Γ_{hu}. If we then decrease Γ_h below Γ_{cu} once flow has begun, the thickness of the layer also decreases but does not vanish until Γ_h reaches a lower critical value Γ_{cd} where the relative motion completely stops. As Γ_h is decreased from above towards Γ_{cd}, grains near the end walls stop moving first, while grains in the middle keep moving. It is perhaps not totally surprising that the initial transition to flow is hysteretic, since the onset of flow is accompanied by dilation and by the breaking of static friction, whereas once flow has begun, dynamical friction is involved, assuming that grains remain in motion and in a dilated state. Some hysteresis was present in the 2D studies by Ristow et al.[6], but the hysteresis in the our 3D experiments is larger, particularly for rougher materials.

The location of Γ_{cu} and Γ_{cd} are reproducible for a given height of a particular material. That is, different ω's or A's yield the same critical Γ_h's for a given material and fill height. Thus, Γ_h is the relevant control parameter, as opposed to say E or some other dynamical measure. The values of Γ_{cu} and Γ_{cd} are also not affected by the total height of the layer, if that quantity is roughly 20 grains or higher.

These transtions values of Γ_h do depend on the physical properties of the material, as shown in Fig 2. For instance, Γ_{cu} and Γ_{cd} increase as the roughness of the granular materials increases. The same is true for the difference ($\Gamma_{cu} - \Gamma_{cd}$), which is higher for rougher granular beds than for smooth ones. This can be attributed to two factors: first, the ability of rough grains to roll is reduced because of the interlocking of grains; and second, the effective macroscopic frictional forces between grains and between grains and walls may also be higher for rough grains. These effects are in principle distinct, although in an experiment, it may be difficult to distinguish between them. It is also possible to map out the full bifurcation diagram, including the unstable states, as shown in Fig. 3. Here, the basis technique that we used involved applying known perturbations to the system, and then observing to see whether the fluid or solid was the end state following the perturbation.

To obtain additional insight into the relative importance of dilatancy and friction, we have carried out two additional experiments. In the first of these, we fluidized the granular bed by passing air through it, using the flow-controlled air supply (see Fig. 1), where the porous base of the cell acts as the gas distributor. Fig 3, which presents Γ_{cu} and Γ_{cd} as functions of the air flow through the bed, shows a very strong dependence of these quantities on the air flow. In particular, a modest air flow reduces the critical Γ's and ultimately effectively removes the hysteresis in the initial transition. Here, ΔP is the pressure difference across the granular layer, and W is its weight per area. A key point is that the measured dilation of the bed due to the air flow is very small; the maximum dilation for these experiments corresponds to less than one granular layer for a bed of 45 layers, i.e. less than 2% dilation. In a second experiment to probe the effects of dilatancy, we placed a small amount of overburden (here a thin piece of plastic of weight comparable to about half a layer of grains). Surprisingly, we found that the hysteresis was lifted, and that the transition to flow occured at Γ_{cd}. This result is particularly noteworthy because we might expect that the addition of weight would delay the transition, rather than enhance it.

An additional point concerns nucleation times. These include the time to nucleate the fluid layer from the solid when Γ_h is stepped slightly past Γ_{cu}, and the time for all of the fluid phase to decay away if Γ_h is stepped slightly below Γ_{hd}. We show results for both processes in Fig. 4. Again, there is a surprise: there is a very well defined power law for each process, with the relaxation time always varying as $\tau = A/\varepsilon$, where $\varepsilon = (\Gamma_h - \Gamma_{ci})/\Gamma_{ci}$, $(i = u, d)$. This set

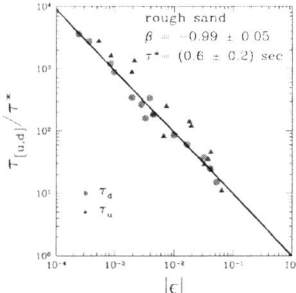

Figure 4: Relaxation times for rough sand vs. $\varepsilon = (\Gamma_{ci} - \Gamma_h)/\Gamma_{ci}$ for the nucleation of the fluidized layer with increasing Γ_h and the vanishing of the fluidized layer with decreasing Γ_h.

of results is striking first because it suggests a universal behavior, and second, because the power of ε is minus one, whereas relaxation for a backward bifurcation occurs on a timescale of $\varepsilon^{-1/2}$. We speculate that one factor of $\varepsilon^{-1/2}$ comes from the fact that a small step past one of the Γ_{ci} occurs only during a small part of the cycle, because of the oscillatory nature of the driving. An additional factor of $\varepsilon^{-1/2}$ must be attributable to another cause, the simplest being that it is an inherent property of the backward bifurcation.

An important issue concerns the sources or "motors" which drive both the convective flow parallel to the shaker plane and the convective flow perpendicular to the shaker plane. It is relatively easy to understand why the layer begins to slosh back and forth above a critical acceleration, but it is much less clear why there should be a transverse flow and the formation of a dome shaped heap. The convective flow parallel to the shaker plane is due to the avalanching of grains at the end walls during the sloshing motion of the liquefied layer. Each half of the shaker cycle grains pile up on one side wall and open a gap near the opposite side wall. When the opening at the wall is sufficiently large, grains near the opening avalanche down to fill part of the gap. This process repeats at an endwall every half cycle. The convective flow transverse to the shaker plane is due to a completely different process. To investigate this cross flow, we have carried out additional experiments in which we only oscillate one of the long side walls (actually, we fix that side wall and oscillate the rest of the cell, but this is not a critical difference at modest Γ). We find that the oscillation of the sidewall at essentially any Γ or A leads to a convective flow where grains fall at the sidewall, and are then pushed inward, and upward within a shearing layer, which is 5-10 grains wide for slow to moderate shearing. It is possible to monitor darker grains along this wall, and we find that the average speed at which grains fall along the shearing wall, v_z, is simply a linear function of the speed, $A\omega$ of the wall relative to the grains. We have also carried out MD simulations[8] that reinforce this interpretation. This kind of shear-induced convection is likely to occur generally when a 3D material is sheared in the presence of a gravitational field.

When the whole layer of material is now shaken, the following scenario describes the onset of the instability and the mechanism for the cross-convective flow. The flow begins when the grains overcome frictional and dilatancy effects. At this point there is a shear flow of surface grains relative to the long walls (i.e. the direction of shaking) which is a maximum in the longitudinal middle of the cell. Near the side walls, there is a strong dilatancy of the moving grains, due to shearing. The highly dilated grains near the side wall are mobile, and grains reaching the top of this region percolate downward. However, grains falling along these walls may eventually participate in a stress chain which pushes grains inwards. Such an effect can also push grains upward towards the free surface, but not into the compacted non-mobile layer at the bottom of the sample.

4 Combined Vertical and Horizontal Shaking

We now turn to studies of combined, independent shaking in both the vertical and horizontal directions. To our knowledge, these studies are the first to address stability and flow of granular materials with this approach. The advantage of this technique is that we can effectively modulate gravity, and probe in greater detail the nature of frictional failure for granular materials.

When we apply both vertical and horizontal sinusoidal vibrations to the bed at the same time, the resulting flow fields depend, for a given material, mainly on Γ_v and Γ_h, and the phase difference Φ between the two motions. (For concreteness, we will refer to $\Phi = 0$ as "in phase" if the maximum (most positive) vertical acceleration (y-direction) occurs at the same time as the maximum (most "rightward") horizontal acceleration (x-direction).)

The possible dynamics become much more complicated than for pure-mode shaking, and we will focus on: 1) in-phase shaking when $\Gamma_v < \Gamma_{vc}$; and 2) the states that occur when $f_h = f_v$, but when $\Phi \neq 0$; and 3) states that occur when $|f_h - f_v| \equiv \Delta f \neq 0$ but small $(f = \omega/2\pi)$, with Γ_v moderate, but not necessarily $\Gamma < \Gamma_{vc}$. In case 3, Φ shifts steadily in time at the rate Δf.

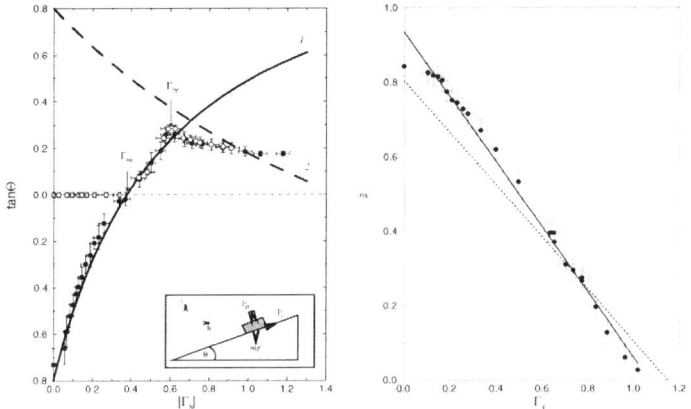

Figure 5: Left: Circles: heap angle, θ of a thick $(5.8cm)$ Ottawa sand bed as a function of Γ_h, for $\Gamma_v = 0.682$ for in-phase ($\Phi = 0$) horizontal and vertical forcing $(f_h = f_v = f = 4.99Hz, A_v = 6.77mm, A_h$ varying), starting from a flat surface. Square symbols correspond to data for the same control parameters, but starting from a heap at the angle of repose. The solid and dashed lines pertain to the model (respectively curve 1 and curve 2) with $\mu = 0.80$ and $\Gamma_{vc} = 1.15$. Inset: sketch of inclined surface that defines the model. Right: Data for Γ_{h1} vs. Γ_v. According to the Coulomb model, these data should have slope $-\mu$ and intercept $\Gamma_{vc} = 1$. Solid curve corresponds to least squares fit to the experimental data $(\Gamma_{vc} = 1.07; \mu = 0.87)$ and the dotted curve corresponds to this model with $\mu = 0.80$ and $\Gamma_{vc} = 1.15$.

In the first set of experiments, we vibrated the system at the same horizontal and vertical frequency with a fixed (and controlled) phase difference $\Phi = 0$. In the second set of experiments, the shaking frequencies ω_h and ω_v differed slightly, so that Φ varied in time. The layer depths were relatively large, roughly 10 cm.

Case 1: $f_v = f_h = f_0 = const.; \Phi = 0; \Gamma_v < \Gamma_{vc}$. We begin by considering what happens if the system is prepared in an initially flat state. The most notable feature of this regime is the spontaneous formation of a new steady (i.e. no flow) state consisting of a heap against the "right" wall, beginning at $\Gamma_h = \Gamma_{h1}$, where Γ_{h1} depends on Γ_v. For example, the circles in Fig. 5 show the height of the heap starting from a flat surface as a function of Γ_h for $\Gamma_v = 0.682$. These data were taken while keeping $f_h = f_v = f = 4.99Hz$, A_v fixed, and by then varying A_h. For $\Gamma_{h1} =$

$0.39 < \Gamma_h < 0.6 = \Gamma_{h2}$, the heap *is* actually the result of a transient, and in the steady state there is no flow until Γ_h exceeds a second threshold, Γ_{h2}, discussed below. More specifically, we observe that just above $\Gamma_{h1}(\Gamma_v)$, a small amount of grains near the top surface shifts towards the side wall. At each value of Γ_h, this shifting of grains along the surface stops when the increased slope is high enough.

If the initial state was a heap at the static repose angle, then for Γ_h increasing from 0, we found the data shown by dark squares in Fig 5. The magnitude of the repose angle decreased with increasing Γ_h, and merged with the data obtained from an initially flat surface at Γ_{h1}.

This process changed qualitatively when Γ_h exceeded some value $\Gamma_{h2}(\Gamma_v)$. At Γ_{h2}, the top layer along the slope became liquefied and moved back and forth under the shaking. This state is the analogue of the state discussed under *pure horizontal shaking*. As a result, there was no further heaping of grains. Any further increase in Γ_h increased the thickness of the liquefied layer along the slope, and resulted in a lower average slope. For example, Fig 5 shows that for $\Gamma_v = 0.682$, θ increases with increasing Γ_h until $\Gamma_{h2} \approx 0.6$, after which the average slope decreases.

A simple Coulomb friction model[12] can capture many of the features observed in these experiments and provide additional insight into pure vertical or horizontal shaking. Good fits to the data based on the model can be obtained for any given property such as Γ_{h1} or Γ_{h2}. However, it is not possible to quantitatively describe all the experimental observations with a single set of model parameters. Thus, more sophisticated models are needed.

We pose the model, which is in essence Mohr-Coulomb failure, in Fig. 5, inset. A block (a model for a surface grain) of static friction coefficient μ is placed on a surface inclined at an angle θ; the surface is subject to vertical and horizontal accelerations, $g\Gamma_v$ and $g\Gamma_h$. For the moment, we envision that these are constant accelerations. Like the experiment, we envision that the signs of the two accelerations are always the same, i.e both positive or both negative. Negative accelerations have a vertical component in the direction of gravity. We then ask at what accelerations does static friction fail.

When $\theta = 0$, i.e. the surface is in the horizontal plane, the block can first begin to slide when $mg\Gamma_h = \mu F_n$, where F_n and F_t are normal and tangential forces. Thus, sliding first occurs when $\Gamma_h = \mu(1-\Gamma_v) \equiv \mu(\Gamma_{vc} - \Gamma_v)$, and the accelerations are negative. In principle $\Gamma_{vc} \equiv 1$, but in order to obtain reasonable fits to the data, we find that slightly higher values may be needed. Henceforth, we identify μ and Γ_{vc} as the two adjustable parameters of the model. We identify the onset of sliding/heaping on a horizontal surface with Γ_{h1} and we examine this prediction in Fig. 5 by considering the data for Γ_{h1} vs. Γ_v. The data indeed show a nearly linear profile whose slope defines μ and whose Γ_v intercept is close to 1. This intercept corresponds to the onset of heaping under purely vertical shaking, which is typically found experimentally to be somewhat larger than 1. Except near $\Gamma_v = 0$, the data of Fig. 5 are consistent with a straight line with intercept Γ_{vc} slightly above 1.0. If we force this intercept to be 1.00, the corresponding μ is $\mu = 1.05$. For example, $\Gamma_{vc} = 1.15$ leads to $\mu = 0.80$. We will see below, that this second set of parameters provides an improved fit to data for Γ_{h2}, although they are clearly not the best fit to the data for Γ_{h1}.

We next turn to the case when $\theta \neq 0$. By the solid and dashed curves in Fig. 5, we show the loci of frictional failure (or first occurance of slipping) vs. $|\Gamma_h|$ for a fixed $|\Gamma_v| \neq 0$; curve 1 is for negative Γ_h and Γ_v and curve 2 is for positive. In the parlance of soil mechanics, these correspond to passive and active failure. Curve 1 is given by

$$\tan \theta = (\Gamma_h - \Gamma_{h1})/(1 - \Gamma_v + \mu\Gamma_h), \tag{3}$$

and curve 2 by

$$\tan \theta = -(\Gamma_h - \mu(1 + \Gamma_v))/(1 + \Gamma_v + \mu\Gamma_h), \tag{4}$$

where we use $\Gamma_{h1} = \mu(\Gamma_{vc} - (\Gamma_v))$. Note that the layer is more unstable when the accelerations are negative, so the curves are not symmetric if $\Gamma_v \neq 0$. In an actual experiment, we shake in phase, so that the instantaneous experimental path is a line and the failure locus a surface in a 3D space of Γ_h, Γ_v and $\tan\theta$.

This model can be generalized in a fairly straight forward way to out of phase shaking. In this case, the easiest approach is to look for the interections of the frictional failure curves of a layer with slope $\tan\theta$ with the trajectory of the shaker in $\Gamma_v - \Gamma_h$ space. Experimentally, the relative phase of the two components of shaking plays an important role. For $\Phi = 0°$ or $\Phi = 180°$, the heap forms next to one or the other of the vertical sidewalls. When Φ is close to $90°$, the heap height is reduced from what occurs for $0°$ or $180°$ and is located near the middle of the cell. In the $90°$ case, the maximum horizontal forcing occurs from either direction when the grains are relatively compacted. The grains moves in circular patterns, and when $0° < \Phi < 90°$, circulation in one sense dominates over the other.

We proceed to further interpretation of this model. As discussed above, a flat surface becomes unstable at Γ_{h1}. But, there exist stable surfaces inclined at a positive θ. These persist up to the point at which the loci for negative and postive acceleration intersect; beyond this, there is slipping in both directions. We identify the intersection of curves 1 and 2 with Γ_{h2}, i.e. the onset of surface flow. We expect that when a surface is prepared at $\theta = 0$ and then Γ_h is increased past Γ_{h1}, grains will flow, transiently, until stability is achieved. Thus, curve 1 between Γ_{h1} and Γ_{h2} defines the slope of the static heap that forms above Γ_{h1} but before persistent flow occurs. For $\Gamma_h < \Gamma_{h2}$, the heap is stable for angles lying both below curve 2 and above curve 1.

We now apply these additional predictions to the experiment. Figure 5 shows the slope, $\tan\theta$, versus Γ_h for Ottawa sand and $\Gamma_v = 0.68$. The solid (dashed) curve in the figure corresponds to curve 1 (2) in Fig. 5 between Γ_{h1} and Γ_{h2}, with $\mu = 0.80$ and $\Gamma_{vc} = 1.15$. As discussed above, we show data (open circles) for the slope vs. Γ_h starting from an initially flat surface, and data (solid squares) where we have prepared the surface at a negative angle corresponding with the angle of repose. In both cases, we have gradually ramped Γ_h from low to high values at fixed Γ_v. The agreement to the simple friction model is generally good for curve 1, although the onset of flow at Γ_{h2} is overpredicted by the model for these parameters. Good agreement can be obtained for this particular data set by adjusting the fit parameters, at the expense of a poorer fit to data for Γ_{h1}. We have also attempted to obtain data for curve 2, but this proved to be problematic because initial transients tended to disrupt the process.

We extend the comparison between the model and experiment in Fig. 6, left parts, where we consider Γ_{h2} and $\tan\theta$ at Γ_{h2}, respectively vs. Γ_v. The solid curves in each part of this figure correspond to values at the intersections of curve 1 and 2 for various Γ_v when $\mu = 0.80$ and $\Gamma_{vc} = 1.15$. Here, the agreement is moderately good, but not perfect. Indeed, these results suggest that μ and/or Γ_{vc} change as the heap is formed from the original state.

Case 3: $\omega_h \neq \omega_v$. Before closing, we explore a related and novel phenomena which occurs when the shaking frequencies differ by a small amount. As noted above, the location of the heap and nature of the flow state depend on the relative phase difference, Φ of the two shaking modes. When the frequencies of vertical and horizontal shaking are slightly different, the phase difference Φ changes periodically with frequency $|f_h - f_v|$. As Φ changes periodically, the location of the heap moves from one wall to the other, and the convective flow field oscillates in time. Fig. 6 shows just the surface of the heap as a function of time for somewhat less than a half cycle of the heap switching. The time between the curves is 2 s and the period of heap motion is 180 s. Although, we consider a case with $\Gamma_v > 1$, the phenomenology is similar for $\Gamma_v < \Gamma_{vc}$. Densely packed curves in this figure show that the heap spends more than half the time near the walls during each cycle. The discussion above indicates why this process occurs. Specifically, the heap forms along the wall which is pushing on the material during the negative vertical acceleration portion of the cycle. The heap remains near that wall until the horizontal forcing switches direction relative to the vertical acceleration.

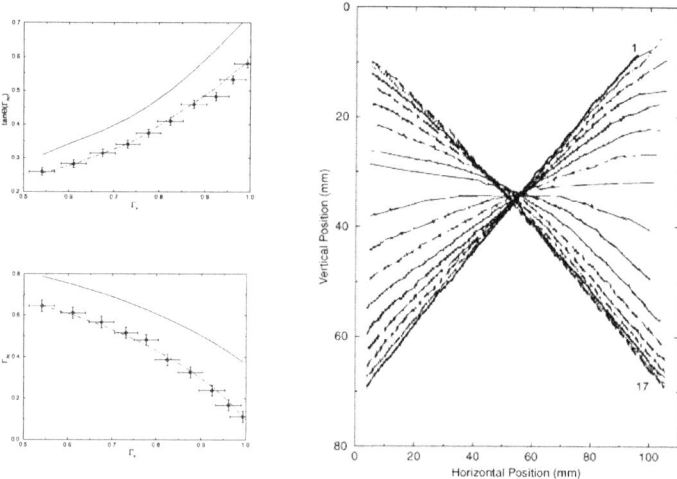

Figure 6: Left: Points: data for bottom) Γ_{h2}, and top) $\theta(\Gamma_{h2})$ vs. Γ_v . Solid curves show predictions of the Coulomb friction model for $\mu = 0.80$ and $\Gamma_{vc} = 1.15$. Dashed curves provide a good fit to the present data, but a worse fit to other data. Right: Shape of the top surface at various times during half a cycle. The time between the curves is 2 s and the period of back and forth heap motion is 180 s.

To conclude, we have characterized the transitions between granular fluid and solid states for pure horizontal shaking and combined vertical-horizontal shaking of granular materials in a 3D geometry. For pure horizontal shaking, and in the absence of additional fluidization by vertical gas flow, this transition is hysteretic. However, by applying a very modest vertical gas flow which creates a dilation in the vertical of ~ 2%, the hysteresis is lifted, and the onset to flow occurs at lower Γ. The hysterisis is also lifted by suppressing the 'sliders', i.e. the free surface motion, with a slight overburden. Interestingly, the nucleation times for the onset of the fluid phase or the disappearance of the fluid phase follow identical power laws in ε, the distance past the bifurcation point: $\tau = B/\varepsilon^{\alpha}$ where B is basically the same for the two cases, and where $\alpha = 1$ within experimental error. These experiments also show that there is convective flow perpendicular to the horizontal shaker motion. The mechanism that drives this convective flow is particularly interesting since it likely occurs in other 3D systems with shearing and gravity. For both horizontal and vertical vibrations, flow properties such as static heaping, and convection depend not only on Γ_v and Γ_h, but also on the phase difference between horizontal and vertical vibrations. The formation of a static heap and the onset to flow are captured semi-quanitatively by a Mohr-Coulomb type friction model. However, these experiments point to the need for a more fundamental and sophisticated picture.

Acknowledgments

This work was supported by NASA under Grant NAG3-1917 and by NSF grants DMR-9802602 and DMS-9803305.

References

[1] H. M. Jaeger, S. R. Nagel, and R. P. Behringer, Rev. Mod. Phys. **68**, 1259 (1996); *Physics Today*, **49**, 32 (1996); See also R. P. Behringer, Nonlinear Science Today, **3**, No. 3, 1-15, (1993); *Granular Matter: An Interdisciplinary Approach*, A. Mehta, Ed. Springer, NY (1994); H. M. Jaeger and S. R. Nagel, Science **255** 1523 (1992); D. Bideau and J. Dodds, *Physics of Granular Media*, Les Houches Series (Nova Science Publishers).

[2] S. Nasuno, A. Kudrolli and J.P. Gollub, Phys. Rev. Lett. **79**, 949 (1997).

[3] P. Evesque, Contemporary Physics, **33**, 245 (1992).

[4] O. Pouliquen M. Nicolas and P. D. Weidman, Phys. Rev. Lett. **79**, 3640 (1997).

[5] K. Liffman, G. Metcalfe, and P. Cleary, Phys. Rev. Lett. **79**, 4574 (1997).

[6] Gerald H. Ristow, Gunther Strassburger, and Ingo Rehberg, Phys. Rev. Lett. **79**, 833 (1997).

[7] M. Medved, D. Dawson, H. Jaeger and S. Nagel, Chaos, **9**, 691 (1999).

[8] S. G. K. Tennakoon, L. Kondic and R. P. Behringer, Europhys. Lett. **45**, 470 (1999).

[9] S. G. K. Tennakoon and R. P. Behringer, Phys. Rev. Lett. **481**, 794 (1998).

[10] H. K. Pak and R. P. Behringer, Phys. Rev. Lett. **71**, 1832 (1993).

[11] E. van Doorn and R. P. Behringer, Phys. Lett. A. **235**, 469 (1997).

[12] M. E. Vavrek and G. W. Baxter, Phys. Rev. E **50**, R3353 (1994).

Mat. Res. Soc. Symp. Proc. Vol. 627 © 2000 Materials Research Society

Flow of Granular Material through Rotating Cylinders: Modelling Transients

Richard J. Spurling, John F. Davidson and David M. Scott
Department of Chemical Engineering, University of Cambridge,
Pembroke Street, Cambridge, CB2 3RA, UK

ABSTRACT

Granular material, fed continuously into the top of a slowly rotating, slightly inclined cylinder, forms a moving bed. Much of the bed rotates with the cylinder in solid body motion. When particles reach the surface of the bed, they move rapidly down it, and are absorbed once more into the solid body motion. Such cylinders are used in calcining, pharmaceutical manufacture, and drying. A steady state transport model, applicable when the bed depth varies slowly along the cylinder, has existed for around 50 years. The bed surface is considered locally flat, and particles in it fall along the line of steepest descent, inclined to the horizontal at the angle of repose. There is reasonable agreement with experiment.

We propose a quasi-steady state dynamical model, in which the steady state model is coupled with a volume balance across an axial element. The model takes the form of a nonlinear diffusion equation which was solved numerically. The parameters of the dynamic model are the dimensions of the cylinder and outlet dam, the inclination of the axis of the cylinder, its rotational speed, the angle of repose of the granular material and its feed volumetric flow rate: the dynamic model has no free parameters. Experiments were conducted using sand, mean particle size 490 μm, in a perspex tube of length 1 m, radius 0.0515 m, lined with sandpaper, with a feed end dam of height 0.029 m, and with no exit dam, or an exit dam of height 0.0105 m. With the system initially in steady state, step changes in feed flow rate, rotational speed or axis inclination were imposed, and the resulting discharge flow rate and bed depth axial profile measured as functions of time. Good agreement is found between model and experiment.

INTRODUCTION

Rotary kilns are widely used in the processing of granular materials in the chemical and metallurgical industries, as mixers, dryers and reactors. A typical kiln used in the sulphate process for the manufacture of TiO_2 is a cylinder 50 m long and 3 m in diameter, with axis inclined at a few degrees to the horizontal, and rotating about its axis at one revolution every 5-10 minutes. Granular material is fed to the upper end of the kiln, and forms a continuously rotating bed resting on the bottom of the kiln, filling each cross section to a level generally less than 30% by volume. The granular material is slowly conveyed along the kiln as a result of the continuous rotation, and the force of gravity down the slope. Some kilns have lifters in the wall, but here attention will be directed to smooth-walled kilns without lifters. Heating can be carried out by flow of hot gas countercurrent to the flow of solids, but this will not be considered here.

The transverse motion of the bed of granular material in such a kiln has been classified by Henein et al. [1]. At very low rotational speeds, slipping may occur when the bed remains at rest and slips relative to the kiln wall. With increased rotational speed, this gives way to slumping (or avalanching), when the bed is continuously lifted up by the kiln wall and periodically falls down the free surface. The period between slumps decreases with increasing rotational speed, eventually producing rolling motion in which there is a layer of continuously falling particles forming a plane free surface. Slumping and rolling motions will be considered here. The main body of the bed rotates with the kiln in solid body rotation, and when grains reach the upper half of the free surface, they fall rapidly down it, each grain following the path of steepest descent. The grains come to rest, relative to the kiln, on the lower half of the slope, and then enter the mass of grains rotating with the kiln in solid body motion. In slumping and rolling motions, the rotational speed is slow enough that gravitational forces dominate over centrifugal, and the bed surface is planar; at higher rotational speeds the bed surface becomes curved, and at very high rotational speeds, centrifuging occurs.

Bulk motion in the steady state can be described by a geometrical model in which the bed surface is treated as locally flat, and particles fall down this flat surface at the angle of steepest descent. It is further assumed that particles spend a negligible amount of time in the fall down the surface of the bed compared to the time they spend in solid body motion. Using these simple ideas, Saeman [2] and Kramers and Croockewit [3] have shown that the volumetric flowrate of granular material, Q, is related to the rotational speed, n, kiln radius, R, bed depth h, the angle of repose, γ, and the angle of inclination of the kiln axis, β, by

$$Q = \frac{4\pi n R^3}{3 \tan \gamma} \left[\frac{\tan \beta}{\cos \gamma} + \frac{dh}{dx} \right] \left[\frac{h}{R} \left(2 - \frac{h}{R} \right) \right]^{3/2} \tag{1}$$

where x is axial distance along the bed, measured from the discharge end. In deriving eq. (1), it has been assumed that the slope of the bed surface relative to the kiln axis, dh/dx and the kiln inclination, β, are small. A number of authors have performed experimental studies validating the accuracy of eq. (1), and reasonable agreement is found, but with systematic deviations when the total inventory is near the limits of its range [3-6]. Spurling et al. [7] have in addition used the geometrical ideas to analyse the orientation of the bed surface in the case $Q = 0$, which is of relevance to back-spill in the feed region of the kiln; the theory is again in good agreement with experiment.

The aim of this work is to analyse transients, which in industrial operation can occur at start-up and shut-down, and for example when there are changes in feed flow rate caused by upstream fluctuations. Experiments on a small scale laboratory apparatus will be described, along with a theoretical analysis which extends the geometrical analysis of the steady state to describe transients. Good agreement between theory and experiment is found. Preliminary results have been given by Spurling et al. [8], and the description of further details is in preparation [9, 10].

EXPERIMENT

The apparatus consisted of a clear Perspex cylinder of length 1 m and internal radius 0.0515 m lined on the inner wall with sandpaper of particle size 127 μm. The cylinder was supported by four rubber rollers and rotated by an electric motor and belt drive up to a maximum rotational speed of 0.125 r.p.s., about 6% of the critical speed $\sqrt{(g/R)}/2\pi$. The cylinder mounting was hinged so that the feed end could be raised above the discharge end, inclining the cylinder axis at an angle of 0-5° to the horizontal. The feed end was fitted with a thin dam of height 0.029 m; at the discharge end there was either no dam or a thin dam of height 0.0105 m.

Granular material was fed from a calibrated vibratory feeder. The outflow at the discharge end of the cylinder was directed into a collection bucket and weighed by an electronic balance, yielding the mass flow rate of the outflow. The granular material used was sieved sand of size range 300-600 μm, and mean particle size 490 μm. The bulk density, measured by packing as loosely as possible in a 500 ml volume, was 1600 kg/m^3; this was the density used to relate mass and volumetric flow rates. The static angle of repose was measured in a horizontal cylinder, giving $\gamma = 32\pm0.5°$. For the experiments described here, the bed was in the rolling mode.

The transient response of the discharge mass flow rate was measured following step changes in (i) feed flow rate, or (ii) rotational speed, or (iii) axis inclination, for a range of experimental parameters. After the step, each experiment was continued with a steady feed rate until the discharge mass flow rate became constant and equal to that of the feed.

A second set of experiments was performed to measure the development of the axial profile of bed depth during the transient response. Experiments were performed as previously, but at a number of times following the step, the feed and rotation were stopped, freezing the bed, and the bed depth was measured. The sandpaper lining of the inner wall was installed as a series of 24 mm axial sections separated by clear gaps, and through these gaps the position of the top of the bed at the wall could be located; the bed surface was assumed to be locally flat. The discharge rate from such an experiment was found to be the same as the discharge rate measured in an otherwise identical continuous experiment, suggesting that the stopping and starting did not affect the response.

DYNAMIC MODEL

It is assumed that changes are slow enough that the steady state expression for Q, eq.(1), can be used, with $(\partial h/\partial x)_t$ replacing dh/dx. The model is written using ϕ defined in Fig. 1, in terms of which $h = R(1 - \cos \phi)$. Then eq. (1) gives

$$\left(\frac{\partial \phi}{\partial x}\right)_t = \frac{3\tan\gamma}{4\pi n} \frac{Q}{R^4 \sin^4\phi} - \frac{\tan\beta}{\cos\gamma} \frac{1}{R\sin\phi} \tag{2}$$

Changes in Q along the kiln are related to the cross sectional area, A, occupied by granular material by a volumetric balance of granular material, where $A = R^2(\phi - \sin\phi \cos\phi)$:

$$\left(\frac{\partial A}{\partial t}\right)_x = \left(\frac{\partial Q}{\partial x}\right)_t \tag{3}$$

Then eqs. (2) and (3) can be written as a nonlinear diffusion equation in ϕ,

$$\sin^2\phi\left(\frac{\partial \phi}{\partial t}\right)_x = \frac{2\pi n R}{3\tan\gamma} \frac{\partial}{\partial x}\left\{\sin^3\phi\left\{R\sin\phi\left(\frac{\partial \phi}{\partial x}\right)_t + \frac{\tan\beta}{\cos\gamma}\right\}\right\} \tag{4}$$

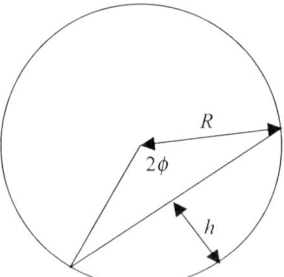

Figure 1. *Cross section of the cylinder and bed; anticlockwise rotation.*

and solved numerically. The initial conditions correspond to the bed depth calculated from the steady state model. The boundary conditions are as follows. At the discharge end, the bed depth is taken to be equal to the height of the discharge dam; this is an approximation as the flow of material over the discharge dam is of finite thickness. At the feed end, the feed flow rate Q_f is specified, and then eq. (2) gives $(\partial\phi/\partial x)_f$ at the feed end.

The parameters of the steady state model are the physical dimensions of the apparatus, the rotational speed, and physical properties of the granular material. Just the same parameters are used in the dynamic model, which has no adjustable parameters.

RESULTS

Experiments were carried out in a variety of configurations, and only a sample of the results is given here. Fig. 2 shows the response of discharge flow rate to a step increase in feed flow rate at $t = 0$. The experiments started in steady state, so the discharge rate at $t = 0$ was the initial feed rate. The experiments continued after the step until the discharge flow rate was again steady, and equal to the feed rate after the step. The results of the model are shown as solid lines. Fig. 3 shows the development of the axial profile of bed depth following a step increase in feed flow rate at $t = 0$. Again, the results of the model are shown as solid lines. Similar experiments were carried out for step decreases in (i) feed flowrate, or (ii) rotation speed, or (iii) cylinder inclination. In all cases, there is good agreement between and experiment and model.

The model has been tested with data from Sai *et al.* [5] and Sriram and Sai [11], who use a cylinder of length 5.9 m and diameter 0.0735 m. Again, the model agrees well with experiment.

CONCLUSIONS

There is good agreement between model and experiment. The model has no adjustable parameters, making it useful for scale-up. To test this further, investigations are underway using cylinders of different sizes.

The model assumes small bed slope, and that grains spend a negligible time falling down the free surface. Further work will investigate the validity of these assumptions.

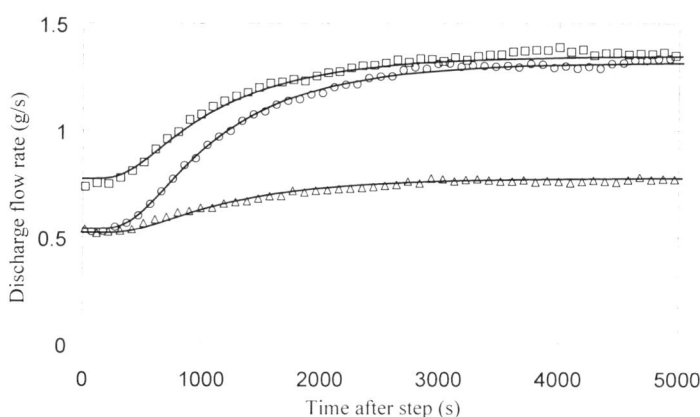

Figure 2. *Discharge flow rate following step increases in feed rate.* $n = 0.0515$ *rps,* $\beta = 1°$, *no discharge dam. The (theoretical) initial mean residence times are 1474 s (squares), 1563 s (circles) and 1570 s (triangles).*

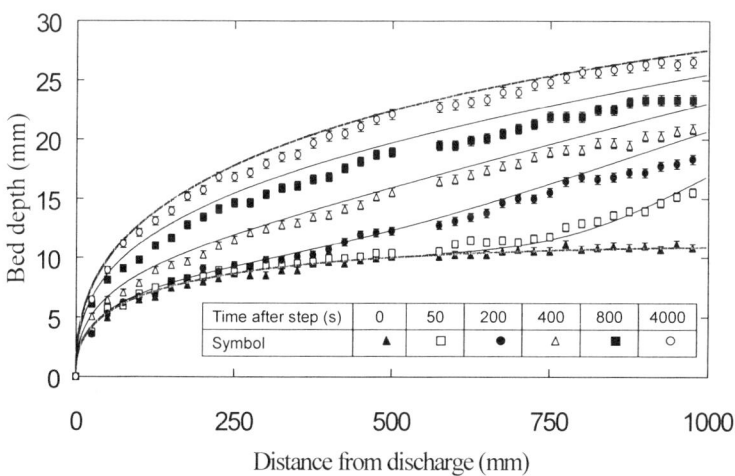

Time after step (s)	0	50	200	400	800	4000
Symbol	▲	□	●	△	■	○

Figure 3. *Development of axial profile of bed depth following a step change of feed rate from 0.625 g/s to 2.45 g/s.* $n = 0.086$ *rps,* $\beta = 1°$, *no discharge dam.*

NOTATION

A area of bed in a cross section of the cylinder (m^2)
h bed depth (m), see Fig. 1
n cylinder rotational speed (rev/s)
Q volumetric flowrate of granular material (m^3/s)
R cylinder radius (m)
t time (s)
x axial distance from cylinder discharge end (m)
β inclination of cylinder to horizontal (rad)
ϕ half-angle subtended by bed at cylinder axis (rad), see Fig. 1
γ angle of repose (rad)

ACKNOWLEDGMENTS

We are grateful to the EPSRC and ICI for a CASE award for RJS, the EPSRC for support through grant GR/M23083, and to Huntsman Tioxide, Billingham, and the PEPT Group at the University of Birmingham for help and advice. DMS is grateful to Cambridge University and the Royal Academy of Engineering for funds to attend the 2000 MRS Spring Meeting.

REFERENCES

1. Henein, H., Brimacombe, J.K. and Watkinson, P., Experimental study of transverse bed motion in rotary kilns, Met. Trans. B **14B**, 191 (1983).
2. Saeman, W.C., Passage of solids through rotary kilns, Chem. Eng. Prog. **47**, 508 (1951).
3. Kramers, H. and Croockewit, P., The passage of granular solids through inclined rotary kilns, Chem. Eng. Sci. **1**, 259 (1952).
4. des Boscs, J., Granular motion in rotating drums, M.Phil. thesis, Faculty of Engineering, University of Birmingham, UK (1998).
5. Sai, P.S.T., Surender, G.D. and Damodaran, A.D., Prediction of axial velocity profiles and solids hold-up in a rotary kiln, The Canadian Journal of Chemical Engineering, **70**, 438 (1992).
6. Lebas, E., Houzelot, J.L., Ablitzer, D. and Hanrot, F., Experimental study of residence time, particle movement and bed depth profile in rotary kilns, The Canadian Journal of Chemical Engineering, **73**, 173 (1995).
7. Spurling, R.J., Davidson, J.F. and Scott, D.M., The no-flow problem for granular material in rotating kilns and dish granulators, Chem. Eng. Sci. **55**, 2303 (2000).
8. Spurling, R.J., Davidson, J.F. and Scott, D.M., The transient response of granular flows in an inclined rotating cylinder, proceedings of Control of Particulate Processes VI, Queensland, Australia (1999).
9. Spurling, R.J., Davidson, J.F. and Scott, D.M., The transient response of granular flows in an inclined rotating cylinder, in preparation.
10. Spurling, R.J., Granular flow in an inclined rotating cylinder, Ph.D. thesis, Department of Chemical Engineering, University of Cambridge, UK, in preparation.
11. Sriram. V. and Sai, P.S.T., Transient response of granular bed motion in rotary kiln, The Canadian Journal of Chemical Engineering, **77**, 597 (1999).

Mat. Res. Soc. Symp. Proc. Vol. 627 © 2000 Materials Research Society

Wavelength Selection in a Vibrated Granular Layer

Eric Clément, Laurent Labous
Laboratoire des Milieux Désordonnés et Hétérogènes – UMR 7603
Université Pierre et Marie Curie- Boîte 86
4, Place Jussieu, F-75252 Paris, France

ABSTRACT

We present a numerical study of a surface instability occurring in a bidimensional vibrated granular layer . The driving mechanism for the formation of stationary waves is closely followed. Two regimes of wavelength selection are identified : a dispersive regime and a saturation regime. For the saturation regime, a connection is established between the pattern formation and an intrinsic instability occurring spontaneously in dissipative gases. We also address the importance of the detailed dissipation laws determining the wavelength values.

INTRODUCTION

In a series of experiments Melo et al.[1] reported a pattern forming instability taking place in a vibrated thin layer of grains (for a recent report see [2] and refs. inside). The apparent phenomenology of the patterns (squares, stripes, hexagons) is strongly reminiscent of the outcome of a parametric instability occurring in vibrated fluid layers called the Faraday instability[3]. Experiments showing parametric surface patterns were also reported in 2D granular layers [4] . Numerical simulations were performed using an event driven algorithm in a 2D[5] and in a 3D geometry[6]. Several theoretical viewpoints were proposed to address the pattern formation issue (see [7] and refs inside) but there is so far no clear vision of the basic mechanisms driving this instability. Here, we report some results on an extensive numerical study we performed using an optimized version of an event-driven algorithm. We address specifically the wavelength selection problem algorithm. In these proceedings we just summarize the principal results. We suggest for more details to look in reference [8] where an extensive report is provided.

NUMERICAL SIMULATIONS

The system we investigate consists of N beads in a container of size L constrained to move in 2D. The bottom plate moves vertically with a trajectory $z(t) = a\sin\omega t$ (a is the amplitude and $f = \omega/2\pi$, the frequency. The lateral boundary conditions are periodic. The collision interactions stem from a collision matrix described in [9] (see refs and details in [6]). The collision parameters are a frontal restitution ε coefficient, a tangential restitution coefficient β (with a maximal value β_0) and a friction coefficient μ. The frontal restitution coefficient is taken to decrease with velocity: $\varepsilon(U) = 1 - \varepsilon_0(U/U^*)^{1/5}$ with U the relative velocity in the normal direction and $U^* = 1$m/s. We use a cut-off velocity ($U_{cut} = 10^{-5}$m/s) for which $\varepsilon = 1$. We use $\varepsilon_0 = 0.4$ (for bead-plate collisions, this coefficient is set equal to 0. The other parameters are $\beta_0 = 0.0$ and $\mu = 0.2$.

RESULTS

Spheres of diameter d are initially packed in a cell with a horizontal width L. We can handle system sizes of the order of 10000 beads. The layer thickness is defined as $H = \sqrt{3/2}N_h$. For a relative acceleration $\Gamma=a\omega^2/g$ situated in a moderate range beyond the threshold $\Gamma=2.5$, an instability occurs and stationary patterns are obtained with a wavelength λ decreasing with frequency (at a constant acceleration). This pattern is made of peaks for which minima and maxima exchange positions at each period of collision with the plate, thus the layer response is $\omega_0=\omega/2$. We monitor in detail the transfer of energy and momentum during the different phases and we observe a non uniform driving, in time and space which is a priori fundamentally different from the driving mechanism in fluids (namely a uniform acceleration modulation). At a constant acceleration, when the frequency is varied, we evidence two different regimes : i) a dispersive regime at low frequencies where the selected wavelength varies with frequency and is close to the dispersion relation observed for gravity waves in a fluid and ii) a saturation regime, at larger frequencies, with a fixed wavelength depending only on the layer height. Now, both regimes are examined independently. The dispersion regime is presented on fig. 1 and the saturation regime on fig. 2.

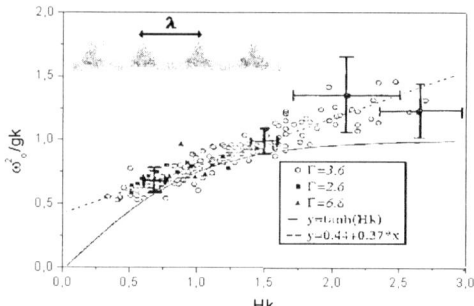

Figure 1. *Selected wavelength in the dispersion regime : $\lambda=2\pi/k$ is displayed in the form: ω_0^2/gk as a function of $Hk=2\pi H/\lambda$; $f_0 =\omega_0/2\pi$ is layer response frequency. $\Gamma =3.6(O)$; $\Gamma =2.6$ (■); $\Gamma =6.8$ (▲) (for many layer heights and bead diameters), the dotted line is the empirical linear best fit (see text); the solid line is the fluid dispersion relation for gravity waves.*

For the dispersion regime (lower frequencies), we plotted the quantity $\omega^2/4gk$ as a function of $Hk=2\pi H/\lambda$ and we observe a collapse of the whole data set around a straight slope bounded by two extreme values : $0.25<Hk<2.8$. The data collapse is interrupted, at high frequencies, for a given set of experiment by the wavelength saturation λ_{sat}, reported on fig2. Note that we found these results to be in agreement with the wavelength selection observed experimentally in 2D[4] and in 3D [1,2]. We notice that the non-saturating regime is consistent with a standard mechanical picture where the average momentum density or the mass flux transferred during the energy input phase ($\rho V_{impact}/T$) is driven by a pressure difference on the scale of a wavelength ($\Delta P/\lambda$). If we estimate that the pressure difference in this regime scales like the peak amplitude (experimentally we measured $p \cong 4a$, see in ref. [4] for a scaling argument) , i.e. $\Delta P \approx gp$, we obtained the balance equation: $\rho a \omega^2 \approx \rho ga/\lambda$. This relation is a dimensional argument which could explain why we observe the limiting law: $\omega^2/4gk$=const at low frequencies. But at larger frequencies (before the saturation occurs) the internal density and pressure waves play an important role, since now, the peak amplitude is reduced and the limiting behavior should corresponds to a limiting velocity of internal density waves caused by the impact with the bottom plate. The pressure scale is now defined by the hydrostatic pressure scale: $\Delta P \approx \rho gH$ and the density waves have a velocity $c \approx (gH)^{1/2}$. Thus in this limit, the selection mechanism would correspond to a dispersion relation $\omega = ck$. As a consequence, we see that the selection mechanism is indeed a complex interplay between the possibilities of global deformation of the granular layer producing gravity restoring forces and the dynamics of internal pressure/density waves driving the flow. The previous argument gives some rational to a selection law of the type :

$$\omega_0^2/gk = A + BHk \qquad (1)$$

with $A=0.44\pm0.02$ and $B=0.37\pm0.02$, that we observe empirically. Note that in this dispersive regime, the exact value of the collision parameters does not matter much as long as there is enough dissipation to be far from the homogeneous fluidization regime. When the dissipation is decreased, the patterns are progressively blurred but the wavelength is unchanged. We also noticed that we need a little bit of rotational friction to stabilize the patterns.

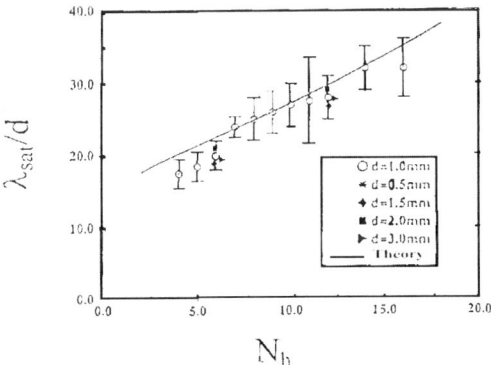

Figure 2. *Selected wavelength in the saturated regime : λ_{sat}/d as a function of the number of layers N_h , for d=.5mm(*), d=1mm(O), d=1.5mm($\blacklozenge$), d=2mm(v), d=3mm($\blacktriangle$); the solid line is the mean field theoretical prediction (see text).*

Now, we focus on the saturating regime obtained at high frequencies. We measure the saturation wavelength for various couples of parameters $\lambda_{sat}(N_h,d)$. From our measurements (see fig.2), we observe a linear increase of this selected length with the number of layers. In parallel we performed a set of simulations where we follow the patterns created by the impact of a moving plate on a layer of grains initially at rest and in the absence of gravity (see fig.3). The impact velocity and the initial separations s_0 between the beads are the typical conditions for the vibrated layer (i.e. U_0= 1m/s, $s_0 \cong .05d$). We calculated the horizontal density distribution. On fig 3b we display the power spectrum of this distribution: $S_k(t)$ for 5 different times after impact. We observe a fast growing wavelength characterized by a wave number : $k_s = 2\pi\backslash\lambda_s$. On fig 3c we report the saturation wavelength $\lambda_{sat}(N_h,d)$ as a function of the selected wave length λ_s and we get with a good precision :

$$\lambda_{sat}(N_h,d) = \lambda_s(N_h,d) \qquad (2)$$

Therefore, the wavelength obtained in the saturation regime for the vibrated layer is equal to the wavelength obtained in a simple collision experiment.

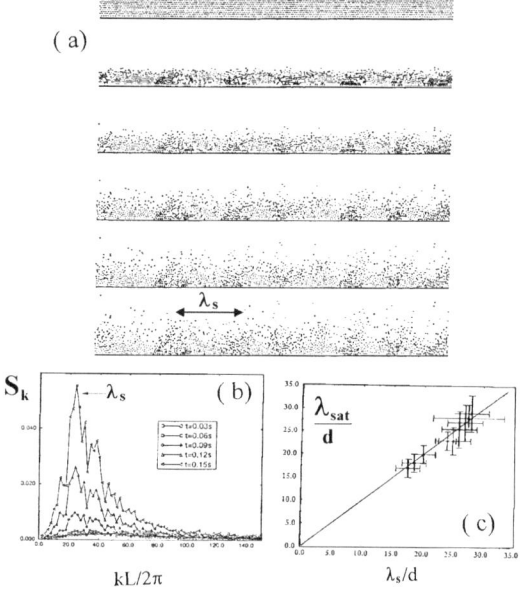

Figure 3. *Pattern formation in a granular layer impacted from the bottom (without gravity). Fig3a : evolution of the grains after the impact. The black circle particles have a horizontal velocity to the right and horizontal dashes to the left. Fig.3b : time evolution of the structure factor $S_k(t)$ after impact. Fig 3c, comparison between the selected wavelength λ_s/d and the saturation wave length λ_{sat}/d, for different layer heights N_h. The straight line is $y=x$.*

Now we relate this pattern selection to the general issue of granular stripe stability. We study the collision of a granular layer with a plate at a velocity U_0, such that the grains dissipate with a *constant* restitution coefficient. The dynamics is followed in detail and we identify two important phases as a result of the impact : i) just after the impact, an upwards compression wave and a downwards dilation wave cross the layer at very large speeds. These waves do not cause global distortion of the layer but are extremely dissipative ; ii) after the passage of those waves an expansion of the layer follows, characterized by a vertical velocity gradient and subsequently, the layer losses contact with the plate and expands in the vacuum. A systematic study of the scaling behavior of the initial temperature before the expansion phase shows that the average temperature T of the layer prior to expansion is:

$$T^{1/2} \approx U/N_h * \exp[-\xi(N_h-1)(1-\varepsilon)] \qquad (3)$$

with $\xi \cong 1.1 \pm 0.1$. During the expansion phase, the layer is unstable and a wavelength λ_s characterizing density modulations in the horizontal direction shows up. We identify two limiting scaling behaviors for the selected wavelength λ_s . Thus we have :

$$\lambda_s/d \approx (1-\varepsilon^2)^{-\beta} , \qquad (4)$$

where the value of the exponent is $\beta=0.44\pm0.08$ in the low dissipation limit and in the high dissipation regime we measure $\beta =1 \pm0.01$. Interestingly relation (4) does not depend on the number of layers N_h .

The same systematic study was performed with a restitution coefficient now *depending* on the collision velocity. We have in the early stage an impact with a layer moving at a constant velocity U_0, and therefore, a typical restitution coefficient $\varepsilon_i(U_0) \cong 1- \varepsilon_0(U_0/U^*)^{1/5}$. But after a large number of collisions between the grains, the average kinetic energy of the layer is decreased tremendously and we get a final restitution coefficient of magnitude: $\varepsilon_f(U_f) \cong 1- \varepsilon_0(U_f/U^*)^{1/5}$ characterized, just before the expansion phase, by a typical collision velocity between the beads U_f. Therefore, using a relation: $T \equiv U_f^2$, with a temperature T determined from the previous study (see equ.(3)), we estimate the final restitution coefficient using the relation : $(1- \varepsilon_f)/(1- \varepsilon_i)= (U_f/U_0)^\alpha$ (for our simulations we used $\alpha=1/5$). Numerically, we determine a scaling relation: $\lambda_s/d \approx (1-\varepsilon_f^2)^{-1/2}$. Putting together these last equations, we get a mean-field theoretical prediction for the saturation wavelength displayed on fig.2. Interestingly, in the limit of weak dissipation, i.e. $\varepsilon_i \rightarrow 1$, we obtain a interesting asymptotic formula for the selected wavelength λ_s for any α and N_h :

$$\lambda_s/d \approx (1-\varepsilon_i)^{-1/2}N_h^{\alpha/2}\exp[-\xi\alpha/2(N_h-1)(1-\varepsilon_i)] \qquad (5)$$

On this equation, we check easily that the exponential growth of the selected length with the number of layers N_h is damped by the weak value of the coefficient $\alpha=1/5$ characterizing the velocity dependence of the binary collision dissipation. Thus, both antagonistic effects give approximately a linear increase of the wavelength. As a consequence, the presence of an intrinsic instability due to the dissipative character of the granular layer prevents the selected wavelength to decrease when the frequency is increased at a value smaller than the intrinsic dissipative length. The vibration only sustains the motion due to this imposed wavelength. Thus, after impact, the regions with larger densities will create larger agitation and thus larger pressures. As a consequence, a horizontal flow develops towards lower pressure regions and the alternative horizontal motion is be sustained at the pace of the vertical impacts with the plate.

CONCLUSION

In conclusion, we present here the results of a numerical study on a pattern forming instability occurring in a 2D vibrated layer of dissipative grains. We focus on the mechanisms leading to the formation of stationary oscillating surface peaks which are separated by a well-defined wavelength $\lambda=2\pi/k$. We identify two distinct regimes. The first regime (dispersive) corresponds to a periodic excitation of the layer where the gravity restoring force plays an important role in competition with internal density and pressure waves created by repeated impacts with the bottom plate. The dispersion relation relating the wavelength and the layer oscillation frequency $f_0=\omega_0/2\pi$, is such that we have in general a relation of the type: $\omega_0^2/gk=O(1)$ with a value smaller for thin channels and larger for thick channels (sizes being compared to the selected wavelength). We propose an empirical relation : $\omega_0^2/gk=A+BHk$, with values of A and B independent on the acceleration. At larger frequencies (acceleration being constant) this dispersion relation is interrupted by a saturation regime where the wavelength is now independent of the frequency. We show how this new wavelength selection mechanism is related to an instability occurring spontaneously in a dissipative gas colliding with a plate. This effect might be linked more generally, to other instabilities found in dissipative gases [10,11]. We also stress on the influence of the detailed microscopic dissipation laws affecting the values of the selected wavelength.

REFERENCES

[1]   F. Melo, P. Ubanhovar and H. Swinney, Phys. Rev. Lett. **72** , 172 (1994); ibid **75**, 3838 (1995).

[2]   P. Umbanhowar, Ph.D. dissertation The University of Texas in Austin (1996) ; P.Umbanhowar and H. Swinney, *preprint* (2000).

[3]   M. Faraday, Philos. Trans. R. Soc. **121**, 299 (1831).

[4]   E. Clément, L.Vanel, J. Duran and J. Rajchenbach, Phys. Rev. E **53**, 2972 (1996).

[5]   S. Luding, E. Clément, J. Rajchenbach and J. Duran, Europhys.Lett, **36**, 247 (1996).

[6]   C.Bizon, M.D.Shatuck,J.B.Swift, W.D.McCormick and H.Swinney,Phys. Rev. Lett. **80**, 57 (1998)

[7]   C. Bizon, M. D. Shattuck, and J. B. Swift, Phys.Rev.E **60**, 7210 (1999).

[8]   L.Labous, Thèse de Doctorat, Université Paris VI (1998); E.Clément, L.Labous *preprint* , submitted to Phys.Rev.E (2000).

[9]   O.Walton et al., J.Rheol. **30**, 949(1983).

[10]   B.Bernu and R.Mazighi, J.Phys.A **23**, 5745 (1990); S.MacNamara, W.R.Young, Phys.Fluid A **5**, 34 (1993).

[11]   I.Goldhirsch and G.Zanetti, Phys.Rev.Lett. **70**, 1619 (1993); S.MacNamara and W.R.Young, Phys.Rev.E **50**, R28 (1994).

Mat. Res. Soc. Symp. Proc. Vol. 627 © 2000 Materials Research Society

Mixing and Segregation Processes in Turbula Blender

Nathalie SOMMIER[1], Patrice PORION[2], and Pierre EVESQUE
Laboratoire de Mécanique MSSM, UMR 8579 CNRS - Ecole Centrale Paris, F-92295 Châtenay-Malabry cedex, France
[1] Laboratoire de Physique Pharmaceutique, UMR 8612 CNRS - Université Paris XI, 92296 Châtenay-Malabry cedex, France.
[2] Centre de Recherche sur la Matière Divisée, UMR 6619 CNRS - Université d'Orléans, F-45071 Orléans cedex 2, France.

ABSTRACT

Magnetic Resonance Imaging (MRI) technique was used to study the mixing and segregation processes of granular materials in a sophisticated tumbling blender (Turbula® mixer) using binary mixtures of sugar beads of different diameters d. Its motion generates mixtures with complex patterns. Effects of some parameters (beads diameter ratio, rotation speed, mixing time) were checked on segregation and mixing processes. We report in this paper, a qualitative and quantitative analysis of these phenomena. A segregation index S was defined to study the homogeneity and the kinetics of the mixing/segregation processes. When the ratio of bead diameters d_{max}/d_{min} is approximately 1, mixing process is observed but segregation occurs as soon as d_{max}/d_{min} is greater than 1.1.

INTRODUCTION

The segregation of granular materials such as grains and powders is a significant problem when processing granular media in the pharmaceutical and food-processing industries. Indeed, heterogeneous mixtures tend to segregate by size, even in situations when one might expect good mixing such as shaking or rotation. Many experiments have been dealing with the segregation occurring in rotating cylinders or in drum mixers. For instance, the well-known axial segregation was already reported in 1939 by Oyama [1]; such axial bands of segregation appear after a few tens of rotation when a binary mixture of different size is rotated in a drum blender and it is now a standard problem in the field of granular-media studies [2-7]. More recently, Magnetic Resonance Imaging (MRI) studies were used to analyze the axial and radial segregation within the bulk of granular media [8] or to study granular flow [9-10] and convection [11-13]. From a more fundamental point of view, the understanding of the basic laws of segregation and mixing appears to be a fundamental challenge of the physics of granular matter, of which very little is known [14]. Furthermore, this topic is most likely related to other fundamental problems of fluid mechanics, such as stirring and turbulence [15], and is coupled to the problem of control and enhancement of the kinematics of chemical reactions [16-17]. In this paper, MRI experiments on mixing and segregation of granular matter performed with a sophisticated tumbling blender called Turbula® mixer (Willy A. Bachofen AG Maschinenfabrik, Basel, Switzerland) are reported. This apparatus is commonly used in chemical and pharmaceutical researches as it is supposed to generate good homogenization of the granular samples due to the intricate motion of the cell (see the patents [18] for a more complete description of the motion). This tumbling blender generates a 3d flow mainly localised at the free surface of the granular sample, but which is made intricate because i) the orientation of the free surface changes continuously in two directions and ii) the center of the cell performs a complex cyclic trajectory. Thus, the motion of the blender imposes an intricate 3d flow that causes the material to be swept by an intense "turbulence", and this apparatus should be more efficient to prevent segregation than classical rotating 2d flow generator devices. Effects of different parameters (bead size ratio R, number of rotation N_R, rotation speed Ω) have been investigated and the following conclusion will be drawn: (1) The segregation process appears as soon as the bead diameter ratio d_{max}/d_{min} is larger than 1.1. (2) Segregation increases with the relative difference of radii $(d_{max}-d_{min})/d_{min}$ till an asymptote is reached quickly $(d_{max}-d_{min})/d_{min}=0.7$. (3) Segregation develops much faster (10 times) than the mixing of identical beads. (4) Mixing and segregation occurs fast (14.4 and 1.4 rotations respectively). (5) The segregation observed can be interpreted as an attractor, near the outer edge, which kept larger particles near the edge.

EXPERIMENTAL METHOD

Sugar beads of different diameters d (NP Pharm, France), ranging from 0.35 mm to 1.1 mm, were used. In order to make them MRI-sensitive, one of the species, called reference beads $(d_{ref}=1 \text{ mm})$, was doped with an organic oil. The beads diameter ratio R, defined as d_{ref}/d, was varied from R=0.9 to R=2.8. A Bruker DSX100 NMR imager/spectrometer operating at 100 MHz was used with a probehead size of the birdcage coil equal to 56 mm (Mini 036 Bruker) and a cylindrical container with dimensions axis Oz, diameter 52 mm, and height 66 mm.

In the first experimental study, the container was at first 2/3 filled with two horizontal layers each of 22 mm height, the bottom of reference beads and the top of the other type. The initial configuration therefore corresponds to a completely segregated sample. A first MRI image of this initial state (t=0) was acquired, and the container then placed in the Turbula® blender which was then run during a given time t with a given rotation speed Ω equal to 22 rpm corresponding to the smaller rotation speed of the blender. After this "mixing" time t_{mix} related to the rotation number N_R through $N_R=\Omega.t_{mix}$, the container was removed and a new NMR image taken, and so on. These 3d MRI experiments were performed using selective excitations and a usual two-dimensional Fourier imaging (spin-warp method) [19]. Labeling Oz the cylinder axis, two different series of 16 images at different depths were obtained, one corresponding to 16 horizontal slices parallel to the xy plane (axial cuts), the other to 16 vertical slices parallel to the zy plane (sagittal cuts). The thickness of a slice (3.5 mm) and the whole 16 slices correspond to the height 56 mm; each 2d image contains 64x64 pixels and corresponds to a field of 60 mm x 60 mm; the spatial resolution is 940 μm. So the voxel size contains about 3-4 grains and the voxel intensity is proportional to the local number of reference beads. Consequently, these slices allow a complete analysis of the spatial concentration of the different grains in the sample. An adequate signal-to-noise ratio requires about 20 minutes acquisition time for the set of sixteen images.

In the second study, the MRI was used to follow the displacement of a single bead (poppy seed) in a media of sugar beads. Indeed, Nakagawa et al. have shown that oil-containing seeds provide sufficient free proton in liquid state to produce a discernable NMR signal [9]. In this case, using a 1d-profile MRI sequence along the 3 principal directions, the poppy seed position r in the sugar bead bulk was determined as function of the number of revolution N_R.

RESULTS AND DISCUSSION

After a sufficient number of rotation, two different steady states were observed depending on whether R=1 or R≠1 (see Figures 1a and 1b). Mixing process was observed if doped and undoped beads were similar but segregation occurred for beads different in size since the larger ones were localised along the inner wall of the container.

Direct eye analysis of the NMR images is able to determine if a sample is segregated or not, but it does not allow a quantitative analysis, which first requires the definition of a segregation index S. It is known that different indexes can be used; the one we have used is based on a statistical approach already described by Lacey [20]. Let n be the number of voxels in the granular sample stemmed from the MRI images (n≈22000), and $x_i(N_R)$ their intensity at a given time N_R ; x_i is proportional to the number of doped beads in voxel i; the segregation index $S(N_R)$ is defined as $S = \sigma/\bar{x}$ where σ and $\bar{x}$ are respectively the standard deviation and the mean value of the distribution $x_i(N_R)$. With this definition, our segregation index $S(N_R)$ spreads from 1 (for complete segregation) to $1/\sqrt{b}$ (for a perfectly homogeneous mixture) for a binary mixture (50-50 %), where b is the mean number of grains per voxel.

Quantitative analysis of mixing/segregation experiments is reported in Figure 2a, which shows the variations of S versus N_R. For mixing (R=1), S decreases from 0.95 to 0.48 when the number of cycles N_R increases from 0 to 30. Owing to the fact that S varies randomly between 0.9 and 0.95 instead of 1 when N_R=0, the uncertainty on S is about 10 %; it is generated at the interface. The steady state value S=0.48 for mixing is obtained already at N_R=30. Turning now to the segregation process (R=2.8), S evolves from S_{max}=0.92 to S_{min}=0.78, the latter being already reached after only 5 cycles. This demonstrates that an efficient segregation process develops much faster than mixing, and confirms the segregation process observed in Figure 1b and the efficiency of the Turbula® blender as segregation and separation device. Characteristic times of the kinematics of mixing and segregation were obtained from a best-fit procedure of the two curves of Figure 2a using an exponential law. For the segregation process, the characteristic time is found to be 1.4 rotations, whereas it is 14.4 rotations for mixing. So, the characteristic time of segregation is much shorter than the mixing one in the Turbula® apparatus.

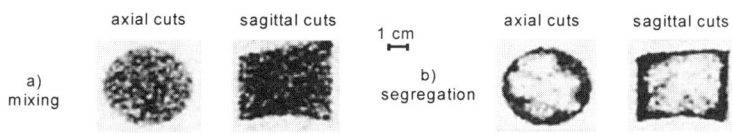

Figures 1. *Stationary pattern observed after 220 rotations at 22 rpm in the middle of the container filled at 2/3 of the height, the larger beads regions appears in dark zone on the picture: a) Mixing for identical beads (R=1), b) Segregation for size-different beads (R=2.8).*

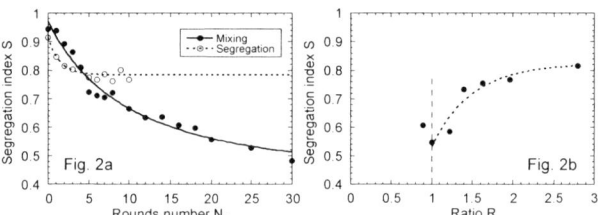

Figures 2. *Segregation index S obtained at large N_R (N_R=220) and Ω=22 rpm, the points are experimental data: a) variation of S plotted versus the number of cycles N_R, for mixing and segregation. The curves represent exponential law fits. b) S for different bead ratio R, the R=0.9 and R=1.2 results are quite similar. Note that segregation is already important at R=1.4.*

The variations of the steady-state segregation index S (i.e. after 220 cycles) is reported in Figure 2b as a function of R. In absence of finite-size effects, one expects $S(R) \approx S(1/R)$. This is indeed approximately observed for the two points at R=0.9 and R=1.2. Furthermore, the samples with R>1 and R<1 exhibit similar segregation patterns with the larger beads located at the edge of the container and the smaller ones in the centre.

In the second experiment study, in order to follow the displacement of a single bead in the granular bulk, we have used a poppy seed (d≈1.1 mm) in a sugar-bead bulk (3 different sizes: 0.35, 1, and 1.7 mm). In Figures 3, for each sugar bead size, we report the renormalized distance D of the poppy seed from the centre of the bulk versus N_R. D is defined as the real distance r from the centre of the bulk divided by the wall distance in this same direction. So, D varies from 0 to 1 when the poppy seed moves from the centre to the edge of the container. We have also reported $D(N_R+1)$ as a function of $D(N_R)$. Figures 3(a) and 3(d) concern the displacement of a poppy seed in small beads (0.35 mm); they show clearly that the edge of the cylinder can be interpreted as an attractor for the segregation process in this 3d-blender. Indeed, when N_R increases, R reaches 1. On the other hand, the Figures 3(c) and 3(f) for large sugar beads show that the poppy seed does not evolve to the external region but it shows that it moves everywhere in the pile. For the intermediate size (Figures 3(b) and 3(e)), the segregation process is still observed but it is attenuated. From the $D(N_R)$ curves, we have quantified the evolution towards the attractor of the dynamic by a characteristic time. We have found respectively 2.3 and 3.6 rounds for the 0.35 and 1 mm sugar beads which confirms our previous qualitative analysis. It is interesting to compare the 0.35 mm result with the other one stemmed from the 3d-MRI macroscopic analysis described above. Indeed, the 3d-MRI analysis of the segregation process gives a characteristic time equal to 1.4 rounds which is closed to 2.3, found for isolated particle.

CONCLUSIONS

This study shows that the segregation process occurs in the Turbula® blender as soon as the bead diameter ratio d_{max}/d_{min} is larger than 1.1, with the larger beads located in the outer part of the sample. The kinetic study shows that there are two characteristic time scales, and that the segregation process (1.4 rotations) is faster than mixing (14.4 rotations). So, segregation is quite efficient, which is relatively surprising from a topology argumentation, since better mixing shall be expected for 3d developed flows [21] instead of 2d one so that one could have expected that mixing be better in Turbula® blender than in a classical drum.

Although the devices and the techniques used are quite different, one may compare the present experimental result with those recently obtained by Dury and Ristow [22] on mixing and segregation using numerical simulations in a 3d rotating cylinder. They observed similar segregation with $d_{max}/d_{min} \rightarrow 0$, with radial segregation scaling linearly with particle size difference and already visible for 10 % difference. We also show that segregation can be interpreted as attractor notion.

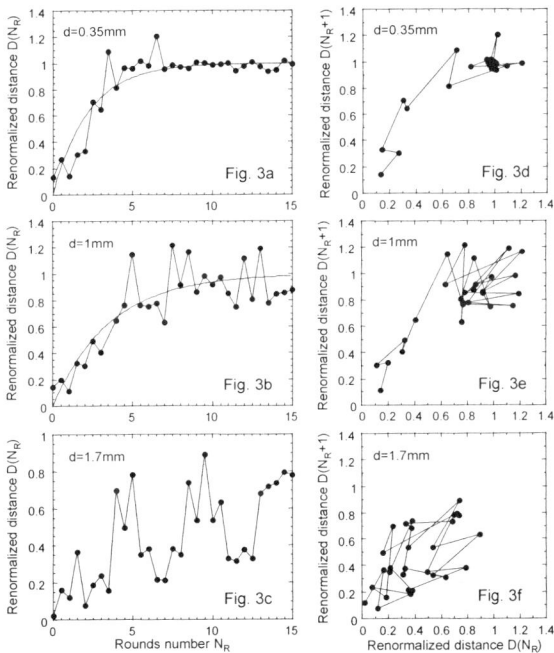

Figures 3. *Evolution of the renormalized distance from the bulk center $D(N_R)$ of a poppy seed inside the granular media plotted versus N_R at $\Omega=22$ rpm: a) 0.35 mm media, b) 1 mm media, and c) 1.7 mm media. On 3a and 3b, the curves represent exponential law fits. Evolution of $D(N_R+1)$ plotted versus $D(N_R)$ for: d) 0.35 mm media, e) 1 mm media, and f) 1.7 mm media.*

ACKNOLEDGMENTS

Profs. G. Couarraze, P. Tchoreloff, Dr. B. Leclerc, L. Pothuaud are thanked for helpful discussions, and M.P. Faugère for her technical assistance. We also acknowledge P. Sanial from the NP Pharm company for supplying the sugar beads samples. P. P. wants to thank Dr. P. Levitz and Prof. H. Van Damme for their encouragement. The Bruker spectrometer and minicomputers used in this study were purchased thanks to grants from CNRS and Région Centre (France). P. E. wants to thank CNES for partial financial support.

REFERENCES

1. Y. Oyona, Bull. Inst. Phys. Chem. Res. (Tokyo) **5**, 600 (1939); Y. Oyona and K. Ayaki, Kagaku Kikai **20**, 6 (1956).
2. M. Donald and B. Roseman, Br. Chem. Eng. **7**, 749 (1962).
3. J. C. Williams, Powder Technol. **15**, 237 (1976).
4. J. Bridgwater, Powder Technol. **15**, 215 (1976).
5. S. D. Gupta, D. V. Khakhar, and S. K. Bhatia, Chem. Eng. Sci. **46**, 1513 (1991).
6. S. B. Savage, in *Disorder and Granular Media*, edited by D. Bideau and A. Hansen (North-Holland, Amsterdam, 1993), p. 255.
7. R. Hogg and D. W. Fuerstenau, Powder Technol. **6**, 139 (1972).
8. K. M. Hill, A. Caprihan, and J. Kakalios, Phys. Rev. Lett. **78**, 50 (1997); Phys. Rev. E **56**, 4386 (1997).
9. M. Nakagawa, S. A. Altobelli, A. Caprihan, E. Fukushima, and E. K. Jeong, Exp. Fluids **16**, 54 (1993).
10. M. Nakagawa, S. A. Altobelli, A. Caprihan, and E. Fukushima, in *Powders and Grains 97*, edited by R. Behringer and J. Jenkins (Balkema, Rotterdam, 1997), pp.447-450.
11. E. E. Ehrichs, H. M. Jaeger, G. S. Karczmar, J. B. Knight, V. Y. Kuperman, and S. R. Nagel, Science **267**, 1632 (1995).
12. V. Y. Kuperman, E. E. Ehrichs, H. M. Jaeger, and G. S. Karczmar, Rev. Sci. Instrum. 66, 4350 (1995).
13. A. Caprihan, E. Fukushima, A. D. Rosato, and M. Kos, Rev. Sci. Instrum. **68**, 4217 (1997).
14. H. M. Jaeger, S. R. Nagel, and R. P. Behringer, Rev. Mod. Phys. **68**, 1259 (1996).
15. H. Aref, J. Fluid Mech. **143**, 1 (1984).
16. J. M. Ottino, *The Kinematics of Mixing: Stretching, Chaos, and Transport* (Cambridge Univ. Press, Cambridge, 1989).
17. J. M. Ottino, Annu. Rev. Fluid Mech. **22**, 207 (1990).
18. P. Schatz, Deutsches Reichspatent Nr. 589 452 in *Der allgemeinen Getriebeklasse* (1933); US Patent Nr. 2 302 804 (1942).
19. P. T. Callaghan, *Principles of Nuclear Magnetic Resonance Microscopy* (Clarendon Press, Oxford, 1991).
20. P. M. C. Lacey, J. Appl. Chem. **4**, 257 (1954).
21. P. Porion, N. Sommier, and P. Evesque, Europhys. Lett. (to be published).
22. C. M. Dury and G. H. Ristow, Phys. Fluids **11**, 1387 (1999).

Stress Distributions

Mat. Res. Soc. Symp. Proc. Vol. 627 © 2000 Materials Research Society

Statistical Mechanics of Stress Transmission in Static Arrays of Rigid Grains

Dmitri V. Grinev and Sam F. Edwards
Cavendish Laboratory, University of Cambridge,
Madingley Road, Cambridge CB3 OHE, U.K.

ABSTRACT

We develop the statistical-mechanical theory that delivers the fundamental equations of stress equilibrium for static arrays of rigid grains. The random geometry of static granular packing composed of rigid cohesionless particles can be visualised as a network of intergranular contacts. The contact network and external loading determine the network of intergranular forces. In general, the contact network can have an arbitrary coordination number varying within the system. It follows then that the network of intergranular forces is indeterminate i.e. the number of unknown forces is larger than the number of Newton's equations of mechanical equilibrium. Thus, in order for the network of intergranular forces to be determined, the number of equations must equal the number of unknowns. We argue that this determines the contact network with a certain fixed coordination number. The complete system of equations for the stress tensor is derived from the equations of intergranular force and torque balance, given the geometric specification of the packing. The granular material fabric gives rise to corrections to the Euler-Cauchy equation that become significant at mesoscopic lengthscales. The stress-geometry equation establishes the relation between various components of the stress tensor, and depends on the topology of the granular array. We show that the macroscopic stress tensor σ_{ij} satisfies the following system of equations

$$\nabla_j \sigma_{ij} + \nabla_j \nabla_k \nabla_m K_{ijkl} \sigma_{lm} + \ldots = g_i$$

$$P_{ijk} \sigma_{jk} + \nabla_j T_{ijkl} \sigma_{kl} + \nabla_j \nabla_l U_{ijkl} \sigma_{km} + \ldots = 0$$

and derive expressions for $K_{ijkl}, P_{ijk}, T_{ijkl}, U_{ijkl}$ in terms of fabric tensors. The problem of incorporating the "no tensile forces" constraint into the formalism is considered. The inability of an intergranular contact to support tensile forces constrains the geometry and the stress state of the contact network. We establish a link between the structural component of the angle of repose and the packing fabric.

INTRODUCTION

In this paper we show that the microstructural origin of granular stress and granular fabric can be incorporated into continuum equations of stress transmission. After presenting a brief characterisation of granular materials and their structure we review basic structural properties and propose a model description of granular packings. In our model the geometric characterisation of disordered granular packings is the key to any qualitative and quantitative understanding of the physics of stress transmission. We argue that it is in principle possible to develop an analytical formalism if one ignores a number of complications that are present in a real packing of particles. The straightforward metric characterisation of microstructure in terms of fabric tensors will allow us to derive fundamental equations of stress transmission.

GEOMETRY AND NEWTON'S LAWS

We model the granular material as an assembly of discrete rigid particles whose interactions with their neighbours are localised at point-like contacts. Therefore the description of the network of intergranular contacts is essential for the understanding of force transmission in granular assemblies.

The mathematical description of geometrical properties of granular packing in our model is straightforward. Let us consider an array of N particles. The reference particle α has the coordination number z and therefore z contact point vectors $\underline{R}^{\alpha\beta*}$. Grain α exerts a force on grain β at a point $\underline{R}^{\alpha\beta*}$. The contact point is a point in a plane whose normal is $\underline{n}^{\alpha\beta}$. The centroid of contacts of particle α is defined as

$$\underline{R}^{\alpha} = \frac{1}{z}\sum_{\beta} \underline{R}^{\alpha\beta*} \tag{1}$$

Vector $\underline{r}^{\alpha\beta}$ joins the centroid with the contact point

$$\underline{R}^{\alpha\beta*} = \underline{R}^{\alpha} + \underline{r}^{\alpha\beta} . \tag{2}$$

So that

$$\sum_{\beta} \underline{r}^{\alpha\beta} = 0, \tag{3}$$

where $\sum_{\beta}$ means summation over the nearest neighbours. We introduce a vector that joins the centroid of grain α with that of grain β

$$\underline{R}^{\alpha\beta} = \underline{R}^{\beta} - \underline{R}^{\alpha} . \tag{4}$$

We describe the fabric (i.e. the arrangement of structural units) in terms of tensors composed of vectors that specify the contact points and the distance between the centroids of contacts of nearest neighbours. Fabric tensors can be defined in different ways. These tensors represent various features of the packing microstructure associated with the type and degree of the anisotropy. Probability distribution $P(\underline{n}^{\alpha\beta})$ of contact normal $\underline{n}^{\alpha\beta}$ can be introduced. For isotropic systems $P(\underline{n}^{\alpha\beta}) = \dfrac{1}{2\pi(d-1)}$ where $d = 2,3$ is the number of dimensions. It is important to distinguish between the so-called inherent or geometrical anisotropy (which is a characteristics of the geometry of packing) and the stress-induced or mechanical one. Anisotropy in granular media has two key sources of origin: deposition history (which is certainly related to the shape of particles and friction coefficient) and applied external stress (which can either cause the rearrangement of grains and/or change the degree of the heterogeneity of the stress distribution). Typical examples of anisotropic phenomena in granular media are arches and force chains [5,6,8,11]. Apparently anisotropy might manifest itself at different length-scales that are the microscopic, mesoscopic and macroscopic scales. The choice of the length-scale depends on several factors. Anisotropy at the microscopic length-scale appears due to the presence of steric correlations whereas at the macroscopic scale boundary conditions and stress at the boundaries are the key factors. The mesoscopic length-scale is the most mysterious one. Possibly the highly correlated local density fluctuations (e.g. due to the presence of polydispersity in the packing and/or fluctuations of the coordination number) give rise to anisotropy at the mesoscopic length-scale. Clearly, given some specified deposition technique, elongated and platy particles have a higher probability of forming anisotropic (i.e. possessing some degree of orientational ordering) packings, in comparison with the isometric grains. However our model allows us to avoid all these complexities of shape, friction and deposition history by having this information incorporated into the set of contact points $\left\{ \underline{R}^{\alpha\beta*} \right\}$. The fabric tensor F_{ij}^{α} for grain α can be introduced as

$$F_{ij}^{\alpha} = \sum_{\beta} R_i^{\alpha\beta} R_j^{\alpha\beta} \ , \tag{5}$$

which characterises the first coordination shell of the grain α. The mesoscopic fabric tensor corresponding to some mesoscopic length-scale can be defined in the same manner. Other types of fabric tensors will also appear in our model.

The choice of the particular type of fabric tensors depends on the mechanical properties under investigation, the corresponding length-scale and the structural features of the packing itself. The macroscopic fabric tensor characterises the packing fabric at the macroscopic length-scale and can be obtained by averaging (5)

$$\left\langle F_{ij} \right\rangle = \frac{1}{N} \sum_{\alpha=1}^{N} \ \sum_{\beta} R_i^{\alpha\beta} R_j^{\alpha\beta} \ . \tag{6}$$

In a static array of grains Newton's equations of intergranular force and torque balance are satisfied. Balance of force around the grain α requires

$$\sum_{\beta} f_i^{\alpha\beta} = g_i^{\alpha} \ , \tag{7}$$

$$f_i^{\alpha\beta} + f_i^{\beta\alpha} = 0 \ , \tag{8}$$

where g_i^{α} is the external force acting on grain α. Further on in this paper $\underline{g}$ is used also for the external forces at the boundaries. The equation of torque balance is

$$\sum_{\beta} \varepsilon_{ikl} f_k^{\alpha\beta} r_l^{\alpha\beta} = c_i^{\alpha} \ . \tag{9}$$

In general, the centroid of the contact points need not coincide with the centroid of the forces e.g. the centre of mass of a solid grain, but we will assume it is so in order to keep the analysis simple. Also we ensure that the microscopic stress tensor is symmetric by putting $c_i^\alpha = 0$.

The microscopic version of force analysis is to determine all of the intergranular forces, given the applied force, torque loadings on each grain and geometric specification of a granular array. The number of unknowns per grain is $\dfrac{zd}{2}$. Required force and torque equations give $d + \dfrac{d(d-1)}{2}$ constraints. The system of equations for the intergranular forces is complete when the coordination number is $z = d + 1$. Hence, for the intergranular forces in the static array to be determined by these equations, the coordination number $z = 3$ in 2-D and $z = 4$ in 3-D is required. Let us clarify what we mean by particles with "perfect" or "infinite" friction. In a static granular array friction appears only when we count the number of intergranular force components. If grains have a zero friction then there are only normal intergranular forces within the material $\underline{f}^{\alpha\beta} = \left| f^{\alpha\beta} \right| \underline{n}^{\alpha\beta}$. Therefore the number of degrees of freedom is $\dfrac{zN}{2}$. The number of equations is $\dfrac{d(d+1)N}{2}$. The system of equations is complete for $z = d(d+1)$. Thus one can see that for arrays of frictionless particles minimal coordination number is higher compared to that for the opposite limit. This observation can be seen as a confirmation of the validity of our analysis of intergranular forces. It is clear that the coordination number z controls the connectivity of granular media. We will assume that z is indeed 3 in 2-D and 4 in 3-D, for this is the simplest situation, and one which is physically possible. We will try to justify the generality of our approach and offer a conjecture for granular arrays with an arbitrary fluctuating coordination number.

STRESS PROBABILITY FUNCTIONAL

Our goal is to determine the macroscopic stress tensor at every point of a granular array, given external loadings and geometric specification. The macroscopic state of stress is a function of the distribution of contact forces. For any aggregate of discrete grains subjected to external loading, the transmission of stress from one point to another can only occur via the intergranular contacts.

Therefore it is clear that the network of contacts determines the distribution of stresses within the granular array. The network of contacts is determined by the deposition history of the sample and the external loading on the boundaries. We define the tensorial force moment

$$S_{ij}^\alpha = \sum_\beta f_i^{\alpha\beta} r_j^{\alpha\beta}, \tag{10}$$

which is the microscopic analogue of the stress tensor. With $c_i^\alpha = 0$ S_{ij}^α will be symmetric. Our goal is to find a complete system of equations for the macroscopic stress tensor $\sigma_{ij}(\underline{r})$, which is supported by the given network of contacts in the state of mechanical equilibrium.

Given an assembly of discrete grains whose geometry is represented by the network of contacts, we will eventually associate a continuous medium to have continuously distributed properties. However such spatial smoothing or averaging will not be accomplished just as a formal procedure but will take into account the presence of underlying microstructure.

To obtain the macroscopic stress tensor from the tensorial force moment to the macroscopic stress tensor we average it over an ensemble of configurations. The macroscopic stress tensor is

$$\sigma_{ij}(\underline{r}) = \Big\langle \sum_{\alpha=1}^{N} S_{ij}^{\alpha} \delta(\underline{r} - \underline{R}^{\alpha}) \Big\rangle . \tag{11}$$

In the simplest cases of isotropic and homogeneous arrays this is not a problem.

The difficulties appear when the packing under consideration is anisotropic and/or inhomogeneous. Within the confines of this paper we explore only the simplest cases.

The number of equations required equals the number of independent components of a symmetric stress tensor $\sigma_{ij} = \sigma_{ji}$ and is $\dfrac{d(d+1)}{2}$. At the same time, the number of equations available is d. These are vector equations of the stress equilibrium

$$\frac{\partial \sigma_{ij}}{\partial r_j} = g_i \tag{12}$$

that have their origin in Newton's second law. Therefore we have to find $\dfrac{d(d-1)}{2}$ equations, which contain the information from Newton's third law, to complete the system of equations, which governs the transmission of stress in a granular array. Thus in 2-D there is one missing equation, and we derive it in terms of the geometry of the system. Let us give a simple illustration of our theory. Suppose that we can ``invert'' $\dfrac{Nd(d+1)}{2}$ equations for microscopic stress tensors and write the intergranular force $\underrightarrow{f}^{\alpha\beta}$ as

$$f_i^{\alpha\beta} = \sum_{\gamma} A_{ikl}^{\alpha\beta\gamma} S_{kl}^{\gamma} . \tag{13}$$

This representation of intergranular force gives a complete system of equations for the microscopic stress tensor field

$$\sum_{\beta,\gamma} A_{ikl}^{\alpha\beta\gamma} S_{kl}^{\gamma} = g_i^{\alpha} \quad , \tag{14}$$

and

$$\sum_{\beta,\gamma} \varepsilon_{ikl} A_{kmn}^{\alpha\beta\gamma} S_{mn}^{\gamma} r_l^{\alpha\beta} = 0 . \tag{15}$$

There are two ways to deal with this system of discrete equations. The first one is to average them and obtain continuous equations for the macroscopic stress tensor. The second way is to solve the system of discrete equations, obtain the microscopic stress tensor field $\{S_{ij}^{\alpha}\}$ and then average it. However the "inversion" procedure is non-trivial for disordered packings (as compared with the ordered ones [14]). We will develop a different formalism, which is based on the concept of probability functionals for the force and stress states.

Given the set of equations for the equilibrium field of intergranular forces we can write the probability functional for this field as

$$
P\{\underline{f}^{\alpha\beta}\} = \tilde{N} \prod_{\alpha,\beta=1}^{N} \delta(\sum_{\beta} \underline{f}^{\alpha\beta} - \underline{g}^{\alpha})\delta(\underline{f}^{\alpha\beta} + \underline{f}^{\beta\alpha})\delta(\sum_{\beta} \underline{f}^{\alpha\beta} \times \underline{r}^{\alpha\beta}),
\tag{16}
$$

where the normalisation is $\tilde{N}$. The probability of finding the set of tensorial force moments $\{S_{ij}^{\alpha}\}$ is

$$
P\{S_{ij}^{\alpha}\} = \tilde{M} \int \prod_{\alpha,\beta=1}^{N} P\{\underline{f}^{\alpha\beta}\}\delta(S_{ij}^{\alpha} - \sum_{\beta} f_i^{\alpha\beta} r_j^{\alpha\beta})D\underline{f}^{\alpha\beta},
\tag{17}
$$

where $\int D\underline{f}^{\alpha\beta}$ implies integration over all point functions $\{\underline{f}^{\alpha\beta}\}$, since all the constraints on the vector field of intergranular forces have been experienced.

The normalisation, $\tilde{M}$, which is a function of a configuration, is given by

$$
\tilde{M}^{-1} = \int \prod_{\alpha,\beta=1}^{N} P\{\underline{f}^{\alpha\beta}\}\delta(S_{ij}^{\alpha} - \sum_{\beta} f_i^{\alpha\beta} r_j^{\alpha\beta})D\underline{f}^{\alpha\beta} DS_{ij}^{\alpha}.
\tag{18}
$$

We assume that the $z = d + 1$ condition means that the integral exists. The algebra that follows, will transform the stress probability functional into the following form

$$
P\{S_{ij}^{\alpha}\} = P\{S_{ij}^{\alpha} | force\}P\{S_{ij}^{\alpha} | geometry\}
\tag{19}
$$

that will incorporate a complete set of equations for N microscopic stress tensors $\{S_{ij}^{\alpha}\}$

$$
P\{S_{ij}^{\alpha}\} = \prod_{\alpha=1}^{N} \delta(\sum_{\beta} K_{ijk}^{\alpha\beta} S_{jk}^{\beta} + g_i^{\alpha})\delta(P_{ijkl}^{\alpha} S_{kl}^{\alpha}),
\tag{20}
$$

where $K_{ijk}^{\alpha\beta}$ and P_{ijkl}^{α} are tensors, which encode the packing geometry.

The delta functions in Eqn. (20) contain the complete system of equations for $\{S_{ij}^{\alpha}\}$. Given this system of equations for microscopic stress tensors we can derive a complete set of equations for a macroscopic stress tensor. Let us briefly discuss the algebra.

We exponentiate the delta functions and thus introduce the set of conjugate fields $\{\varsigma_{ij}^{\alpha}\}$, $\{\gamma_i^{\alpha}\}$, $\{\lambda_i^{\alpha}\}$ and $\{\eta_i^{\alpha\beta}\}$

$$P\{S_{ij}^\alpha\} = \tilde{M}\int \prod_{\alpha,\beta=1}^{N} \exp(iA)D\underline{f}^{\alpha\beta} D\underline{\underline{\varsigma}}^\alpha D\underline{\lambda}^\alpha D\underline{\eta}^{\alpha\beta} ,\tag{21}$$

where A is

$$A = \sum_{\alpha=1}^{N} \varsigma_{ij}^\alpha\left(S_{ij}^\alpha - \sum_\beta f_i^{\alpha\beta} r_j^{\alpha\beta}\right) + \gamma_i^\alpha\left(\sum_\beta f_i^{\alpha\beta} - g_i^\alpha\right) + \eta_i^{\alpha\beta}\left(f_i^{\alpha\beta} + f_i^{\beta\alpha}\right) + \lambda_i^\alpha\left(\sum_\beta \varepsilon_{ikl} f_k^{\alpha\beta} r_l^{\alpha\beta}\right). \tag{22}$$

The $\underline{\lambda}^\alpha$ field term gives the symmetry of S_{ij}^α. After integrating out the fields $\underline{f}^{\alpha\beta}$ and $\underline{\eta}^{\alpha\beta}$ we find the following linear equation for the conjugate fields

$$\varsigma_{ij}^\alpha r_j^{\alpha\beta} - \gamma_i^\alpha = \varsigma_{ij}^\beta r_j^{\beta\alpha} - \gamma_i^\beta . \tag{23}$$

The idea of the conjugate field method is to use these equations for the ς_{ij}^α field in the stress probability functional and derive the complete system of equations for the microscopic stress tensor. The general solution of the above equation is a sum of the $\varsigma_{ij}^{\alpha 0}$ field which is the particular solution and depends on γ_i^α, and $\varsigma_{ij}^{\alpha*}$ which is the complimentary function

$$\varsigma_{ij}^\alpha = \varsigma_{ij}^{\alpha 0} + \varsigma_{ij}^{\alpha*}. \tag{24}$$

Let us clarify this statement by considering the following equation

$$\underline{a}\cdot\underline{x} = b, \tag{25}$$

where vector $\underline{x}$ is the unknown variable. The general solution of this equation is

$$\underline{x} = \frac{\underline{a}}{|\underline{a}|^2}b + \underline{a}\times\underline{c} \tag{26}$$

Vector $\underline{a}\times\underline{c}$ is the analogue of our tensorial $\varsigma_{ij}^{\alpha*}$ field. The $\varsigma_{ij}^{\alpha*}$ field satisfies the set of $\frac{zdN}{2}$ equations

$$\varsigma_{ij}^{\alpha*} r_j^{\alpha\beta} - \varsigma_{ij}^{\beta*} r_j^{\beta\alpha} = 0. \tag{27}$$

The fact that there is a split of ς_{ij}^α into $\varsigma_{ij}^{\alpha 0}$ and $\varsigma_{ij}^{\alpha*}$ justifies the form of Eqns. (19-20).

It is easy to see that the system of coupled discrete equations (27) has its origin in the Newton third law (10) and contain information about the transmission of a tensor through the contact network.

If we introduce the fabric tensor F_{ij}^α and its inverse $M_{ij}^\alpha = \left(F^\alpha\right)_{ij}^{-1}$ we can rewrite the Eqn. (27) in the following form

$$\varsigma_{ij}^\alpha = M_{jl}^\alpha \sum_\beta R_l^{\alpha\beta}\left(\gamma_i^\alpha - \gamma_i^\beta\right) + M_{jl}^\alpha \sum_\beta R_l^{\alpha\beta} r_k^{\beta\alpha}\left(\varsigma_{ik}^\beta - \varsigma_{ik}^\alpha\right). \tag{28}$$

We can iterate this equation, and by doing so, one can relate the ς_{ij}^α field to the subset of ς_{ij}^δ fields corresponding to the next nearest neighbours of the reference grain α. The second iteration propagates this equation further on within the network of contacts.

Let us derive the set of Nd equations for $\left\{S_{ij}^\alpha\right\}$ that contains information from Newton's second law and the geometry of the contact network

$$P\left\{S_{ij}^\alpha \big| force\right\} = \prod_{\alpha=1}^N \delta\left(\sum_\beta K_{ijk}^{\alpha\beta} S_{jk}^\beta + g_i^\alpha\right). \tag{29}$$

Let us consider the field $\varsigma_{ij}^{\alpha 0}$ that satisfies the following equation

$$\varsigma_{ij}^{\alpha 0} = M_{jl}^\alpha \sum_\beta R_l^{\alpha\beta}\left(\gamma_i^\alpha - \gamma_i^\beta\right) + M_{jl}^\alpha \sum_\beta R_l^{\alpha\beta} r_k^{\beta\alpha}\left(\varsigma_{ik}^{\beta 0} - \varsigma_{ik}^{\alpha 0}\right). \tag{30}$$

We start our approximation procedure by ignoring the second term. In general one can develop a more advanced theory which takes into account correlations due to the presence of the next nearest neighbours but at this point we aim to consider the simplest case. We now integrate out the $\left\{\gamma_i^\alpha\right\}$ field, which gives us the stress-force equation. We obtain

$$\sum_\beta S_{ij}^\alpha M_{jl}^\alpha R_l^{\alpha\beta} - S_{ij}^\beta M_{jl}^\beta R_l^{\beta\alpha} = g_i^\alpha. \tag{31}$$

Let us derive d equations for the macroscopic stress tensor $\sigma_{ij}(\underline{r})$. We expand β quantities about α quantities, multiply the set of discrete equations by $\delta\left(\underline{r} - \underline{R}^\alpha\right)$ and sum them over the whole system. Thus we obtain d equations for the macroscopic stress tensor $\sigma_{ij}(\underline{r})$

$$\frac{\partial \sigma_{ij}}{\partial r_j} + \frac{\partial^3 L_{klmj}\sigma_{ij}}{\partial r_k \partial r_l \partial r_m} + \dots = g_i, \tag{32}$$

where $K_{klmj} = \left\langle R_k^{\alpha\beta} R_l^{\alpha\beta} R_m^{\alpha\beta} R_j^{\alpha\beta}\right\rangle$ and gives the correction to the standard equation of stress equilibrium at the length-scale which is small compared to the size of the system. These corrections correspond to the presence of the second, third etc. nearest neighbours and topological correlations and must vanish at the macroscopic length-scale. These corrections become significant when $\underline{k} \to \dfrac{n\pi}{a}$, $n \succ 3$ is some integer number, and a is a typical particle diameter. Let us recall that we have introduced three basic length-scales, namely: macroscopic $\underline{k} \to 0$, mesoscopic $\underline{k} \to \dfrac{n\pi}{a}$ and microscopic $\underline{k} \to \dfrac{\pi}{a}$. Our formalism allows one to study different types of stress tensor, which correspond to these length-scales. Averaging procedure can be revisited for different length-scales. For example it would be interesting to study the patterns of the mesoscopic stress tensor $S_{ij}^{meso}\left(\underline{R}^{\delta}, ...\underline{R}^{\alpha}, ...\underline{R}^{\gamma}\right)$, which is defined, say, for a cluster of particles carrying high forces. Mathematics for studying mesoscopic stress states is, perhaps, complex. However, one can always investigate propagation of stress within the framework of our model by means of computer simulations. Throughout this paper we assumed that the granular aggregate under consideration is homogeneous and that the average density of the packing is constant. We have also used the simplest averaging procedure. One can think of more sophisticated schemes of averaging (i.e. involving some weight functions which explicitly characterise the microstructure). In order to obtain macroscopic equations one can also use the Fourier transform. However if the packing is disordered one encounters certain difficulties with the inverse transform. Due to the lack of space we will not discuss this here (for details see [14]).

So far the well-known equations have been derived by using the conjugate fields to transform the information from Newton's second law into Nd constraints on $\left\{S_{ij}^{\alpha}\right\}$. However we still have unused information from Newton's third law. By integrating out the $\varsigma_{ij}^{\alpha*}$ fields we will obtain the missing $\dfrac{Nd(d-1)}{2}$ equations that close the system

$$P\left\{S_{ij}^{\alpha}\big|geometry\right\} = \prod_{\alpha=1}^{N} \delta(P_{ijkl}^{\alpha} S_{kl}^{\alpha}) \ . \tag{33}$$

We start with

$$P\left\{S_{ij}^{\alpha}\big|geometry\right\} = \int \prod_{\alpha,\beta=1}^{N} \exp\left(i\sum_{\alpha=1}^{N} \varsigma_{ij}^{\alpha*} S_{ij}^{\alpha} \right) \delta\left(\varsigma_{ij}^{\alpha*} r_{j}^{\alpha\beta} - \varsigma_{ij}^{\beta*} r_{j}^{\beta\alpha}\right) D\varsigma_{ij}^{\alpha*} \ , \tag{34}$$

since

$$\varsigma_{ij}^{\alpha*} r_{j}^{\alpha\beta} - \varsigma_{ij}^{\beta*} r_{j}^{\beta\alpha} = 0 \tag{35}$$

follows from Eqns. (23-27). If we count degrees of freedom in the equations satisfied by $\left\{\varsigma_{ij}^{\alpha*}\right\}$, it could be shown (one has to use the relation $\sum_{\beta} \underline{r}^{\alpha\beta} = 0$) that there are only d equations for every label α. Hence dN constraints give $\dfrac{Nd(d-1)}{2}$ equations for $\left\{S_{ij}^{\alpha}\right\}$ in (34). Using $\underline{R}^{\alpha\beta}$ and $\underline{Q}^{\alpha\beta} = \underline{r}^{\alpha\beta} + \underline{r}^{\beta\alpha}$, we reduce Eqn. (35) to the correct number of equations by projecting, into the two scalar equations

$$\varsigma_{ij}^{\alpha*} \sum_{\beta} R_i^{\alpha\beta} R_j^{\alpha\beta} + \sum_{\beta} \left(\varsigma_{ij}^{\alpha*} - \varsigma_{ij}^{\beta*} \right) r_j^{\beta\alpha} R_i^{\alpha\beta} = 0 , \tag{36a}$$

$$\varsigma_{ij}^{\alpha*} \sum_{\beta} R_j^{\alpha\beta} Q_i^{\alpha\beta} + \sum_{\beta} \left(\varsigma_{ij}^{\alpha*} - \varsigma_{ij}^{\beta*} \right) r_j^{\beta\alpha} Q_i^{\alpha\beta} = 0 , \tag{36b}$$

and assuming as before that $\varsigma_{ij}^{\alpha*} - \varsigma_{ij}^{\beta*}$ gives rise to gradient terms we can exponentiate (36a,36b) by parametric variables ϕ^{α} and ψ^{α}

$$P\left\{ S_{ij}^{\alpha} \middle| geometry \right\} = \int \prod_{\alpha=1}^{N} \exp \left\{ i \sum_{\alpha=1}^{N} \varsigma_{ij}^{\alpha*} \left(S_{ij}^{\alpha} - \phi^{\alpha} F_{ij}^{\alpha} - \psi^{\alpha} G_{iij}^{\alpha} \right) \right\} D\varsigma_{ij}^{\alpha*} D\phi^{\alpha} D\psi^{\alpha} , \tag{37}$$

where F_{ij}^{α} is given by (5) and

$$G_{ij}^{\alpha} = \frac{1}{2} \left(\sum_{\beta} R_i^{\alpha\beta} Q_j^{\alpha\beta} + R_j^{\alpha\beta} Q_i^{\alpha\beta} \right). \tag{38}$$

We would like to note that the system under consideration is disordered and therefore $\underline{Q}^{\alpha\beta} \neq 0$ (whereas for a honeycomb periodic array of monodisperse spheres $\underline{Q}^{\alpha\beta} = 0$). After integrating out the ς_{ij}^{α}, ϕ^{α} and ψ^{α} fields, we find the following constraint on the $\left\{ S_{ij}^{\alpha} \right\}$

$$\begin{vmatrix} S_{11}^{\alpha} & S_{12}^{\alpha} & S_{22}^{\alpha} \\ F_{11}^{\alpha} & F_{12}^{\alpha} & F_{22}^{\alpha} \\ G_{11}^{\alpha} & G_{12}^{\alpha} & G_{22}^{\alpha} \end{vmatrix} = 0 . \tag{39}$$

So far we have derived a complete system of equations for a set of microscopic stress tensors $\left\{ S_{ij}^{\alpha} \right\}$ in two dimensions. In 3-D we need three stress-geometry equations. These can be easily derived. However an additional dimension requires a third vector $\underline{P}^{\alpha\beta} = \underline{Q}^{\alpha\beta} \times \underline{R}^{\alpha\beta}$ in addition to $\underline{Q}^{\alpha\beta}$ and $\underline{R}^{\alpha\beta}$. It is easy to see from Eqn. (35) that three fabric tensors are also required in 3-D. However due to the lack of space the detailed analysis of 3-D problem will be reported elsewhere [14]. Let us concentrate on the detailed analysis of a 2-D case. It follows, that F_{ij}^{α} and G_{ij}^{α} can be written as

$$F_{ij}^{\alpha} = \begin{pmatrix} 1 & 0 \\ 0 & 1 \end{pmatrix} , \tag{40a}$$

$$G_{ij}^{\alpha} = \begin{pmatrix} \sin\theta^{\alpha} & \cos\theta^{\alpha} \\ \cos\theta^{\alpha} & -\sin\theta^{\alpha} \end{pmatrix}. \tag{40b}$$

Thus we have the system of $\dfrac{Nd(d+1)}{2}$ discrete linear equations for $\{S_{ij}^{\alpha}\}$

$$\sum_{\beta} S_{ij}^{\alpha} M_{jk}^{\alpha} R_k^{\alpha\beta} - S_{ij}^{\beta} M_{jk}^{\beta} R_k^{\beta\alpha} = g_i^{\alpha} , \tag{41}$$

$$S_{11}^{\alpha} - S_{22}^{\alpha} = 2S_{12}^{\alpha} \tan\theta^{\alpha} . \tag{42}$$

Let us consider the simplest analysis of these equations. In order to obtain continuum equations for macroscopic stress tensor one has to average discrete equations. However, in this problem, averaging is subtle. Let us consider the simplest averaging procedure. One can try out the simplest averaging procedure and replace every quantity by its average over the whole system. If the system is strongly anisotropic (i.e. there exists a preferred direction characterised by some angle φ) and $\tan\theta^{\alpha}$ has an average value $\tan\varphi$, then this simple averaging works and (42) becomes

$$\sigma_{11} - \sigma_{22} = 2\sigma_{12} \tan\varphi , \tag{43}$$

where φ can be considered as an angle of repose. This equation is known as the Fixed Principal Axes model [9], and has been used with notable effect to solve the problem of the stress distribution in sand piles. If the packing is isotropic the naive method of averaging has the following problem. If one averages Eqn. (42) the macroscopic stress-geometry equation is

$$\sigma_{11} - \sigma_{22} = 0 . \tag{44}$$

However, one can rewrite (42) as

$$\left(S_{11}^{\alpha} - S_{22}^{\alpha} \right)\cot\theta^{\alpha} = 2S_{12}^{\alpha} , \tag{45}$$

and average it in the same way. In this case one has

$$\sigma_{12} = 0 . \tag{46}$$

Are equations (44) and (46) equivalent to each other? In other words, do they give the same macroscopic stress state? If the packing under consideration is isotropic then the macroscopic stress state is characterised by some diagonal stress tensor $\sigma_{ij} = \sigma\delta_{ij}$, where σ is determined by the external force at the boundaries. This macroscopic stress tensor is rotationally invariant. One can write down any stress tensor σ_{ij} in terms of its eigenvalues σ_1, σ_2 and the angle of rotation θ

$$\sigma_{ij} = \begin{pmatrix} \sigma_1 \cos^2\theta + \sigma_2 \sin^2\theta & (\sigma_1 - \sigma_2)\sin\theta\cos\theta \\ (\sigma_1 - \sigma_2)\sin\theta\cos\theta & \sigma_2 \cos^2\theta + \sigma_1 \sin^2\theta \end{pmatrix}. \tag{47}$$

One can see that in the case of isotropic packings (i.e. for some arbitrary angle θ) Eqns. (44) and (46) give the same macroscopic stress state with $\sigma_1 = \sigma_2$. Let us give a probabilistic interpretation of the Eqn. (42b). Thus, if we are given S_{12}^α, the probability of finding $S_{11}^\alpha - S_{22}^\alpha$ is

$$P\left\{ S_{11}^\alpha - S_{22}^\alpha \,\middle|\, S_{12}^\alpha \right\} = \frac{2}{\pi} \frac{\left| S_{12}^\alpha \right|}{\left(S_{11}^\alpha - S_{22}^\alpha \right)^2 + \left(S_{12}^\alpha \right)^2} \,. \tag{48}$$

The mean values of $\left(S_{11}^\alpha - S_{22}^\alpha \right)$ and of S_{12}^α are zero, hence we predict, rather obviously, that $\sigma_{ij} = \sigma\delta_{ij}$. However, notice that we are able to predict the fluctuations away from this "granulostatic" stress state, and would do more on correlations if one could find a pathway to measure them.

FRICTION AND TENSILE FORCES

We have shown that the presence or absence of friction in our model affects the values of the coordination number z. In particular we have seen that z is higher for packings of frictionless particles (where only normal components of intergranular force are nonzero) compared with packings where intergranular forces have nonzero tangential component. These are very simple arguments for limiting cases that give an insight into a very difficult problem of the distribution of contact forces. In general friction coefficient is finite. Its value influences the response of a granular material to external perturbations (i.e. compression or shear) that cause internal rearrangements of particles within the packing. In this paper we do not consider such cases and concentrate on static arrays with fixed geometry. It is clear that in our model no contact within the static packing can support tensile forces. The presence of tensile forces would cause internal rearrangement. The constraint that prohibits the presence of tensile forces within the packing can be written in terms of intergranular forces and contact normals

$$\underrightarrow{f}^{\alpha\beta} \cdot \underrightarrow{n}^{\alpha\beta} > 0 \tag{49}$$

This inequality can be transformed (using the conjugate fields method) into

$$S_{ij}^\alpha M_{jk}^\alpha R_k^{\alpha\beta} n_i^{\alpha\beta} > 0 \tag{50}$$

One can see that for a given set of microscopic stresses $\left\{ S_{ij}^\alpha \right\}$ there exist constraints on permitted geometric configurations. This inequality can be coarse-grained and inequalities for the macroscopic stress tensor $Det\sigma_{ij} > 0$ and $Tr\sigma_{ij} > 0$ can be obtained. Both inequalities have a meaning that eigenvalues of the macroscopic stress tensor have the same sign and that macroscopic tensile stresses cannot be supported by a static granular array composed of cohesionless particles. One can also think of packings where fabric and distribution of intergranular forces allow the existence of limiting equilibrium zones. Then inequality (50) becomes an equation

$$S_{ij}^{\alpha} M_{jk}^{\alpha} R_k^{\alpha\beta} n_i^{\alpha\beta} = 0 . \tag{51}$$

This equation constrains the ratio of eigenvalues of S_{ij}^{α}. We can use our simple procedure to average this equation

$$\sum_{\alpha=1}^{N} \sum_{\beta} S_{ik}^{\alpha} \delta(\underline{r} - \underline{R}^{\alpha}) M_{kl}^{\alpha} R_l^{\alpha\beta} n_i^{\alpha\beta} = 0 , \tag{52}$$

which can be written as $\sigma_{ik} L_{ki} = 0$, where $L_{ki} = \left\langle M_{kl}^{\alpha} R_l^{\alpha\beta} n_i^{\alpha\beta} \right\rangle$. It is tempting to establish a link between the inequality (52) and traditional phenomenological equations like the Mohr-Coulomb criterion [1-3]. The Mohr-Coulomb criterion can be written in terms of the principal stresses σ_1 and σ_2

$$\sigma_1 - \sigma_2 = 2\sqrt{\sigma_1 \sigma_2} \tan\varphi . \tag{53}$$

 In general the angle of repose is not equal to the angle of internal friction [1]. In principle the angle of repose φ depends on the fabric of the packing. For anisotropic or/and inhomogeneous packings φ might be a complex quantity. Equation (52) can be reduced to a condition on the ratio of the principal stresses $\dfrac{\sigma_1}{\sigma_2} = f(\theta)$, where $f(\theta)$ is some function of the packing fabric

$$f(\theta) = \frac{\left\langle \underline{\underline{R}}(\theta) \left\langle M_{kl}^{\alpha} R_l^{\alpha\beta} n_i^{\alpha\beta} \right\rangle \underline{\underline{R}}^{-1}(\theta) \right\rangle_{22}}{\left\langle \underline{\underline{R}}(\theta) \left\langle M_{kl}^{\alpha} R_l^{\alpha\beta} n_i^{\alpha\beta} \right\rangle \underline{\underline{R}}^{-1}(\theta) \right\rangle_{11}} , \tag{54}$$

where $\underline{\underline{R}}(\theta)$ and $\underline{\underline{R}}^{-1}(\theta)$ are rotation matrices. From Eqn. (52) and Eqn.(54) we obtain $(f(\theta)-1)^2 = f(\theta)\tan^2\varphi$.

Thus we have obtained (in a rather crude approximation) a relation between the phenomenological parameter of the Mohr-Coulomb theory and the fabric of the contact network.

DISCUSSION

The objective of this paper is to investigate the fundamental mechanisms that govern transmission of stress in static granular arrays. Due to its importance in civil and chemical engineering, the problem of prediction and measurement of stress distribution in granular media has a long history [1-3]. There has been a recent upsurge in interest in the problem of stress transmission as a topic in physics [4-6]. Stress patterns in static granular materials exhibit interesting and unusual features. For instance, the vertical pressure distribution in a conical sand pile often has a minimum underneath the apex [5-7]. Photoelastic and sound transmission experiments show that stresses in static and quasi-static granular media concentrate along well-defined paths [8,11]. A large amount of force is transmitted via these paths or "chains" as can be seen in computer simulations [12,13]. Local constitutive relations between stress tensor components (known as oriented stress linearity models) have been postulated to account for the pressure "dip" in conical sand piles [9]. A vector model, which relates the minimum in the stress distribution to the presence of force lines and formation of arches has been proposed [10]. Experiments and computer simulations give clear evidence that the distribution of intergranular forces is highly heterogeneous [6,8,11-13,15]. Both experimental measurements and numerical modelling indicate that the stress fluctuations in granular media can be as large as the average stress itself, even on scales much larger than the typical grain size [5,6,11]. The intergranular contact forces determine the bulk mechanical properties (e.g. the load bearing capability of the aggregate). Recent experiments [8,15] and computer simulations [5,6,12,13] show that the distribution of forces, where it is found that forces above the mean decay exponentially, is a robust property of static granular media. It is clear that we need a fundamental and predictive theory of stress propagation. In order to do this we have proposed a model description of granular materials. The purpose of this model description is to single out and make available for intensive investigation those structural characteristics of granular packings that are of the most importance in the problem of stress transmission. We have stated that it is possible to idealise granular media as arrays of rigid cohesionless particles that interact with their nearest neighbours via point-like contacts. The geometry of such packings is then determined by sets of vectors that specify the position of contact points of all particles. We have started with the observation that the system of Newton's equations of intergranular force and torque balance is not, in general, complete. In other words, for an arbitrary value of the coordination number the number of unknown intergranular forces does not equal the number of available equations of balance. We resolve this difficulty by choosing the contact network with a certain fixed coordination number. This allows us to develop the statistical-mechanical formalism that delivers the complete system of fundamental equations governing the equilibrium stress state.

We have derived the complete set of equations for the stress tensor. The stress-force equation is the Euler-Cauchy equation. The granular nature of the packing fabric gives rise to gradient corrections that become significant at mesoscopic length-scales

$$\nabla_j \sigma_{ij} + \nabla_j \nabla_k \nabla_m K_{ijkl} \sigma_{lm} + \ldots = g_i. \tag{55}$$

The stress-geometry equation is the analogue of constitutive relation in theory of elastoplastic granular media. Within the framework of our model it relates the stress tensor components with mathematical characteristics of the packing fabric. It has its origin in Newton's third law and contains the information about the contact network

$$P_{ijk} \sigma_{jk} + \nabla_j T_{ijkl} \sigma_{kl} + \nabla_j \nabla_l U_{ijkl} \sigma_{km} + \ldots = 0. \tag{56}$$

In order to solve these equations one has to know the geometric quantities P_{ijk}, T_{ijkl}, U_{ijkl} etc. These can be obtained from experimental measurements of correlation functions. We see that geometry of disorder imposes non-trivial physics of stress transmission dominated by topological correlations at different length-scales and governed by equations of balance. This makes the averaging procedure to be non-trivial for granular media. In general the averaging formalism depends on the packing fabric [14].

In our model contacts between particles cannot support tensile forces. We have incorporated this constraint into our formalism. This provides us with a deep insight into the fundamental physics of Mohr-Coulomb theory of limiting stress states and the microstructural origin of its phenomenological parameters such as the angle of repose. Though our derivation is fairly crude it clearly shows that one can develop advanced models within the framework of our formalism. Let us repeat that all simplifications we have made so far can be easily controlled. Our aim is to establish a theory that explains the macroscopic stress behaviour of large static packings of rigid particles in terms of their statistical micromechanics. Our model achieves this goal by providing the simplest, however sufficiently realistic approach of tackling this problem.

CONCLUSIONS

Let us briefly list the possible directions of future research, which appear to be both important and interesting. As a further development of our theory of stress transmission it would be interesting to calculate the stress correlation and response functions and study the issue of non-trivial stress correlations induced by the deposition history of the material.
The striking feature of particulate systems is the presence of strong fluctuations in stress. Photoelastic visualisation experiments show that stresses in static granular media concentrate along certain well-defined paths. A large amount of force is transmitted via these stress paths within the material. Computer simulations also give clear evidence that the distribution of intergranular forces is highly heterogeneous. Both experimental measurements and simulations indicate that the stress fluctuations in granular media can be as large as the average stress itself, even on scales much larger than the typical grain size. The issues of the onset of rigidity, fragility of stress bearing paths and their relationship to physics of the jamming transition are still a subject of controversy in the current literature and therefore should be investigated by means of theory and simulation. It is necessary to obtain a proper theoretical understanding of factors and underlying physical mechanisms that determine the size and spatial structure of these stress paths and govern the statistical distribution of the stress fluctuations. The length of stress chains may give some insight into the effects of spatial correlations. On of the statistical quantities that characterise the stress chain pattern in the sample is the distribution of chain lengths. An improved statistical measure might be a two-variable distribution of chain lengths for particles lying along straight lines and having force within a given range. Theoretical understanding of such heterogeneous distribution of stress will give deep insight into the physics of jamming in dense flows of pastes, powders and colloids. The challenge is to comprehend relationships between stress and geometry at different length scales. This has led already to a well-developed theory of ``homogenisation" when the scales are well separated. When the scales are separated but still comparable, there is a need for a micromechanical rationale for including scale effects in macroscopic models. These issues present fascinating scientific challenges and are of real importance. We believe that the new fundamental knowledge will in time impact upon understanding materials of industrial importance such as colloids, foams and soils. The future development of the statistical-mechanical theory of stress transmission in granular media must be based on experiments that give definite and reliable results. The main goal of such experiments should be to give insights, check theoretical results and determine the limits of their applicability.

ACKNOWLEDGEMENTS

We wish to acknowledge the financial support of ROPA grant from EPSRC (UK), the Leverhulme Foundation, Shell Research and Technology Centre (Amsterdam) and of Gonville and Caius College (Cambridge).

REFERENCES

1. J. Feda, *Mechanics of Particulate Materials: The Principles* (Elsevier, 1982).
2. D. M. Wood, *Soil Behaviour and Critical State Soil Mechanics* (Cambridge University Press, 1990).
3. R. M. Nedderman, *Statics and Kinematics of Granular Materials* (Cambridge University Press, 1992).
4. S. F. Edwards, R. B. S. Oakeshott, *Physica* **D38**, 88 (1989).
5. D. Bideau, A. Hansen (eds.), *Disorder and Granular Media* (Elsevier, 1993).
   H. J. Herrmann, J. -P. Hovi and S. Luding (eds.), *Physics of Dry Granular Media*, (Kluwer Academic, 1998).
6. Focus issue: Granular materials, *Chaos* **9**, 3 (1999).
7. R. Brockbank, J. M. Huntley and R. C. Ball, *J. de Physique II (France)* **7**, 1521 (1997).
8. C. H. Liu *et al.*, *Science* **269**, 513 (1995); C. Liu, S. R. Nagel, *Phys. Rev.* **B48**, 15646 (1993); X. Jia, C. Caroli, B. Velicky, *Phys. Rev. Lett.* **82**, 1863 (1999).
9. J. P. Wittmer, P. Claudin, M. E. Cates, *J. de Physique I (France)*, **7**, 39 (1997).
10. S. F. Edwards and C. C. Mounfield, *Physica* **A226**, 1 (1996).
11. B. Miller, C. O'Hern and R. P. Behringer, *Phys. Rev. Lett.* **77**, 3110 (1996).
12. C. Thornton, *KONA: Powder and Particle*, **15**, 81, (1997).
13. F. Radjai *et al.*, *Phys. Rev. Lett.* **80**, 61 (1998).
14. R. C. Ball and D. V. Grinev, http://xxx.lanl.gov, *cond-mat/9810124*; S. F. Edwards and D.V. Grinev, *Phys. Rev. Lett.* **82**, 5397 (1999); D.V. Grinev and S. F. Edwards, preprint, to be submitted to *Phys. Rev. E*.
15. D. M. Mueth, H. M. Jaeger and S. R. Nagel, *Phys. Rev. E* **57**, 3164 (1998).
16. M. E. Cates, J. P. Wittmer, P. Claudin, and J-P. Bouchaud, *Phys. Rev. Lett.* **81**, 1841 (1998).

Mat. Res. Soc. Symp. Proc. Vol 627 © 2000 Materials Research Society

Mechanical response of a static granular piling

Guillaume Reydellet, Loic Vanel, Eric Clément
Laboratoire des Milieux Désordonnés et Hétérogènes – UMR 7603
Université Pierre et Marie Curie Boîte 86
4, Place Jussieu, F-75252 Paris

ABSTRACT

We present the results of different experiments which aim is to investigate the response of a static granular assemblies to various local perturbations. We design a sensitive experimental method to probe the response of a localized stress at the top of a granular piling. The piling is an horizontal 3D granular layer confined in a box and the spatial distribution of loads at the bottom is monitored. This analysis defines some crucial experimental test to inform a currently debated issue on the mechanical status of static granular assemblies.

INTRODUCTION

Understanding the exact mechanical status of static or quasi-static granular assemblies is still an open and debated issue (see for example [1] and refs inside). Models typically used in soil mechanics assume an elasto-plastic behavior for a granular material subjected to applied loads [2,3]. Relations between components of the stress tensor and deformations are determined empirically from triaxial test experiments. These relations may be quite complex if they are to reproduce the behavior of "real" granular assemblies (piece-wise, non-linear, anisotropic...) and many empirical constant may enter in the description. But overall, below the plastic threshold, the constitutive relations between stresses and deformations give to the system of partial differential equations (PDE) describing static equilibrium, the structure of elasticity (elliptic equations). Beyond the plastic threshold, the system is described using an adaptation of the Coulomb plasticity theory and we have a situation where a propagative character for the PDE (hyperbolic equations) will appear in regions experiencing yield[3].

An alternative theoretical approach was propose recently and is based on a microscopic viewpoint that incorporates a propagative character for the contact forces between the grains [4]. When extended to the continuum limit, this model predicts simple relations between components of the stress tensor, and a set of hyperbolic PDEs is obtained to solve for the distribution of stresses. A particular example of such a relation is called the Oriented Stress Linearity model (OSL) [5]. These models are dramatically different in character (below the plasticity threshold). The difference in the two approaches is manifested in the nature of the boundary conditions. In a propagative (hyperbolic) model, stresses must be specified on just a part of the boundaries, (for example at the free surface) and propagate in the bulk. The equation solutions and the boundary reflection properties determine stresses on the other boundaries. For an elliptic equation, stresses and/or strains must be specified at all boundaries [4,6,7] and influence the whole solution in the bulk.

Recently several reproducible experiments were performed in a sand-pile (see[8] and references therein) and on a granular column [9]. It was shown that an OSL modeling is indeed able to reproduce the observed phenomenology. In essence, these models contain a propagation angle for the stress components that can be modified by the local value of the shearing stress.

From this viewpoint, it is this effect that can lead to a pressure dip in the middle of a sand-pile that is built by successive avalanches or that can yield a flat stress distribution in the center of a sand-pile that is created by an homogeneous rain [8]. Also, at the bottom of a granular column, a pressure overshoot for an overload at the top, as observed experimentally[10] is explained by arguments based on reflections at the wall. Nevertheless, the agreement between recent experiments and the hyperbolic models does not rule out elliptic models. Hence, there is a considerable need for tests at the most fundamental level in order to probe the mechanical behavior of a granular assembly. Here, we present an experiment method suited to probe the response of static granular assemblies to a local stress perturbation. The layer under test consists of a horizontal 3D granular assembly confined in a box. The spatial distribution of stress responses at the bottom is monitored.

THE EXPERIMENTAL SET-UP

Already at the most basic level, measuring meaningful stresses in granular assemblies is a non-trivial question. Generally, experimentalists are confronted to three fundamental problems i) the response of the probe depends on the history of the preparation as evidenced on sand pile (which might be a physically relevant issue, ii) the probe surface has a limited number of contacts with the granular medium which is at the origin of an inherent fluctuations scale (which importance should decrease when the probe size increases), iii) the physical characteristics of the probe itself may have an influence on the measurements (large deformations of the probe may change drastically the local force distributions). To illustrate this difficulty we show the response of two $1cm^2$ size stress probes made of a thin metal membrane of thickness e=0.1mm and welded on a cylinder. The stiffness of both membranes was measured to differ by a factor 2.

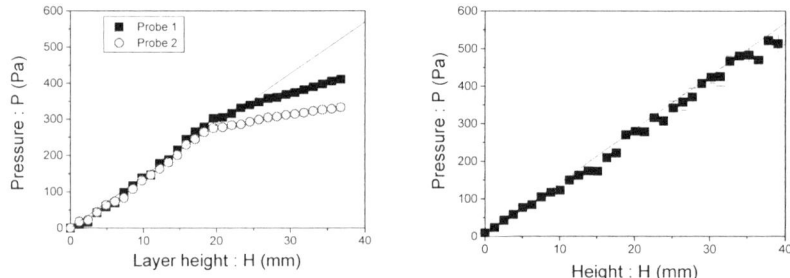

Figure1 *Pressure response of a $1cm^2$ stress probe to a granular layer of height H (see text for procedures). The grains are d=1mm glass beads. The solid line is the hydrostatic curve. Fig1a response to progressive layering; probe 1 is stiffer than probe 2. Fig2b average response after vibration for each layer deposit.*

The deformations of the membrane (estimated to be smaller than 1μ) are monitored using a sensitive capacitive technique. We present the results of two different preparation techniques. On fig 1a, we show the response of two probes with different stiffness for an increasing height H. The grains (d = 1mm glass spheres) are simply poured creating small height increases of about 1mm each time. The straight line corresponds to the relation P=ρgH. But due some localized vaults effects (and possibly to the preparation method as well), we evidence a systematic change of slope for heights H*≅15-20d, maybe corresponding to a vault effect around the probe. Note that this relation was tested on several granular materials with a size ranging between d=0.3mm and d=3mm. Deviations from a simple hydrostatic linear relation are more pronounced for the thinner membranes (we tested thickness between 20 and 200 μm). On Fig.1b, we present results obtained in a similar way but after each height increase, the pile is slightly vibrated. This procedure produces some fluctuations from one step to the other but in the average, the hydrostatic relation is well recovered (with a precision of 5%, which is the typical uncertainty of our packing fraction determination). As a consequence, for all the experiments we describe in the following, we use the second procedure.

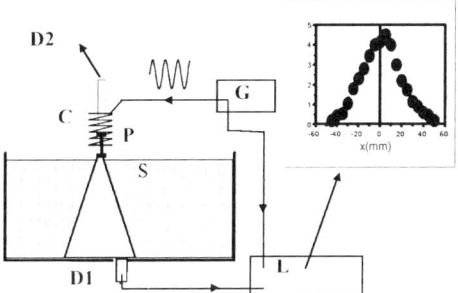

Figure 2 Schematic of the experimental set-up. G is the sine generator driving the coil C and the stress source S through the piston P. The piston fixed position is monitored with the probe D2. the stress response is monitored using the probe D1 which signal is sent to the phase lock-in amplifier L which source is G. The out-put of L is monitored as a function of the relative horizontal position x between D1 and P.

Furthermore, probing the response of a granular static piling to a localized perturbation is a priori a difficult issue since the perturbing stress must not create plastic reorganization of the grains. Therefore, weak perturbations must be used but on the other hand, the spatial response is hard to monitor since noise due to finite size of the probes is likely to hide the desired signal. To overcome this difficulty, we use a modulated force created at the bottom of the pile by a piston of mass M=5g on a surface of A=1cm². The stress modulation is achieved using a periodic magnetic field that is created by an electric current in a coil driven by a low frequency generator (see G on fig.2). We impose on the free surface a local stress source (see S on fig2) of value Σ_0=Mg/A. In the piston, a permanent magnet is inserted and thus the modulation of the magnetic field creates a stress modulation : $\Sigma = \Sigma_0 (1 + \alpha I \sin \omega t)$, where I is the current intensity α is a constant and f=ω/2π is the frequency of modulation, here f=40Hz.. The signal of the stress probe D1 is then directed to a lock-in amplifier L synchronized by the generator exciting the source. Moreover, a sensitive displacement probe (see D2 on fig.2) monitors the piston position to check that no plastic yield occurs during the data collection. The relative horizontal position between the piston P and the probe D1 is varied. The signal hence obtained, is the convolution of the mechanical response function (the Green's function) of the piling by the width of the source and the width of the probe.

RESULTS

We tested that, in the limit where no "sinking" of the piston inside the pile is observed, the response is linear in the value of the imposed stress. We found also (see fig 3) that the value of the slope relating the applied force to the observed stress may depend strongly on configurations. This is evidenced in fig.3 where we display for three different pilings obtained in the same conditions, the response amplitude after detection by the lock-in amplifier as a function of the force modulation amplitude F.

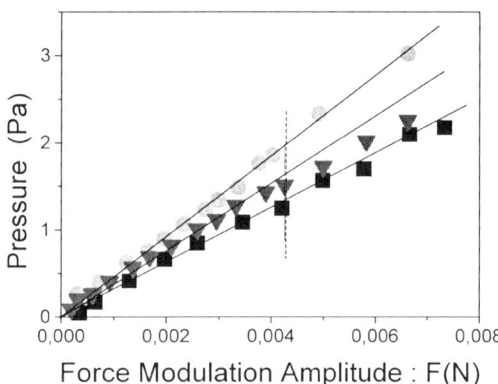

Figure 3 *Response of the stress probe as a function of the modulation force F, for a probe situated at the direct vertical position H = 27mm below the piston. The response is given for three different independent experiments and for the same preparation. The vertical dashed line defines a limit where some yield was observed.*

A typical pressure distribution is displayed in fig4. It was obtained for d=1mm glass beads at a depth of H=27mm. Therefore we are in a position to test the various theoretical propositions by measuring the value of the Green's function for different layer heights and for different grains sizes.

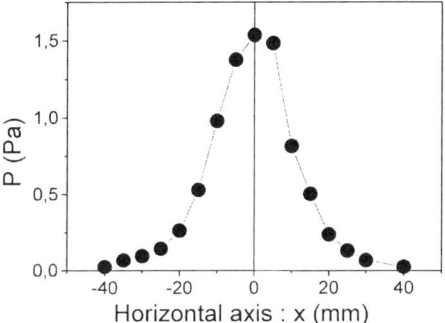

Figure 4 *Horizontal distribution of pressures at the bottom of a granular layer at vertical depth H=27mm for d = 1mm beads.*

The shape of the mechanical response (the Green's function) does not show the two side bumps as predicted by the hyperbolic models. We are now able to offer a complete study of this Green's function with depth, for various granular materials and for different piling procedures. These systematic results will be reported in a future contribution [11].

CONCLUSIONS

In conclusion, we describe an experimental set-up which is suited to obtain the spatial distribution of stresses in response to a localized pressure at the surface of a granular layer. First, we address the fundamental question of stress measurement in a granular assembly. We show that the results may depend strongly on the preparation method and eventually, on some mechanical characteristics of the probe such as stiffness. We define a procedure to prepare the granular assembly in reproducible conditions. Then, we describe in details an experimental technique designed to achieve very sensitive stress measurements using a force modulation as a source and a lock-in detection. Importantly, we show that the response is *linear* with the amplitude of the applied force which validates the experimental approach. We also exhibit, for the first time, the spatial distribution of pressures at the bottom a pile. The distribution shape shows a single peak and thus, has the qualitative form of an elastic response. In the future, measurements will be systematically obtained at various depths and for different packing procedures. Then, the experimental results should be compared precisely to the outcome of the different theories available and provide a fundamental test to understand more deeply the mechanical status of static granular assemblies[11].

ACKNOWLEDGMENT
We acknowledge many discussions and interactions with Prof. R.P.Behringer and the financial support of the PICS-CNRS n° 563.

REFERENCES

[1]   *Physics of Dry Granular Media*, eds H.J.Herrmann, J.P. Hovi, S.Luding. NATO ASI series, Kluver, Amsterdam (1998) ; E. Clément, Current Opinion in Rheology and Interface Science **4**, 294 (1999).

[2]   R.A.Schoefield and P.Wroth, Critical State in Soil Mechanics, McGraw Hill (1968)

[3]   R.M Nedderman, *Statics and Kinematics of Granular Materials*, Cambridge University Press (1992)..

[4]   P.Claudin, M.E Cates, J.Wittmer, J.P Bouchaud, Phys.Rev.E, **57**, 4441 (1998).

[5]   M.E Cates, J.Wittmer, J.P Bouchaud et P.Claudin, Phil.Trans.Roy.Soc.London.A, **356**, 2535 (1998).

[6]   F.Cantelaube and J.D Goddard in *Physics of Dry Granular Media*, eds H.J.Herrmann, J.P. Hovi, S.Luding. NATO ASI series, Kluver, Amsterdam (1998).

[7]   S.B.Savage in *Physics of Dry Granular Media*, eds H.J.Herrmann, eds J.P. Hovi, S.Luding. NATO ASI series, Kluver, Amsterdam (1998).

[8]   L.Vanel, D.Howell, D.Clark, R.P.Behringer and E.Clément Phys.Rev.E **60**, R5040 (1999).

[9]   L.Vanel and E.Clément, Eur.Phys.J B, **11**, 525 (1999).

[10]   L.Vanel, P.Claudin, J.P Bouchaud, M.E. Cates, E. Clément et J.Wittmer, Phys.Rev.Lett **84**, 1439 (2000).

[11]   G.Reydellet and E.Clément , *submitted* (2000).

Mat. Res. Soc. Symp. Proc. Vol 627 © 2000 Materials Research Society

Persistence of Granular Structure during Die Compaction of Ceramic Powders

William J. Walker, Jr.
New York State Center for Advanced Ceramic Technology at Alfred University,
Alfred, NY 14802, U.S.A.

ABSTRACT

Glass spheres were used as a model system to investigate granule failure during die compaction. Stresses within an assembly of spheres follow a network of pathways. When the spheres are of uniform composition, the magnitude of the stresses within a pair of contacting granules is a function of the locally transmitted stress and the diameter of the two spheres. Results obtained using glass spheres demonstrated the statistical nature of granule failure during compaction, with some granules failing at very low applied pressures while others (~40% by volume) persist at even the highest applied loads. Within a distribution of granule sizes, those granules with smaller diameter were seen to have a higher probability of failure at low pressure than were larger granules. These results are consistent with those observed during die compaction of granulated alumina powder.

INTRODUCTION

Die compaction of granulated powder is a common forming process used in the ceramics industry. Granulation of fine powders is necessary to produce the free flowing feed material required for die filling when high-speed presses are used. However, it is desirable to eliminate all artifacts of the granules during the compaction process in order to produce defect-free sintered components. In this work, statistical analysis of the fragmentation of glass beads during die compaction was used as a model granular system to achieving a better understanding of persistent granular structures in compacts.

During compaction of granulated powder, the density increase that results from applied pressure occurs in three stages [1]. Stage I consists of granule rearrangement at low pressures resulting in a small increase in density of the granular assembly. Above an apparent yield pressure P_y, which marks the onset of Stage II, the interstitial pores between granules (intergranular pores) are reduced in size as the granules break down or deform, causing a linear increase in density with log (compaction pressure). In Stage III, the intergranular pores are mostly eliminated, and particle rearrangement within the granules causes increased densification at high pressures. The types of granule-related defects that may persist after compaction include persistent intergranular pores and poorly joined interfaces between granules.

Stress transmission during compaction has been modeled by a number of researchers using a continuum approach [1]. This method provides a good description of the pressure gradients that occur within compacts and the resulting density gradients that are observed. A discrete particle approach provides better insight into density gradients on a smaller scale. Granule deformation is also better understood using a discrete particle approach. Computer simulation [2] and a photoelastic disk method [3] are methods that have been used to model stress transmission through a packed bed of discrete granules. Internal stress distributions in photoelastic materials are easily observed as color changes when viewed under polarized light. Thus, stress distributions through a two-dimensional cell packed with photoelastic disks can be directly observed. Results from both computer simulations and photoelastic disk studies show that stress is transmitted through a granular material along a branched network of pathways which concentrate stress in some disks and bypass others. Flow occurs from sliding of block-like regions along slip planes.

THEORY

Below some critical level of stress, granules are elastic spheres. Above that level, granules will yield, either by fracture or deformation. In the elastic range, stresses resulting from the contact of two spheres follows Hertzian behavior, and are a function of the transmitted load and the diameter of each sphere [4]. Therefore, a model of elastic spheres at small strains can be used to describe compaction of an assembly of granules at pressures near P_y, before significant deformation of granules has occurred. For point loading of two elastic spheres with diameters d_1 and d_2, it can be shown that the maximum stress is given by

$$\sigma_{max} = C \left(\frac{P}{K_d^{\,2}} \right)^{1/3}$$ (1)

where

$$K_d = \frac{d_1 d_2}{d_1 + d_2} \, .$$ (2)

The constant C is equal to 0.616 for compressive stresses transmitted between loading points, and C is approximately equal to 0.0819 and 0.205 for tensile and shear stresses respectively, which occur near the point of contact.

For an assembly of contacting spheres of uniform size, maximum stresses inside each sphere are proportional to the cube root of the applied stress, assuming that force is uniformly transmitted through each sphere. Since the force through an assembly of spheres is not uniformly transmitted, but instead follows a network of pathways, some spheres will experience much higher levels of stress than others. For spheres of non-uniform size, the maximum stress will depend on the sizes of both spheres in each contacting pair and the locally transmitted load.

Consider a short chain of spheres of varying size that is transmitting a load through a small portion of a large assembly. Within this chain, the smaller spheres have a higher state of stress and will fail at lower loads than larger spheres. As individual granules fail, either from fracture or deformation, the network of stress pathways rearranges. In the case of brittle granules, fragmentation increases the packing efficiency of the assembly and stress continues to be transmitted via point contacts between elastic particles, now either granules or granule fragments. The model can be continually applied to an ever-changing size distribution of elastic particles. In the case of plastic granules, the density of the assembly increases as granules deform. Deformation will allow contact between larger numbers of adjacent spheres, increasing the complexity of the network of stress pathways to include more and more granules, while at the same time, maintaining existing pathways. Additionally, the stress will be transmitted through larger contact areas, which will lessen the local stresses that drive further deformation. Smaller granules, because of their smaller cross sections, will tend to experience higher levels of stress and deform more rapidly than larger granules. In both the cases of brittle and plastic granules, small spheres will be more likely to fail at lower loads and large spheres will be more likely to persist intact after compaction. This model assumes spheres of uniform strength.

EXPERIMENT

Submicron alumina powder (A16-SG, Alcoa, Bauxite, AR)) was granulated by spray-drying an aqueous suspension of alumina which contained (based on the weight of the dry powder) 0.75% ammonium polyacrylate deflocculant (Darvan 821A, R.T. Vanderbilt, Norwalk, CT) and 3.0% polyethylene glycol binder (Carbowax Compound 20M, Union Carbide, Danbury, CT). The mean granule size of the spray-dried powder was 90 μm. The spray-dried powder was compacted in a steel die to 51 MPa and sintered in air to 1650°C.

Glass beads (3M Company, St. Paul, MN) with mean diameter of 85.5 μm were compacted in a steel die at several pressures. The die had a cylindrical cavity with diameter 12.7 mm, and was fabricated in two halves that could be separated to facilitate removal of the specimens without an ejection step. After compaction, the die was opened and the glass beads were dispersed in water using ultrasound. Size distributions of the compacted beads were measured using light scattering (Microtrac 9200, Leeds and Northrop, North Wales, PA) and plotted as a function of compaction pressure.

RESULTS

Figure 1 shows fracture surfaces of alumina compacts. In Figure 1(a), the fracture path of an unfired compact follows the surfaces of some large granules while evidence of small granules is obliterated. In Figure 1(b), granule-related structure is evident after sintering. The poorly joined interfaces in the unfired compact persist as macro-scale defects in the sintered microstructure.

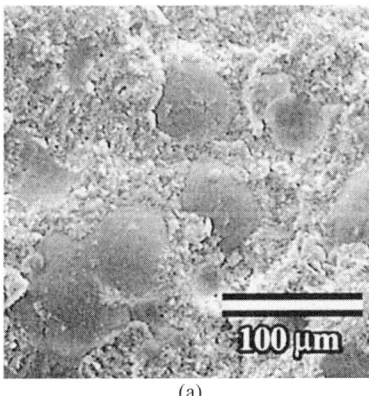
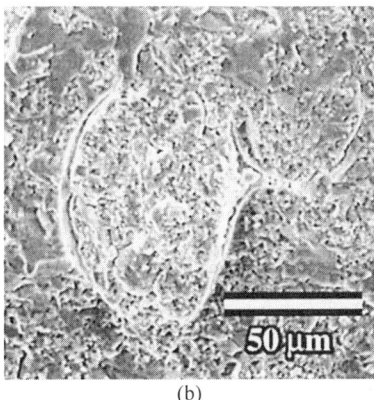

(a)   (b)

Figure 1. Fracture surfaces of (a) unfired and (b) fired alumina compact showing artifacts of spray-dried granular material.

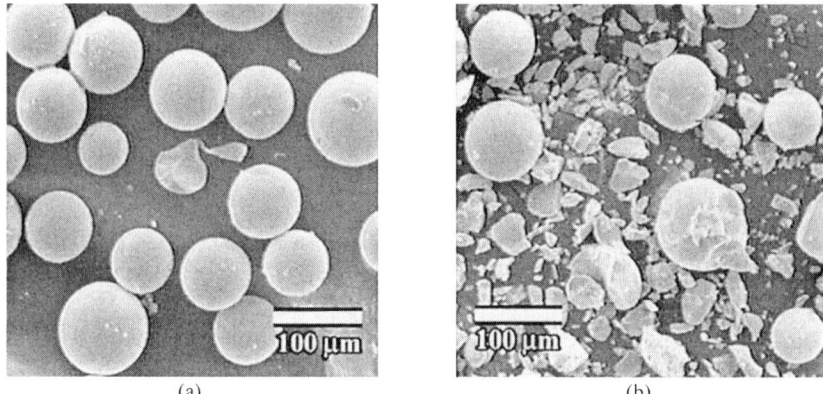

(a)   (b)

***Figure 2**. Glass beads compacted to (a) 53 MPa and (b) 420 MPa.*

When the glass beads were compacted, some beads were observed to fracture at pressures as low as 53 MPa, as is shown in Figure 2(a). As pressure was increased, larger numbers of fragments were observed, but some large spheres remained intact at the highest compaction pressures tested, as is shown in Figure 2(b). With the large number of spheres ($\sim 10^9/cm^3$) that are present in a single compact, simple observation does not give an accurate indication of trends involving so many particles. Particle size analysis provides a statistical method.

The particle size distribution of the uncompacted glass beads was log normal. The size distribution remained constant at low pressures. Changes in size distribution near P_y have been previously reported [5]. At a compaction pressure of 60 MPa, which is about $P_y/3$, the size distribution began to change as glass beads fractured in response to the applied pressure. The median size and the fine end of size distribution both began to decrease at about 60 MPa ($\sim P_y/3$). The coarse end of size distribution remained constant until a point near P_y (~ 150 MPa). As pressure was increased above P_y, the entire distribution shifted toward smaller sizes as more spheres became fractured. The increase in fines in the size distribution at compaction pressures below the point where the coarse end of the distribution begins to change indicates that the largest spheres are more likely to remain intact at low compaction pressures, verifying the behavior predicted by the model.

Figure 3 shows the change in particle size distribution of the glass beads at different compaction pressures. As pressure was increased, the number of particles in the 70 to 110 μm size range decreased and a tail of fine particles in the 10 to 60 μm range increased as more beads became fractured. The size distribution was bimodal with the coarse fraction consisting of largely unfractured beads and the fine fraction consisting of fractured fragments. This is illustrated in Figure 4, and is consistent with microscopic observation (Figure 2). In Figure 5, the amount of material in each mode is plotted with respect to compaction pressure. The amount of material in each fraction changes rapidly between pressures of about 75 and 400 MPa. At higher pressures, the rate of change decreases. Above about 500 MPa the amount of material in each fraction becomes constant with respect to compaction pressure, with the size distribution

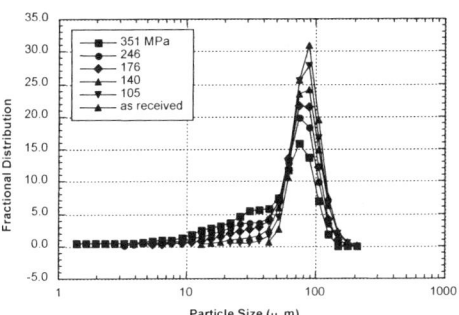

Figure 3. *Change in size distributions of glass beads with respect to compaction pressure.*

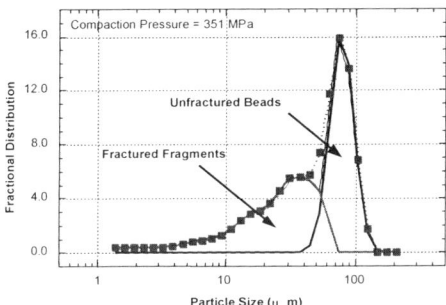

Figure 4. *Bimodal size distribution of glass beads consists of unfractured beads and fractured fragments.*

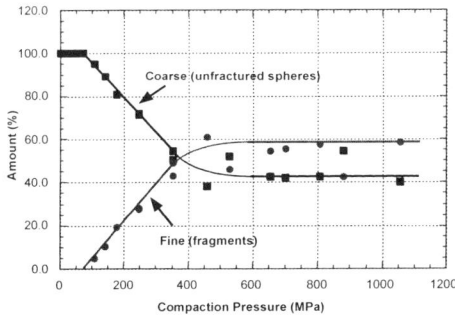

Figure 5. *Change in coarse and fine fractions of size distribution during compaction of glass beads.*

consisting of about 40% coarse and 60% fine. Thus, the increase in density during compaction is a result of fragmentation of 60% of the material while about 40% of the volume of the compact consists of granules which are relatively unchanged from the precursor material. These results are consistent with the behavior observed using granulated alumina, however, in the case of alumina, it was not possible to quantitatively describe the degree to which the precursor granular structure persisted after compaction.

CONCLUSIONS

During die compaction of granulated ceramic powders, the load is transmitted along a branched network of contacting granules, concentrating the stress in some granules while others are bypassed. As the applied load is increased, granules fail by deformation or fracture, the network of stress pathways rearranges to some extent, and the packing density of the granular assembly increases. However, evidence of the granular precursor material can be seen to persist in the microstructure of unfired and fired ceramics prepared by die compaction. Glass spheres were used as a model granular system to investigate the degree to which the granular structure persists during die compaction.

The state of stress within any individual granule depends on contact stresses from the locally transmitted load. The magnitude of the load, the diameter of the sphere, and the diameter of the neighboring spheres which transmit the load are all factors determining the state of stress within a sphere. Within a single chain of contacting spheres, those with smaller diameter experience higher levels of stress and will thus fail at lower loads. Within an assembly of a large number of spheres, a general trend observed is that smaller granules fail at lower compaction loads.

The size distribution of the glass spheres changed with applied compaction load as the spheres failed due to fracture. As the compaction load approached the apparent yield pressure of the granular assembly, a tail of fines was seen in the particle size distribution before any change is observed in the coarse fraction, indicating that fines are formed due to fragmentation of smaller spheres in the initial sample. As pressure was increased, the size distribution became bimodal, consisting of unfractured spheres in the coarse fraction, and fractured fragments in the fine fraction. The amount of material in each fraction changed rapidly during the early stages of compaction, but became constant at higher loads, with about 40% of the volume consisting of unfractured spheres.

REFERENCES

1. J.S. Reed, *Principles of Ceramic Processing*, 2nd ed. (John Wiley and Sons, Inc., New York, 1995), pp. 425-33.
2. P.A. Cundall, A. Drescher and O.D.L. Strack, in *IUTAM Symposium on Deformation and Failure of Granular Materials*, edited by P.A. Vermeer and H.J. Luger (A.A. Balkema, Rotterdam, 1982), pp. 355-70.
3. A. Drescher and G. de Josselin de Jong, J. Mech. Phys. Sol., **20**, 337-51 (1972).
4. R.J. Roark and W.C. Young, *Formulas for Stress and Strain*, 5th ed. (McGraw-Hill Book Company, New York, 1975), pp. 513-29.
5. W.J. Walker, Jr. and J.S. Reed, Am. Ceram. Soc. Bull., **78** (6), 53-57, (1999).

Mat.Res.Soc.Symp.Proc.Vol.627 2000 Materials Research Society

Theory of Stress Distribution in Granular Materials: the Memory Formalism

V.M. Kenkre

Center for Advanced Studies and the Department of Physics and Astronomy
University of New Mexico, Albuquerque, NM 87131, USA

Abstract

A theoretical approach to the description of stress distribution in granular compacts is presented on the basis of a memory function formalism. Experiments which have motivated the approach are mentioned. The formalism is shown to provide an explanation of observed features of stress distribution in compacts, and to lead to existing theories in extreme limits, thereby providing a unification of the theories. The memory functions are shown to be intimately related to characteristic spatial correlations in the granular system and are discussed on the basis of stochastic considerations.

1 Introduction: Experimental Motivation and the Memory Approach

Major areas of current research in granular materials include avalanches, patterns in flow, segregation, sound propagation, and spatial distribution of stress[1, 2, 3, 4, 5, 6, 7]. This article deals with the last of these. The study of stress distribution in static piles of granular material is characterized by undisputable importance from the applications point of view, enormous difficulty in the clear construction of theories as well as in experimental measurement of relevant observables, and, currently, by an unfortunate absence of communication between various groups working in the field. The importance of the field stems, e.g., from the requirement of understanding and control of stress distribution in pre-sintering compacts in almost any manufacturing situation. The difficulty in theory arises from the complexity of the system, involving as it does, friction, as well differing shapes and sizes of the granular particles: for instance, almost nothing is known definitively about the so-called constitutive relations among the stresses. The difficulty in experiment lies in the design of direct probes of stress in the *bulk* of the granular material: it is relatively easy to measure stress at the surfaces of a compact but values of stress in the interior must be often deduced from density distributions or other indirect observations[1].

The focus of the present article is a method we have developed recently[8, 9, 10] for the theoretical description of stress distribution on the basis of what is called a memory formalism. The original motivation for the investigation was provided by reported observations of curious features such as spatial oscillations in stress down the center line in compacts. These features are apparent in recent experimental results [11] as well as in data that have been available in the literature for many years [12, 13, 14, 15]. Experimental information about the distribution of stress in a powder compact has been difficult to obtain unambiguously. Observations have employed, in some cases, direct measurement of the stress with the use of sensors or strain gauges [14, 16] within, or at the edge of, a compact to measure the forces that evolve during pressing. Other cases have involved indirect deduction of the stress distribution from the density distribution within the compact. The first approach suffers from a lack of accuracy and the second from the need for specific assumptions of a local stress-density relation at every point in the compact[17, 18, 19, 20]. Nevertheless, it is quite clear that characteristic unexplained features such as the non monotonic variation of the stress with depth along the centerline of the compact emerge regularly (but not universally), and that a theoretical description of these features is not trivial. Indeed, Aydin et al. [11] have referred to the

---

[1]The third current characteristic, the absence of communication between different sets of workers, is difficult to account for, but could be arising from the fact that the background of the investigators is rather varied, ranging from engineering through physics to applied mathematics.

failure of existing theories to account for the oscillatory behavior. The reader is referred to ref. [9] for a detailed description of the experimental background.

The method of approach we have developed [8, 9, 10] is based on two ingredients: (i) the $t - z$ transformation which singles out one spatial direction in the granular material and treats it for the purpose of description as if it were time, and (ii) a spatially non-local formalism which employs integrodifferential equations of the Volterra type incorporating memory functions which characterize spatial correlations in the granular material. The $t - z$ transformation has appeared in investigations earlier than ours, notably in the work of Bouchaud, Cates and collaborators[7]. The transformation simplifies the mathematical treatment considerably, provides physical intuition based on knowledge of initial value problems in other fields, but, as a result of accompanying approximation procedures, forces certain limitations on the applicability of the entire formalism. The memory formalism[8, 9, 10] has a great deal of analytic power, particularly for the unification of disparate approaches and for the description of intermediate cases, and was suggested by work in the rather distant area of exciton transport in molecular aggregates [21]. It can be viewed as arising from a mathematical generalization of constitutive relations such as those employed in ref.([7]) but, as will be remarked below, is best looked upon as arising from the stochastic properties (spatial correlations) of the granular system.

The first step in our description is, thus, the choice of the z direction as the direction of gravity and/or of the applied stress, along with an attempt to write a closed evolution equation for the scalar field σ_{zz} which represents the zz-component of the stress tensor. The second step, and indeed the characteristic ingredient of our approach, is the use of an 'evolution equation' which is non-local in z :

$$\frac{\partial \sigma(x,y,z)}{\partial z} = D \int_0^z dz' \, \phi(z-z') \left[\frac{\partial^2 \sigma(x,y,z')}{\partial x^2} + \frac{\partial^2 \sigma(x,y,z')}{\partial y^2} \right] \tag{1}$$

The bridge function ϕ which connects the derivatives of the stress at various depths z is the *memory function*, and is a measure of important spatial correlations of the granular material which arise from the granularity (variations in shape and size of the grains) and other properties such as friction [22]. Those properties also determine the value of D. For the sake of simplicity we will refrain from discussing in this paper starting points more general than (1) in which the depth coordinate and the other spatial coordinates are intermingled in the memory description.

The three succeeding sections of this article deal, respectively, with how an equation such as (1) helps in the unification of diverse existing approaches to stress distribution, how they lead to the understanding of observed features such as spatial oscillations of the stress in compacts, and how the memory functions are related to properties of the compact.

2 How Memories Help I: Unification of Existing Approaches

Three particular cases of (1) deserve mention. In the first, we take the memory function to be independent of z: $\phi(z) = c^2/D$. Equation (1) reduces, then, to the wave equation

$$\frac{\partial^2 \sigma_{zz}(x,y,z)}{\partial z^2} = c^2 \nabla^2 \sigma_{zz}(x,y,z) = c^2 \left[\frac{\partial^2 \sigma_{zz}(x,y,z)}{\partial x^2} + \frac{\partial^2 \sigma_{zz}(x,y,z)}{\partial y^2} \right], \tag{2}$$

and thus to the starting point of the analysis of Bouchaud et al. [7]. The parameter c denotes what may be termed the wave speed, which is directly related to the slope of the so-called light cones. This perfect memory situation represents the fact that the stress applied on one particle is transmitted along the lines of contact between particles and there is no loss of information about the original strength and direction of the applied force.

In the second case, we take the memory function to be decaying so rapidly with depth that it may be replaced by a δ-function: $\phi(z) = \delta(z)$. Equation (1) yields the diffusion equation

$$\frac{\partial \sigma_{zz}(x,y,z)}{\partial z} = D \nabla^2 \sigma_{zz}(x,y,z) = D \left[\frac{\partial^2 \sigma_{zz}(x,y,z)}{\partial x^2} + \frac{\partial^2 \sigma_{zz}(x,y,z)}{\partial y^2} \right]. \tag{3}$$

It is possible to show in detail, as we have done elsewhere [8] that the above equation is identical to a simplified version [2] of the starting point of the analysis of Liu et al. [23].

---

[2] The simplification consists in assuming a lack of dependence of D on z as well as on x, y.

The diffusive limit of the evolution has been used in the past for developing mean field treatments [23] and addressing the magnitude distribution of the stresses rather than their spatial variation. The wave limit has been discussed primarily via ray tracing arguments [7] in what may be termed the geometrical limit of the wave equation. Our own approach has been quite different. We have obtained actual solutions of these equations for the propagators (Greens functions) through explicit initial value and boundary value treatments and, with their help, attempted to address the *spatial* distribution of stress in granular systems. Our especial emphasis has been to present an intermediate starting point which combines the physics inherent in the extreme limits of wave-like and diffusive behavior and is capable of describing the *entire* range in between. Therefore, we focus attention on memory functions which are neither constant nor have infinitely fast decay. A simple intermediate situation is the exponential $\phi(z) = \alpha \exp(-\alpha z)$. In the respective limits of small and large α (the latter limit being actually $\alpha \to \infty$, $c \to \infty$, $c^2/\alpha = const.$), the wave and the diffusive case emerge trivially. The intermediate case gives the telegraphers equation

$$\frac{\partial^2 \sigma_{zz}(x, y, z)}{\partial z^2} + \alpha \frac{\partial \sigma_{zz}(x, y, z)}{\partial z} = c^2 \nabla^2 \sigma_{zz}(x, y, z) \qquad (4)$$

with $D = c^2/\alpha$.

Whereas the wave limit of Bouchaud et al. [7] corresponds to identical, frictionless spherical particles arrayed in a perfectly ordered lattice, the intermediate situation above describes a more realistic granular system in which random shaped particles of random sizes are packed in a random arrangement. For the sake of simplicity, we will consider here only a two-dimensional system and thus use, instead of (4), the equation

$$\frac{\partial^2 \sigma_{zz}(x, z)}{\partial z^2} + \alpha \frac{\partial \sigma_{zz}(x, z)}{\partial z} = c^2 \frac{\partial^2 \sigma_{zz}(x, z)}{\partial x^2}. \qquad (5)$$

The easy unification of the extreme limits provided by our memory approach may be appreciated either directly as explained above or through explicit solutions such as those of (5). Take the applied stress $\sigma_{zz}(x, 0)$ at the 'surface' $z = 0$ to be a delta function $\delta(x)$. The solution of (5) is then given by

$$\sigma_{zz}(x, z) = e^{-\alpha z/2} \left[\frac{\delta(x + cz) + \delta(x - cz)}{2} + T \right], \qquad (6)$$

where the term T vanishes identically for $cz \le x$, and equals, for $cz \ge x$,

$$T = \left(\frac{\alpha}{4c} \right) \left\{ I_0 \left(\frac{\alpha}{2c} \sqrt{c^2 z^2 - x^2} \right) + \frac{cz}{\sqrt{c^2 z^2 - x^2}} I_1 \left(\frac{\alpha}{2c} \sqrt{c^2 z^2 - x^2} \right). \right\} \qquad (7)$$

the I's being modified Bessel functions. In the limit $\alpha = 0$,

$$\sigma_{zz}(x, z) = (1/2) [\delta(x + cz) + \delta(x - cz)] \qquad (8)$$

as in ref. [7] and we immediately recover the pheonomenon of 'light cones'. Our solution shows that, in addition, there is a nonvanishing stress distribution *within* the light cones. This stress is given by our term T. In the limit which reduces our theory to the opposite extreme of Liu et al. [23], the light cones spread out to coincide with the surface $z = 0$, and the entire region experiences stress:

$$\sigma_{zz}(x, z) = \frac{e^{-x^2/4Dz}}{(4\pi Dz)^{1/2}}. \qquad (9)$$

It is also possible to analyze with the help of these solutions the well-known 'burial problem', i.e., the question of where one should bury oneself under a sandpile to minimize the stress. One way of addressing the problem is to consider, via the $t - z$ transformation, stresses arising from gravity forces to be applied to circular regions of radii increasing continuously from zero to a maximum at different times (depths), and to sum all the contributions, taking boundary contributions to be relatively negligible. The key quantity to analyze is, thus,

$$Q(x, z) = \int_0^z dz_1 \int_{-vz_1}^{vz_1} dx_1 \psi(x, x_1, z, z_1) \qquad (10)$$

where ψ is the propagator (Greens function) and v describes the slope of the sandpile. The propagator is easily obtained from the memory functions, an explicit example being (6). The vanishing of $\frac{\partial Q(x,z)}{\partial x}$

at a given value of the depth z locates the extremum and the sign of the second derivative identifies the extremum as minimum or maximum. Unlike simple ray tracing arguments, which cannot address the fact that stress extrema appear under the apex in some sandpiles but not in others, the present memory analysis has the potential to show how factors such as the extent of coherence (the value of c/α in our telegraphers equation above) influence the extremum.

3 How Memories help II: Eigenvalue Problem for Compaction in Dies

We now return to the primary problem which motivated the memory approach: spatial oscillations of stress in compacts. The memory approach addresses this issue by developing an eigenvalue analysis of (1) in the compact. Details of the theory may be found in [8] and applications to experiment in [9]. Under the assumption that the extent in the z-direction is large, (5) can be solved through the application of the method of separation of variables:

$$\sigma_{zz}(x,z) = \sum_k \left(A_k \cos kx + B_k \sin kx \right) g_k(z) \tag{11}$$

$$g_k(z) = e^{-\frac{\alpha}{2}z} \left[\cosh \Omega_k z + \frac{\alpha}{2\Omega_k} \sinh \Omega_k z \right], \qquad \Omega_k = \sqrt{\alpha^2/4 - c^2 k^2}. \tag{12}$$

First, let us take the stress to be vanishing on the boundaries of the die which we assume to extend from $x = -L/2$ to $x = L/2$. This is an artificial boundary condition which we consider only for illustrative purposes [8]. If a constant punch pressure p_0 is applied across the top surface of the compact, the center line stress can be evaluated exactly in the Laplace domain as

$$\widetilde{\sigma_{zz}}(0,\varepsilon)/p_0 = \frac{1}{\varepsilon} \left[1 - \mathrm{sech} \left(\frac{L}{2c} \sqrt{\varepsilon^2 + \varepsilon\alpha} \right) \right] \tag{13}$$

where tildes denote the Laplace transform, and ε is the Laplace variable. In the wave limit $\alpha = 0$, the inversion is easy and gives the center line stress as a square wave $W(z)$ along the z coordinate. It is constant at the applied value p_0 for $0 < z < L/2c$, flips to $-p_0$ for $L/2c < z < 3L/2c$, flips back to p_0 for $3L/2c < z < 5L/2c$, and continues alternating in this fashion. In the diffusive limit, the center line stress distribution is given by

$$\sigma_{zz}(0,z)/p_0 = 2 \int_0^{1/2} d\nu \, \theta_1 \left(\nu \left| \frac{4Dz}{L^2} \right. \right), \tag{14}$$

where θ_1 is the elliptic theta-function of the first kind. The general expression for the intermediate region is

$$\sigma_{zz}(0,z)/p_0 = 1 + \int_0^z du \, e^{-(\alpha/2)u} \left[M(u) + (\alpha/2) \int_0^u ds \, I_1(s) M \left(\sqrt{u^2 - s^2} \right) \right] \tag{15}$$

where I_1 is the modified Bessel function and $M(z)$, the derivative $\frac{dW(z)}{dz}$ of the square wave $W(z)$ described above, can be expressed as an infinite sum of δ functions centered at multiples of $L/2c$.

This illustrative analysis shows oscillations in the center line stress but contains unphysical elements which arise from the vanishing boundary conditions at the die walls because of the wave element in the evolution. Realistic considerations involve a decrease of the stress at the pipe walls with increasing depth, and have been treated in [9]. In that treatment, $\sigma_{zz}(\pm L/2, z)$ is not taken to vanish, but rather to be a given function $h(z)$ of the depth:

$$\sigma_{zz}(x,z)\Big|_{x=\pm L/2} = p_0 h(z), \tag{16}$$

where p_0 is the average value of the applied stress at the top surface. The function $h(z)$ is taken directly from experiment. The solution of (5) with such initial and boundary conditions presents an unusual boundary value problem which is analogous to propagation problems in which the boundary condition is dependent on time [24]. We have provided a complete solution in ref. [9] which may be summarized as follows.

In a manner analogous to that used in the treatment of Thompson [18], the applied stress at $z = 0$ is taken to have a parabolic dependence,

$$\sigma(x, 0) = p_0(c_0 + (1 - c_0)\frac{12x^2}{L^2}), \tag{17}$$

with $c_0 = \sigma(0,0)/p_0$ to ensure that the integrated applied pressure is equal to p_0. A typical set of observations taken from Duwez and Zwell [16] is found to be compatible with

$$h(z) = \beta + [(3 - 2c_0) - \beta]e^{-\gamma z}. \tag{18}$$

With the definition

$$a_m = \frac{4(-1)^m}{\pi(2m + 1)}, \qquad m = 0, 1, 2, \ldots \tag{19}$$

the restriction $k = (2m + 1)\frac{\pi}{L}$, and

$$A_k = p_0 a_m \left[c_0 + 3(1 - c_0)\left(\frac{k^2 L^2 - 8}{k^2 L^2} \right) \right], \tag{20}$$

the expression for the stress extended to realistic initial and boundary conditions is found to be

$$\begin{aligned}
\sigma_{zz}(x, z) &= p_0\beta + \sum_k (A_k - p_0 a_m \beta)g_k(z)\cos kx \\
&+ p_0((3 - 2c_0) - \beta)\sum_k \frac{c^2 k^2 a_m}{(\gamma - \alpha/2)^2 - \Omega_k^2} \\
&\times \left\{ e^{-\gamma z} - g_k(z) + \frac{\gamma}{\Omega_k}e^{-\frac{\alpha}{2}z}\sinh\Omega_k z \right\}\cos kx
\end{aligned} \tag{21}$$

Thus, with given distributions of stress along the top surface and the side walls, explicit solutions are found for the stress in the interior and compared successfully to experiment. Oscillations down the center line emerge naturally but not always, the factor governing their appearance being the ratio c/α. Closed contours signifying true wavelike behavior appear in some cases but not in others, also depending on the value of c/α. Practical matters such as the effect of lubrication of the walls and of changing the profile of the applied stress at the top of the compact can be addressed [9].

Careful analysis of the question of whether the diffusive limit alone would suffice to describe the observed stress distribution results in an unequivocal answer in the context of the experiments reported in refs. [11, 14, 15]. The wave ingredient of the telegrapher's equation is found to be essential to explain some of the data (as in uranium dioxide) where oscillations are clearly visible. Furthermore, even for cases which exhibit no such oscillations (as in magnesium carbonate and alumina), a careful analysis based on our predictions lead to the conclusion that the diffusive limit is inadequate for the experiments of refs. [11, 14, 15].

4   Where do the Memories Originate ?   Stochastic Considerations

Having understood how memories help in understanding experiment and in unifying diverse approaches to the calculation of stress distributions, it is necessary to understand how the memories arise. One way is to obtain them phenomenologically through more or less suggestive arguments involving generalizations of previous constitutive relations. Such arguments have been provided in ref. [8] but give only a mathematical justification with little physical content. To understand the physical origin of the memories, consider (for simplicity) a two-dimensional granular compact(z along the vertical and x along the horizontal) consisting of weightless circular discs of a given radius arranged in perfect order. Let a vertical force be applied to the top of one of the discs lying on the top layer of the compact. It is trivial to show, on the basis of Newtonian laws of statics, that the consequent force distribution, equivalently stress distribution, is down two lines in the compact, representative of what has been called [7, 8] light cones. Viewed through the $t - z$ transformation, the representative point in the one-dimensional space of x travels ballistically with constant speed which we will call c. Consider next a more realistic situation. The array is now not perfectly periodic, there being irregularities stemming from changes in shape and size of the discs and/or presence of friction. The speed c will change from location to location, and the path of the representative point will be jagged: the speed c will become a *stochastic* variable. Restricting attention to its z-variation only, we write

$$\frac{dx}{dz} = c(z),\tag{22}$$

with $c(z)$ a given stochastic process. Defining a Liouville density for the process and averaging over all realizations of the stochastic process, it is possible to obtain[25] a variety of evolution equations for the averaged probability density $P(x, z)$ according to the particular stochastic characteristics of the process $c(z)$. In other words, the particular irregularities arising from the shape and size changes in the discs, or from their roughness, are reflected in $c(z)$ and thereby in the evolution of $P(x, z)$. The latter quantity, involving as it does an average over various realizations (the jagged paths) of the stochastic process, can be shown to correspond to the probable value of the stress, equivalently to the probability density of the stochastic process.

One simple example of the stochastic process is one in which $c(z)$ is stationary and Gaussian, with zero mean and a correlation function Δ :

$$\langle c(z)c(z_1)\rangle = \Delta(z - z_1).\tag{23}$$

It leads straightforwardly[25] to

$$\frac{\partial}{\partial t} P(x, z) = D(z)\frac{\partial^2}{\partial x^2} P(x, z),\tag{24}$$

where the depth-dependent diffusion constant $D(z)$ is given as

$$D(z) = \int_0^z dz_1 \Delta(z - z_1).\tag{25}$$

We observe that the depth-dependence of $D(z)$ arises from a direct integration of the correlation function $\Delta(z)$. If the correlation function decays extremely rapidly signifying that the stochastic process corresponds to a perfect random walk, the stress evolution equation is a simple diffusion equation as in (3). Our analysis thus provides an explicit derivation from stochastic considerations of the full (not simplified to a constant D) equation of Liu et al.[23], and clarifies the validity of that equation.

On the other hand, if the stochastic process is a random telegraph, with an exponential correlation, it is also possible[25] to show that, to a good approximation, the stress evolution equation is the telegrapher's equation (4). Generally, the complexities and irregularities of the grain-grain interactions in the compact will influence the details of the stochastic process and thereby the memory function. Computer simulations have been begun to obtain the spatial correlations inherent in $c(z)$ and thence the memory functions. Such simulations, along with attempts to measure the correlations experimentally through scattering experiments, give an *a priori* predictive character to the theory of stress distribution that we have developed. The memory functions are seen, in this manner, to be not merely a phenomenological construct but calculable in principle from microscopic considerations regarding the physical characteristics of the granular system.

There is yet another source of memory functions which has been described in ref. [10] in greater detail than possible here (because of space restrictions). It arises from an effective medium theory of the granularity of the material. The granularity demands that one replace x by a discrete index m and the randomness of shapes and sizes of the particles demands that the rates in the evolution equation be random functions. Even if we start from a diffusive (but discrete) extreme represented, e.g., by

$$\frac{dP_m(z)}{dz} = F_{m+1,m}[P_{m+1}(z) - P_m(z)] + F_{m,m-1}[P_{m-1}(z) - P_m(z)]\tag{26}$$

where P denotes the z-component of the stress, and m is the discrete index representing the horizontal x (or y) coordinate, the randomness of the rates leads to a memory function that arises from *disorder*:

$$\frac{dP_m(z)}{dz} = \int_0^z dz' \mathbf{F}(z - z')[P_{m+1}(z') + P_{m-1}(z') - 2P_m(z')]\tag{27}$$

Here $\mathbf{F}(z)$ is obtained through a mean field argument from the random distribution $\rho(F)$ of the rates. Equation (27) is evidently equivalent to (3) in the continuum limit, $\mathbf{F}(z)$ being proportional to $D\phi(z)$.

The memory functions which arise from such effective medium considerations are characterized by a sum of two parts with differing decay constants, and to stress distributions different from those predicted by a diffusion or telegrapher's equation [10].

5 Conclusions

The formalism of memory functions for the description of stress distribution described in the present paper has achieved unification of diverse approaches such as those applicable in the extreme diffusive and wave limits, treatment of the entire range in between, and explanation of observed features such as oscillations in stress distribution. The memory function may be computed from given stochastic properties of the granular system arising from the varying shapes and sizes of the grains and from the grain-grain interaction. Information about these stochastic properties themselves may be obtained in principle from a combination of scattering experiments and computer simulations. The memory formalism has also been extended [26] to include nonlinearities of the kind relevant to reaction diffusion systems.

Among shortcomings of this approach in its present stage are the assumption that the present does not influence the past (in the sense of the $t - z$ transformation) which means that stresses at smaller depths are considered as not influenced by stresses at larger ones. This assumption is not always valid as the stochastic paths representing the variable $c(z)$ can in some cases turn upwards in a granular system. Indeed, stress distribution cannot be looked upon universally as an initial value problem. This is a difficulty shared by the extreme approaches of Liu et al.[23] and of Bouchaud and Cates [7] as well as by our intermediate formalism. Related to this problem is the evident restriction that the stress analysis presented above for dies be used only in long pipes or media without a bottom. Termination in the z direction as in a compact introduces 'boundary conditions in time' which appear difficult to treat from evolution equations. In the true time evolution situation, we predict behavior at a later time, given spatial boundary conditions for all time and an initial condition. The incorporation of a 'final' condition, i.e., a boundary condition at large values of time seems difficult to implement. Another notable absence from the formalism is the inclusion of features peculiar to the granular system such as isostaticity [27]. This last is a very important matter which, it is hoped, will be incorporated in the analysis at a future time. Indeed, at the present stage, our formalism is too simplistic to address severe complexities peculiar to granular matter such as the dependence of stresses on the history of how the granular system is constructed [3, 28, 29].

6 Acknowledgments

It is a pleasure to thank my collaborators, in particular Alan Hurd, Marek Kuś and Joseph Scott. This work was supported in part by Sandia National Laboratories, a Lockheed-Martin Company, under U.S. Department of Energy contract DE-AC04-94AL85000.

References

[1] H. M. Jaeger, S. R. Nagel, and R. P. Behringer, Rev. Mod. Phys. **68**, 1259 (1996); Physics Today **49**, 32 (1996).

[2] A. Mehta, ed. *Granular Matter: An Interdisciplinary Approach* (Springer Verlag, New York, 1994).

[3] P. C. de Gennes, Rev. Mod. Phys. **71**, S374 (1999).

[4] L. Kadanoff, Rev. Mod. Phys. **71**, 435 (1999).

[5] Jacques Duran, *Sands, Powders and Grains* (Springer 2000).

[6] S. F. Edwards and R. B. S. Oakeshott, Physica D **38**, 88 (1989).

[7] J. P. Bouchaud, M. E. Cates, and P. Claudin, J. Phys. I France **5**, 639 (1995).

[8] V. M. Kenkre, J. E. Scott, E. A. Pease, and A. J. Hurd, Phys. Rev. E **57**, 5841 (1998).

[9] J. E. Scott, V. M. Kenkre, and A. J. Hurd, Phys. Rev. E **57**, 5850 (1998).

[10] V. M. Kenkre, Granular Materials, **XX**, XXXX (2000), to be published (Proceedings of a Consortium of the Americas for Interdisciplinary Science Workshop held in Bariloche, Argentina (2000 March)).

[11] I. Aydin, B. J. Briscoe, and K. Y. Sanliturk, Computational Materials Science **3**, 55 (1994); Powder Technology **89**, 239 (1996).

[12] R. Kamm, M. A. Steinberg, and J. Wulff, Trans. AIME **171**, 439 (1947); **180**, 694 (1949).

[13] G. C. Kuczynski and I. Zaplatynsky, Trans. AIME **206**, 215 (1956).

[14] D. Train, Trans. Instn. Chem. Engrs. **35**, 258 (1957).

[15] H. M. Macleod and K. Marshall, Powder Technology **16**, 107 (1977).

[16] P. Duwez and L. Zwell, Trans. AIME **185**, 137 (1949).

[17] R. P. Seelig, Trans. AIME **171**, 506 (1947).

[18] R. A. Thompson, Ceramic Bulletin **60**, 237 (1981).

[19] A. R. Cooper and L. E. Eaton, J. Am. Ceram. Soc. **45**, 97 (1962).

[20] V. M. Kenkre, M. R. Endicott, S. J. Glass, and A. J. Hurd, J. Am. Ceram. Soc. **79**, 3045 (1996).

[21] See, e.g., V. M. Kenkre, in *Energy Transfer Processes in Condensed Matter, ed. B. Di Bartolo (Plenum Press, New York, 1984), pp. 205 249.*

[22] H. A. Janssen, Z. Ver. Dt. Ing. **39**, 1045 (1895).

[23] C. -h. Liu, S. R. Nagel, D. A. Schecter, S. N. Coppersmith, S. Majumdar, O. Narayan, T. A. Witten, Science **269**, 513 (1995).

[24] S. J. Farlow, *Partial Differential Equations for Scientists and Engineers* (Dover Publications, Inc., New York, 1982), pp. 64 73.

[25] V. M. Kenkre, M. Kuś and A. Hurd, preprint.

[26] K. Manne, A. J. Hurd, and V. M. Kenkre, Phys. Rev. E **61**, 4177 (2000).

[27] C. F. Moukarzel, Phys. Rev. Lett. **81**, 1634 (1998).

[28] J. Goddard in *Physics of Granular Media, Proc. of NATO Institute*, ed. H. Herrmann, J. Hovi and S. Luding (Luwer Academic, Dordrecht, 1998).

[29] S. B. Savage in *Powders and Grains*, ed. R. Behringer and J. Jenkins (Balkema, Rotterdam, 1997).

Mat. Res. Soc. Symp. Proc. Vol. 627 © 2000 Materials Research Society

A New Simple Non Linear Modelling of the Quasi Statics of Granular Media: predictions, comparisons with experiments

Pierre Evesque
Lab. MSSMat, UMR-8579 CNRS, Ecole centrale de Paris, F-92295 Châtenay-Malabry, France

ABSTRACT

First, a non linear incremental modelling is proposed to describe rheological behaviour of granular material under different simple (i.e. triaxial-, oedometric-, undrained-) stress-strain paths. Validity of isotropic-response assumption is demonstrated whatever the stress ratio as far as deformation range remains small ($\epsilon_1 < 5\%$). This contradicts some recent hypothesis made on the evolution of contact distribution during anisotropic loading.

INTRODUCTION

It has been known from long that granular media can be changed of shape at will, just by changing the stress field [1-7]. For instance, figure 1 reports typical vertical stress $\sigma_1 = q + \sigma_2$ variations and volume ($\epsilon_v = \delta v / v$) changes under constant radial stress $\sigma_2 = \sigma_3$ as a function of the vertical strain ($\epsilon_1 = -\delta h / h$), i.e. when decreasing the height h of the sample. Each of the three series of two curves correspond either to the same granular medium packed at the same initial density but compressed under three different values of σ_3. It is known also that the same three different behaviours can be obtained under the same value of σ_3 with the same sand packed at three different densities.

The aim of this paper is to show that one can predict the main rheological behaviours of granular media, using these three curves only. Furthermore, as it was shown by Rowe [4] that the

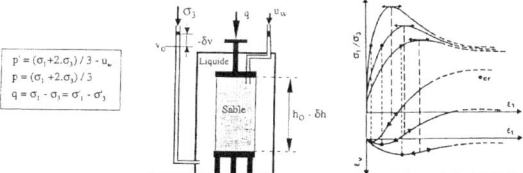

Figure 1: *Typical triaxial cell results at $\sigma_2 = \sigma_3 = c^{ste}$. The state obtained at large deformation is called the "critical state". It is characterised by a single ratio $q/\sigma_3 = M$, (or equivalently by a single $q/p = M'$) and a specific volume v_c which depends on p only; v_c varies with p according to $v_c = v_{co} - \lambda \ln(p/p_o)$, with p_o, v_{co} and λ being three constants for a sand. M & M' are found to independent on the initial volume v_i and on p; M obeys $M = 2\sin\varphi/(1-\sin\varphi)$ and $M' = 3(M-1)/M$, with φ being the fiction angle. The state with no volume change at short deformation is called the characteristic state[7]; its stress obeys $q = M'p$ or $q = M\sigma_3$.*

experimental change of volume $D=-\partial\varepsilon_v/\partial\varepsilon_1$ of figure 1 curve depends only on the friction angle φ of the material and on the stress ratio σ_1/σ_2:

$$\sigma_1/\sigma_2 = (1 - \partial\varepsilon_v/\partial\varepsilon_1)_{\sigma2=cste} \tan^2(\pi/4+\varphi/2) \qquad (1)$$

the experimental determination of q/σ_2 vs. ε_1 is enough to determine the volume change so that there is no real need to determine the volume variation experimentally. This will allow to present a very simple model which catch most of the main features of the rheology of granular matter; it uses isotropic-response assumption. The validity of the model will be discussed using experimental data.

THE MODEL

As a matter of fact, in general, the experimental behaviour of figure 1 are used in soil mechanics to fit a model based on one (or more than one) plastic mechanisms and to predict the behaviour of granular medium using a computer code, since the modelling is rather complicated. This paper is aimed at proposing a much simpler alternative approach, which is based on simple incremental modelling. Such modelling assumes that stress increment $\delta\sigma$ generates a strain increment $\delta\varepsilon$; since the response to stress increment does not depend on the speed of increase of σ, i.e. quasi static regime, the response is incrementally linear per zone [5,6]. However, in order to allow hysteresis the response to $\delta\sigma$ shall be different from the one to $-\delta\sigma$, so that two different zones have to be used where the incremental responses are different. However, only compression case are considered here ($\varepsilon_1>0$), so that the response in a single zone is needed. At last, since figure 1 does not give much detail on the rheology, it is not needed to assume complexe response so that this one is assumed more or less isotropic. In this case figure 1 curves shall be fitted using two parameters, i.e. (i) a pseudo Young modulus $E=1/C_0$ and (ii) a pseudo Poisson coefficient v :

$$\begin{pmatrix} \delta\varepsilon_1 \\ \delta\varepsilon_2 \\ \delta\varepsilon_3 \end{pmatrix} = -C_0 \begin{pmatrix} 1 & -v & -v \\ -v & 1 & -v \\ -v & -v & 1 \end{pmatrix} \begin{pmatrix} \delta\sigma_1 \\ \delta\sigma_2 \\ \delta\sigma_3 \end{pmatrix} \qquad (2)$$

Equations 1 and 2 impose $-D= (\partial\varepsilon_v/\partial\varepsilon_1)_{\sigma2=cste}=1-2v$ and hence:

$$v=(\sigma_1/\sigma_3) \tan^2(\pi/4-\varphi/2)/2 \qquad (3)$$

This modelling will be used to predict behaviours and they will be compared to experimental behaviours

UNDRAINED BEHAVIOUR: ($v=c^{ste}$)

Undrained tests means $\delta\varepsilon_v=0$ and $v=c^{ste}$. So main rheology is given by the evolution of q and p. Typical experimental results of variations of q vs. p are reported in figures 2a & 2b for "loose" and "dense" sands respectively. In the "dense" case, the trajectory starts vertical, i.e. $p=c^{ste}$, then it turns right on the $q=M'p$ line when it reaches it; it ends at the "critical state", i.e; for $p_c/p_o= (v_{co}-v_c)/\lambda$. The trajectory of loose sand starts also vertical, but it turns left much before reaching

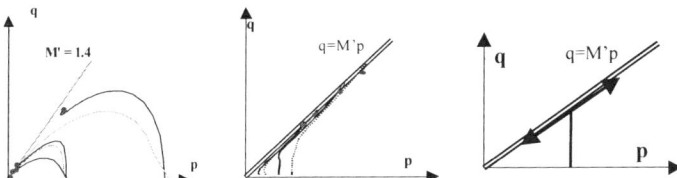

Figure 2: *the trajectory in the (p,q) plane corresponding to undrained (v=c^ste) path: a & b are experimental results for loose (a, left) and dense (b, middle) packings. (c, right) is the predicted behaviour from the modelling.*

the q=M'p line; it ends on this line at the "critical state". On the other hand, the model, i.e. equation 2, with condition $\delta\varepsilon_v=0$ implies:

$$(1-2v)(\delta\sigma_1+\delta\sigma_2+\delta\sigma_3)=0 \qquad (4)$$

This implies either $v=1/2$ or $p=(\sigma_1+\sigma_2+\sigma_3)/3=c^{ste}$, (where p is the mean stress). In turn, $v=1/2$ implies $\sigma_1/\sigma_3 =\tan^2(\pi/4+\varphi/2)$ or, what is the same, q=M'p, or $q=M\sigma_3$. This leads to predict the following behaviour: starting from an isotropic stress state ($\sigma_1=\sigma_2=\sigma_3$), equation 3 imposes $v\neq1/2$ so that p shall be c^{ste} at the beginning of the compression. So the trajectory starts vertically in the (p,q) plane; this trajectory remains always possible (as far as the response remains isotropic); however, when it meets the q=M'p line a second solution occurs since $\delta v=0$ whatever $\delta\sigma_1$, $\delta\sigma_2$, $\delta\sigma_3$, under this stress ratio, since $v=1/2$. So, the q=M'p line becomes a possible trajectory.

So when q/p reaches the value M', the trajectory can undertake a bifurcation and uses the q=M'p line either towards the left or towards the right. It can be shown [8] that this solution costs less energy than the other. This is not a demonstration, but it justifies that the trajectory undergoes a bifurcation at the point when the $p=c^{ste}$ line meets the q=M'p line. This bifurcation is of the "trans" kind. The q=M'p line was called by some authors the characteristic state [7]. Furthermore, soil mechanics teaches that specific volume v_c of granular matter in the critical density depends on the mean stress p according [1, 2] to $v_c=v_{co} -\lambda \ln(p/p_o)$, where v_{co}, p_o & λ are parameters.

Hence, the present model predicts that the trajectory shall turn left on the q=M'p line if the initial density $d_o=1/v_o$ is denser than the critical density at the given initial p_1, i.e. $v_o>v_c(p_1)$, since the mean stress p shall becomes smaller than the initial one p_1; it predicts also that it turns right on the contrary $v_o<v_c(p_1)$. This is sketched on figure 2c.

Comparison with experimental data shows that the modelling is correct for $v_o<v_c(p_1)$; however one observes the trajectory deviates from the prediction for $v_o>v_c(p_1)$: it turns on left much before the q=M'p line; but the start and the end of the trajectory are correct.

Indeed, it is possible to use an anisotropic modelling [8] to describe the more complex behaviour of the "loose"-sand case, by introducing a second pseudo Poisson coefficient in equation 2; however, one can remark also that the predicted trajectory which turns left at the q=M'p point is an unstable one, so that the intersection point between the vertical line and the q=M'p line is a repulsive point in the direction q=M'p. This may explain the enhancement of the non predicted behaviour .

Anyhow, this demonstrate that isotropic response is mainly correct for dense pile, whatever the stress field ratio is; but it is not valid for loose pile. This result may have some implication on analysis of contact forces such as those ones done in [9]: It is not obvious that stress increase generates anisotropy by its own; but instead, it shall be coupled to deformation to change really the real contact distribution. This is hence in contradiction with hypothesis in [9].

OEDEMETRIC BEHAVIOUR: ($R=c^{ste}$; $\varepsilon_2=\varepsilon_3=0$) The Jaky constant

In the general case of isotropic response, the condition $d\varepsilon_2=d\varepsilon_3=0$ combined with equation 2 imposes:

$$-vd\sigma_1+d\sigma_2(1-v)=0 \qquad\qquad (5)$$

Where the pseudo Poisson coefficient v varies with σ_1/σ_2 according to the Rowe's law [4], i.e. equation 1. So replacing v by its value from equation 1 leads to the differential equation:

$$d\sigma_2/\,d\sigma_1=v/(1-v)=[2(1+M)\,\sigma_2-\sigma_1]/\sigma_1 \qquad\qquad (6)$$

At this stage, it is worth recalling that equation 6 governs only the case of compaction, i.e. $\varepsilon_1>0$, since Rowe's relation combined with equation 2 is only valid in this case due to the character of non reversibility of equation 2 when changing the sign of ε_1, (see section on Modelling).

Anyhow, equation 6 can be integrated, at least numerically; it is found that the ratio σ_2/σ_1 tends towards an asymptotic value at large σ_1 when increasing progressively σ_1. It is labelled K_o in general in soil mechanics. The predicted value of K_o depends on M and hence on φ, at large stress σ_1, since the exact asymptotic value can be found analytically since it obeys $d\sigma_2/d\sigma_1= \sigma_2/\sigma_1=K_o$ which has as unique solution [9]:

$$K_o= (\sigma_3/\sigma_1)_{oed} = [1+(9+8M)^{1/2}]/[4(1+M)] \qquad\qquad (7)$$

$$= (1/4)[(1-\sin\varphi)/(1+\sin\varphi)][1+\{[9+7\sin\varphi]/[1-\sin\varphi]\}^{1/2}] \qquad\qquad (7)$$

Indeed, one knows a best fit of the experimental dependence of K_o vs. φ; it is called the Jaky constant [3], labelled K_{Jaky}, which value varies with φ as $K_{Jaky}=1-\sin(\varphi)$. Figure 3 compares these two variations, i.e. K_o from equation 7 and K_{Jaky} . They exhibit a *good correlation*, since it is better than 5% in the real experimental range.

One can discuss also the validity of the isotropic approximation: first of all, one remarks that deformation remains always small in an oedometric test, even when the sand sample is loose. One can also incorporate the undrained results obtained in the previous section to strengthen the approach: Indeed, according to figure 2 results, the rheology of dense sand can always be described within the isotropic assumption. However, some deviation from isotropic response can be observed in figure 2a in the case of loose sand; but it concerns only q/p ratios larger than 0.7 which corresponds to σ_3/σ_1 larger than 0.5 and to values larger than expected value of K_o. So, this confirms the validity of the isotropic approximation used here.

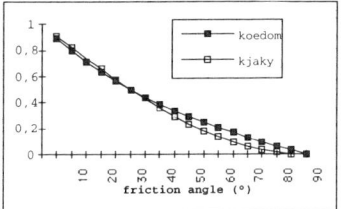 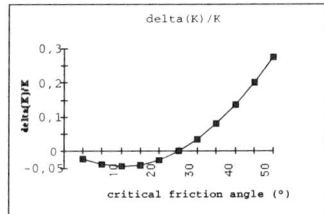

Figure 3: *Comparison between the oedometric modulus* $K_{oedometric}$ *as calculated from triaxial test result, i.e.* $K_{oedometric} = \sigma_3/\sigma_1 = [1+(8M+9)^{1/2}]/[4M+4]$ *, and the experimental best fit of Jaky, i.e.* $k_{Jaky} = 1-\sin\varphi$ *, vs. the friction angle* φ *of perfect plasticity. Left: the two dependencies. Right: their relative difference.*

The exact asymptotic ratio $K_o = (\sigma_2/\sigma_1)_{oed}$ depends on the exact dependence of ν upon (σ_2/σ_1). It might occur that the Rowe's law is not followed exactly by the sample, but only approximately; in this case, the exact asymptotic value $K_o = (\sigma_2/\sigma_1)_{oed}$ will change ; however, such a ratio $(\sigma_2/\sigma_1)_{oed}$ will exist as far as it exists some relationship $\nu = g(\sigma_2/\sigma_1)$ between ν and (σ_2/σ_1); this asymptotic value is obtained by integrating equation 5 or solving equation 6.

CONCLUSION

This paper proposes a simple non linear incremental modelling of the macroscopic rheological behaviour of granular matter in the quasi-static regime. It is based on classical experimental behaviours, well admitted by soil mechanics, at least as a first approximation. It assumes an isotropic behaviour for sake of simplicity. It turns out that that this assumption remains valid from experimental data since deformation is small enough, i.e. $\varepsilon_1 < 5\%$; this condition was always satisfied in the present investigated paths.

Non linear behaviour was taken into account (i) via a strain-dependent pseudo Young modulus which shall be fitted to figure 1 curves and (ii) via a pseudo Poisson coefficient ν; this second parameter can also be fitted from the volume variation given by figure 1 curves; however, as one knows a good estimate which links the volume variation and the stress field, i.e. the so-called Rowe's law, one can use it and use equation 1 as a state equation for ν.

Furthermore, as we investigated only paths obeying some strain invariance, such as constant-volume or constant-radius paths, the modelling predicts that results do not depend on the evolution of the pseudo Young modulus, but do depend on the pseudo Poisson coefficient ν, the evolution of which is controlled by equation (1). This one does not contain any adjustable parameter once the solid friction φ is known. So, the present model does not use any other adjustable parameter than φ in the investigated paths.

Concerning oedometric paths (radius=c^{ste}), one finds that the proposed modelling is able to explain satisfactorily them, whatever the initial density of the pile; especially, it is able to explain the value of the oedemetric ratio at large stress, i.e. the so-called K_o constant, and its variation with φ; it seems that it is the first time that such an explanation is proposed.

Concerning undrained paths ($v=c^{ste}$), the calculated trajectory follows mainly the experimental one in the case of dense sand; in particular it predicts a bifurcation of the trajectory when the $p=c^{ste}$ line meets the $q=M'p$ line. However, some deviation is observed for loose sands since experimental data exhibits a smoothening of the bifurcation. This may be generated simply by the fact that the trajectory becomes unstable after it passed the bifurcation point; it can also be described using a non isotropic response, since deformation becomes larger than $\varepsilon_1 > 5\%$ in this case.

The validity of the isotropic modelling casts a serious doubt on recent assumptions at the microscopic level concerning the contact distribution; these assumptions suppose two series of distinct contacts, the ones which bear the isotropic load, and the ones which bear the anisotropic part. If these assumptions were true, they will predict an anisotropic response as soon as the load is anisotropic. This is not observed experimentally: anisotropic response is generated only after large deformation, and the same stress field can generate an isotropic response or an anisotropic one depending on the material hystory; in other words, from experiments, only deformation coupled with anisotropic load allows to change the contact distribution and the isotropy; this is at least what shall be concluded from direct interpretation of classical results of soil mechanics.

ACKNOLEDGMENTS

This work has benefited from discussions with A. Modaressi, P. G. de Gennes, J. Biarez, F. Darve and M.P. Luong. CNES is thanked for partial funding.

REFERENCES

1. J.H. Atkinson & P.L. Bransby, *The Mechanics of soils*, Mac Graw Hill, Cambridge Un. Press (1977)
2. K.H. Roscoe, A.N. Schofield & C.P. Wroth, " On the yielding of soil ", *Geotechnique* **8**, 22-53, (1958)
3. Terzaghi K., *Theoretical Soil Mechanics*, John Wiley and sons, New York, twelfth printing, (1965)
4. P.W. Rowe, "The stress dilatancy relation for static equilibrium of an assembly of particles in contact", *Proc. Roy. Soc. Lndn* **A269**, 500-527, (1962)
5. F. Darve, L'Ecriture incrémentale des lois rhéologiques et les grandes classes de lois de comportement, *Manuel de Rhéologie des géomatériaux*, ed. F. Darve, presses des Ponts & Chaussées, Paris, 129-152, (1987)
6. B. Loret, Application de la théorie des multimécanismes à l'étude du comportement des sols, *Manuel de Rhéologie des Géomatériaux*, Presses de l'ENPC, Paris, pp. 189-214, (1987)
7. M.P. Luong, *C.R. Acad. Sc. Paris*, **287**, Série II, pp. 305-7 , (1978)
8. P. Evesque, Poudres & Grains **9**, 1-12, (1999) ; also at http://prunier.mss.ecp.fr/poudres&grains
9. F. Radjai, D. Wolf, M. Jean & J.J. Moreau, "Bimodal character of stress transmission in granular packing", *Phys. Rev. Lett.* **90**, 61, (1998)
10. P. Evesque, (1997), *J. de Phys. I France* **7**, 1501-1512, (1997)) ; and see http://prunier.mss.ecp.fr/poudres&grain

From Avalanches
to Sandcastles

Mat. Res. Soc. Symp. Proc. Vol 627 © 2000 Materials Research Society

Humidity-Induced Cohesion Effects in Granular Media

Nathalie Fraysse[1], Luc Petit
Laboratoire de Physique de la Matière Condensée, UMR 6622 CNRS - Université de Nice, Parc Valrose, 06 108 Nice cedex 2, France.
[1]fraysse@unice.fr

ABSTRACT

Experiments were performed under accurately-controlled humidity conditions in order to quantify effects induced by humidity on granular materials. Measurements of the maximal stability angle of a pile made of small glass beads are reported as a function of the relative vapor pressure in the cell, up to close to saturation. The comparison of the results obtained with fluids differing in their molecular interactions with glass, namely water and heptane, shows that the wetting properties of the interstitial liquid on the grains have a strong influence on the cohesion of the non-saturated granular medium. This suggests that gravimetric experiments which could indirectly give information on the size of the capillary bridges that form between grains should be useful to understand the close connection that exists, through interparticle forces, between microscopic properties such as wetting properties and surface roughness of the grains, and global-scale properties of the pile, as its stability and flowability.

INTRODUCTION

The wide variety of behaviors observed for dry granular media according to various external constraints originates from two main interparticle forces that are the solid friction and an elastic repulsion. If now considering wet granular media, capillary forces and possibly viscous forces have also to be taken into account; at the global scale, cohesion and lubrication effects, which are totally absent in dry sand, may appear as the result of these additional local interactions between grains. Such humidity-induced effects come into play in many applications that involve granular materials. The presence of an interstitial liquid can have various causes; the atmospheric humidity is the most obvious and maybe the most common one as, in usual practice, grains are most of the time stored and handled in the open air. Note that the consequences of humidity effects are negative in most of industrial processes (*e.g.* blocked flows from hoppers or silos) but they might be positive: For instance it is conceivable that humidity effects oppose the annoying segregation phenomenon. Therefore it is of great importance to precisely know and to fully understand humidity effects on granular media, from a practical point of view as well as from a fundamental one. In this paper we present experiments that we have performed under accurately-controlled humidity conditions in order to quantify such moisture-induced effects. We report the quantitative measurements of the maximum angle of stability of a pile, as a function of the relative vapor pressure, up to close to saturation. We also investigate the influence of the wetting properties of the interstitial liquid on the grains on the stability of the pile.

EXPERIMENT

One distinctive feature of granular materials that is well known to be strongly affected by humidity is their ability to form piles that are stable up to an inclination of the free surface to the horizontal called "maximal stability angle", θ_m. Any attempt to increase the slope of the pile beyond this value causes an avalanche of grains. The slope of the pile then relaxes to a smaller value called "angle of repose", θ_r. We have focused on the maximum angle of stability which characterizes a critical state of the pile and which is a purely static property of the granular material. On the contrary, the angle of repose is the result of a dynamic process the analysis of which is very complex, even for dry granular media; the three flowing regimes and the phenomenon of clumping observed by Tegzes et al. [1] depending on the liquid content have shown that it is even more complex for wet granular media.

In order to measure the maximum angle of stability, we used the so-called rotating drum method [2]. The granular sample fills 50% volume of a cylindrical cell (length 10 cm, diameter 10 cm, axis horizontal) which can be slowly rotated around its axis. Figure 1 shows a schematic view through the glass end-face of the cell, as well as the angles θ_m and θ_r that we measure using this method.

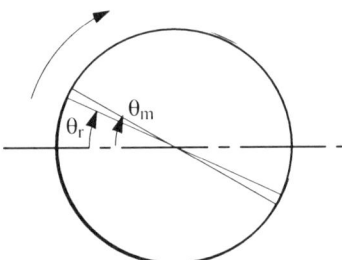

Figure 1. *The rotating drum method. θ_m is the maximum angle of stability and θ_r is the angle of repose.*

The humidity degree of the granular sample naturally appears as an experimental parameter that is essential to control for any quantitative study of the effects of humidity on the behavior of a granular material. At the same time, parameters that have proved important in the physics of dry granular media such as packing density and the history of construction of the pile should not be ignored. The experimental work we have carried out on wet granular media rests on a method of humidification from an under-saturated vapor, according to a procedure that ensures a rigorous control of both humidity and packing density. Very schematically, vacuum is made in the rotating drum, and then the vapor (of water or any other volatile liquid) is injected slowly. Mechanical pressures equilibrate in the cell and throughout the granular medium almost instantaneously and the porous medium constituted by the packing of grains is totally and uniformly invaded by the vapor; at thermodynamic equilibrium of the solid, liquid and vapor phases, the liquid phase, which results from adsorption and capillary condensation, is thus homogeneously distributed throughout the granular sample. Besides, this process of humidification does not perturb the initial packing built with dry grains (unless rearrangements occur as a result of capillary forces); even when the resulting cohesion of the wet granular sample is high, the packing density thus obtained is uniform and reproducible to a good accuracy, which is certainly not true when stirring of the granular material is required in order to get a uniform distribution of liquid.

Note that no dependence of the maximum angle of stability on the waiting time [3] was observed in our experiments. More details about the experimental procedure can be found in [4, 5].

RESULTS AND DISCUSSION

The curves on Figure 2 show the variation of the maximal stability angle that we measured for monodisperse (diameter 200-250 µm) glass beads humidified respectively from water vapor and heptane vapor, as a function of the relative vapor pressure. Humidifying the glass beads with heptane does not significantly change the maximum stability angle up to a value of about 0.9 for P_v/P_{sat} i.e. until very close to saturation; at relative vapor pressures larger than 0.90, a sudden and strong increase in θ_m is observed. On the contrary, humidifying the glass beads with water leads to a gradual increase in θ_m with the percentage of relative humidity.

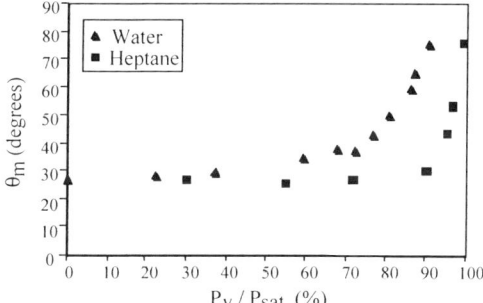

Figure 2. *The maximum angle of stability θ_m as a function of the relative vapor pressure of water (▲) and n-heptane (■). P_v is the vapor pressure directly measured in the cell; P_{sat} (T) is the saturated vapor pressure at the cell temperature T. The relative vapor pressure P_v/P_{sat} (T) expressed in percent is strictly equivalent to the percentage of relative humidity which is commonly used to characterize the atmospheric degree of humidity, noted %RH.*

The comparison of the θ_m curves on Figure 2 shows that the wetting properties of the fluid on the grains have a strong influence on the variation of the cohesion of the granular medium with its degree of humidity. More specifically, both water and heptane completely wet glass at the macroscopic level (contact angle equal to zero) but their physico-chemical interactions with glass are strongly different: Water, which incidentally is of obvious practical interest, leads to highly complex molecular interactions with glass; n-heptane is a light saturated alkane which leads to simple, Van der Waals-type, interactions with glass. Consequently, the local spatial distributions of these liquids in the granular sample made of glass beads are expected to differ significantly. This local, microscopic aspect appears to strongly influence a macroscopic behavior of the granular material such as the stability of a pile.

Two models were proposed in order to analyze the stability of a pile of cohesive grains. Using geometric arguments, Albert *et al.* [6] developed a surface model that analyzes the mechanical stability of a sphere at the surface of the pile. Halsey and Levine [7] applied a modified Mohr-Coulomb criterion for local failure to the cohesive material, in the framework of continuum mechanics. However, whatever model one chooses, a question soon arises: What is the exact value of the cohesive force that exists between the grains due to the liquid bridges that form at their contact zones? The behaviors predicted by the models drastically depend on the answer to this question, in agreement with the strong influence we experimentally observed of the microscopic wetting properties, and more generally of the local spatial distribution of liquid. Direct investigations at the scale of the microscopic roughness of the beads (a few tenths of μm), or at the scale of the average thickness of the liquid films involved, have proved to be difficult [1, 8]. We are currently undertaking new gravimetric measurements under the same rigorously-controlled humidity conditions as in the experiments reported in this paper; our aim is to indirectly get information on the volume of the liquid bridges, which is the theoretically relevant quantity, as well as on the role that surface roughness plays on the capillary bridges formation.

CONCLUSION

In this paper we have reported experimental results obtained under accurately-controlled humidity conditions on the maximal stability angle of a pile made of small glass beads humidified with fluids differing in their molecular interactions with glass. Our study shows that a close connection exists, through interparticle forces, between microscopic properties such as wetting properties and surface roughness of the grains, and global-scale properties of the pile, as its stability and flowability. Information at the local scale is clearly needed in order to better understand the large-scale effects induced by humidity on granular materials.

ACKNOWLEDGMENTS

One of the authors (N. Fraysse) wishes to thank Dr. D. Chatain for the SEM pictures of the glass beads surface she provided us with and for her help regarding the gravimetric measurements. We also would like to thank Dr. N. Olivi-Tran for discussions.

REFERENCES

1. P. Tegzes, R. Albert, M. Paskvan, A.-L. Barabasi, T. Vicsek, P. Schiffer, "Liquid-induced transitions in granular media", *Phys. Rev. E*, **60**, 5823 (1999)
2. R.L. Brown, J.C. Richards, *Principles of Powder Mechanics*, 1st edn. (Pergamon Press, Oxford, 1970), p. 24-26.
3. L. Bocquet, E. Charlaix, S. Ciliberto, J. Crassous, "Moisture-induced ageing in granular media", *Nature*, **396**, 735 (1998).

4. N. Fraysse, H. Thomé, L. Petit, "Effects of humidity on quasi-static behavior of granular media", in *Proceedings of the 3rd International Conference on Powders & Grains, Durham, 1997,* edited by R.P. Behringer and J.T. Jenkins (Balkema, Rotterdam, 1997), p.147.
5. N. Fraysse, H. Thomé, L. Petit, "Humidity effects on the stability of a sandpile", *Eur. Phys. J. B*, **11**, 615 (1999)
6. R. Albert, I. Albert, D.J. Hornbaker, P. Schiffer, A.-L. Barabasi, "Maximum angle of stability in wet and dry spherical granular media", *Phys. Rev. E*, **56**, R6271 (1997).
7. T.C. Halsey, A.J. Levine, "How sandcastles fall", *Phys. Rev. Lett.*, **80**, 3141 (1998).
8. T.G. Mason, A.J. Levine, D. Ertas, T.C. Halsey, "The critical angle of wet sand piles", *Phys. Rev. E*, **60**, R5044 (1999).

Mat. Res. Soc. Symp. Proc. Vol. 627 © 2000 Materials Research Society

DEM APPLICATION TO MIXING AND SEGREGATION MODEL IN INDUSTRIAL BLENDING SYSTEM

KENJI YAMANE
Quality Control Department
Taiho Pharmaceutical Co., Ltd.
Tokusima 770-0194
Japan

ABSTRACT

To predict the motion of powders and grains is important in pharmaceutical industries. Many pharmaceutical engineers have studied granular flows related to powder mixing. In this study, DEM (Discrete Element Method) approach is presented as an industrial application to investigate the behavior of granular flows. The granular motion in a rotating cylinder was focused on the basic study of DEM for industrial application. Rotating cylinder is a fundamental system for commercial blenders widely used in many industrial process. In addition, segregation of particles in a rotating cylinder is very interesting phenomena. Not only industrial engineers but also physicists research this segregation mechanism. DEM simulation showed radial segregation of two different size particles in a rotating cylinder. From the viewpoint of calculated granular temperature, radial segregation system was analyzed. Particle migration in axial direction, which is the source for axial segregation, was also shown by DEM simulation.

INTRODUCTION

Engineers have been interested in the complicated granular flows. Mixing of granular has been studied based on flow mechanism of granular by many engineers. Because the mixing process is a important process in the industries where powders and grains are handled. Specially, on the process of powder mixing of pharmaceutical manufacturing, it is important to assure the blend uniformity of mixtures. Segregation should be avoided on blending process in order to maintain the quality of drug products. While, the phenomena of segregation has captured physicists for long time. The original study was performed by Oyama. In 1939, Oyama [1] first reported that rotating cylinders partially filled with a mixture of granular media may serve to segregate the individual species into bands along the rotational axis. This phenomenon is termed axial segregation. Since Oyama's experiments, a number of studies have shown that mixtures of granular media exhibit a wide range of axial segregation (see, for example [2]-[7]). It has been found from recent studies that the reversible axial segregation of different-sized granular media in a drum mixer is related to variations of the dynamic angle of repose [8], but it is also clear that subsurface effects are critical, because axial bands that do not extend to the surface have been observed within the bulk. In order to study the dynamics beneath the surface involved in the segregation process, DEM simulation was used in the present study to demonstrate granular flow patterns within a drum mixer. Specially, On binary-disperse particles, migration to axial direction was simulated. In this paper, the study of segregation mechanism was conducted by DEM simulation.

CALCULATION THEORY

DEM simulation was explained in many articles. In this paper, only brief explanation is described. Further instruction of DEM simulation is described in the references [9] [10]. Equations for translational and rotational motion for a spherical particle, in the absence of interstitial fluids, are

$$\ddot{\vec{r}} = \frac{\vec{f}}{m} + \vec{g} \quad (1)$$

and

$$\dot{\vec{\omega}} = \frac{\vec{T}}{I} \quad (2)$$

where $\vec{r}$ is the position vector of the particle center, m is the particle mass, $\vec{f}$ is the sum of all contact forces, $\vec{g}$ is the gravitational acceleration, $\vec{\omega}$ is the angular velocity, $\vec{T}$ is the net torque due to contact forces, I is the moment of inertia, and ($\bullet$) indicates the time derivative. Contact forces between spherical particles are modeled by springs, dash-pots, and a friction slider, as originally considered by Cundall and Strack [11].

The parameters special to DEM are as follows. Each time step was 10^{-4} s while Young's modulus and Poisson ratio were 1.0×10^5 N/m^2 and 0.3, respectively. The diameter of cylinder was 69 mm and periodic boundary conditions in the longitudinal direction were used to alleviate the severe computing demands of 3D simulations. Therefore, the simulation takes place in a 15mm long, half-filled cylinder with two periodic end boundaries. About 1000 particles of 4 mm and about 2000 particles of 2 mm were used for the calculation.

RESULTS AND DISCUSSIONS

Radial segregation

The initial conditions of DEM simulation are as follows; large and small particles are separated as shown in Figure 1(a). Next the drum is made rotate with high rotation speed Ω =170 rpm in order to make mixture condition. This mixture is not segregated shown in Fig. 1(b). Next the rotation speed Ω =17 rpm was chosen. This speed corresponds to a region with continuous surface. After some revolutions a clearly segregated core composed of small particles is observed as shown in Fig. 1(c).

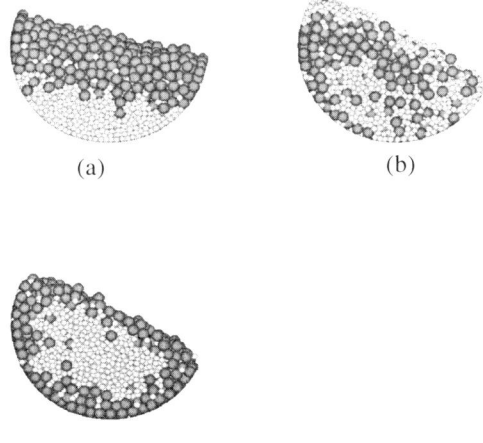

(a) (b)

(c)

Figure 1. *(a) initial condition, (b) mixture condition, (c) radial segregation*

In order to investigate the segregation condition, the author propose further parameter. The parameter that can be computed from the simulation is the fluctuation of velocity defined as

$$< v >= \sqrt{\frac{1}{m} \sum_{i=1}^{m} (v_i - \bar{v})^2} \ . \ (10)$$

where m is the average number of particles in the voxel, v_i is the velocity of i-th particle, and $\bar{v}$ is the average velocity of the particles in the voxel. This is a parameter that is proportional to the granular temperature which is used in kinetic theory treatments of such problems [12][10]. The calculated granular temperature at rotation speed of 30 rpm is shown in Figure 2.

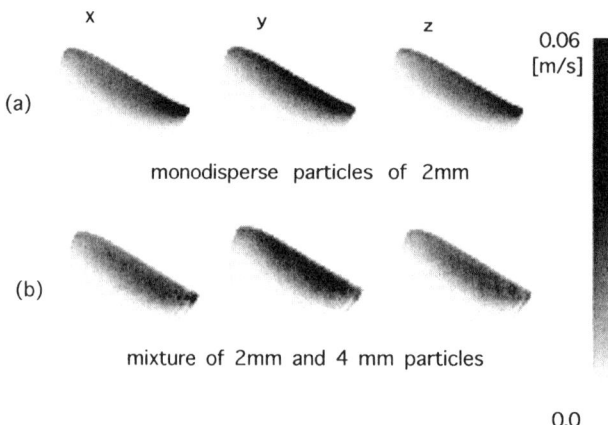

Figure 2. *Fluctuation of velocity components*

Migration to axial direction

Furthermore, as strange phenomena of particles including two different size, axial segregation was well known. In order to show the particle migration to axial direction, DEM simulation was conducted with following calculation conditions. The simulation takes place in a 150mm long, half-filled cylinder of 20 mm diameter with two periodic end boundaries. About 2000 particles of 4 mm and about 7000 particles of 2 mm were used for the calculation. Figure 2 shows particles migration to axial direction. On initial condition, all small particles were set into right side of cylinder, while large particles were set into left side. A few seconds later, particles started to move to assembly of another size particles.

Time= 0.00[sec]

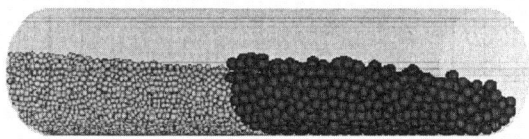

Time= 2.00[sec]

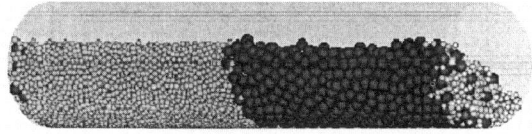

Time= 4.00[sec]

Figure 3. Snapshot for particle migration to axial direction

CONCLUSIONS

Radial segregation was simulated by DEM. In short time, the radial segregation was observed by computer simulation. The migration to axial direction of granular was also shown. The migration is considered as a preliminary stage of axial segregation status. Computer simulation provided information about the movement to axial direction.

REFERENCES

1. Oyama, Y. (1939). *Bull.Inst.Phys.Chem.Rep.* (Tokyo), **18**, 600 (in Japanese).

2. Nakagawa, M. (1994). Axial segregation of granular flows in a horizontal rotating cylinder. *Chem. Eng. Sci.* **49**, 2540-2544

3. Hill, K.M., A. Caprihan, and J. Kakalios (1997). Bulk segregation in rotated granular material measured by Magnetic Resonance Imaging. Phys. Rev. Lett., **78**, 50-53.

4. Donald, M.B., and B. Roseman (1962). Mechanisms in a horizontal drum mixer. *British Chem. Eng.* **7**, 749-753; Effects of varying the operating conditions of a horizontal drum mixer. *Brit. Chem. Eng.* **7**, 823-827.

5. Savage, S.B (1993). Banding or pattern formation in horizontal drum mixers. In D. Bideau and A. Hansen (eds.) *Disorder and Granular Media*: 255-285. Amsterdam: North-Holland.

6. Bridgwater, J., N.W. Sharpe, and D.C. Stocker (1969). Particle mixing by percolation. *Trans. Inst. Chem. Eng.* **47**, T114-T119.

7. Zik, O., D. Levine, S.G. Lipson, S. Shtrikman, and J. Stavans (1994). Rotationally induced segregation of granular materials. *Phys. Rev. Lett.* **73**, 644-647.

8. Hill, K.M. and J. Kakalios (1995). Reversible axial segregation of rotating granular media. *Phys. Rev. E* **52**, 4393-4400.

9. Tsuji, Y., T. Tanaka, and T. Ishida (1992). Lagrangian numerical simulation of plug flow of cohesionless particles in a horizontal pipe. *Powder Technology* **71**, 239-250.

10. Yamane, K., M. Nakagawa, S.A. Altobelli, T. Tanaka and Y. Tsuji (1998). Steady particulate flows in a horizontal rotating cylinder. Phys. Fluids **10**,1419-1427.

11. Cundall, P.A. and O.D. Strack (1976). A discrete numerical model for granular assemblies. Geotechnique **12**, 47-65.

12. Jenkins, J.T. and D. M. Hanes (1993). The balance of momentum and energy at an interface between colliding and freely flying grains in a rapid granular flow, Phys. Fluids A **5** (3), 781-783.

Mat. Res. Soc. Symp. Proc. Vol 627 © 2000 Materials Research Society

An Experimental Study of the Fluctuations in Granular Drag

István Albert[1], Pál Tegzes[1,2], Réka Albert[1], John Sample[1],
Albert-László Barabási[1], Tamás Vicsek[2], B. Kahng[1,3] and Peter Schiffer[1*]

[1]Department of Physics, University of Notre Dame, Notre Dame, IN 46556
[2]Department of Biological Physics, Eötvös University, Budapest 1117, Hungary
[3]Dept. of Physics, Konhuk University, Seoul 143-701 Korea

ABSTRACT

We study fluctuations in the drag force resisting the motion of an object being pulled through a dense spherical granular medium. These fluctuations are stick-slip in nature due to the jamming and reorganization of the grains. The fluctuations in the force are periodic at small depths, but they become "stepped" at large depths. We interpret this transition as a consequence of the long-range nature of the force chains.

INTRODUCTION

Materials in granular form are composed of many solid particles that interact only through contact forces. In a granular pile, the strain resulting from the grains' weight combines with the randomness in their packing to constrain the motion of individual grains. This frustration leads to the phenomenon of "jamming" which also characterizes a variety of other frustrated physical systems [1,2]. Not surprisingly, this frustration of local motion causes the dynamic properties of granular materials to be quite different from those of liquids [3]. The effects of this jamming are also manifested in static properties, leading to inhomogeneous stress propagation through force chains of strained grains [4] and arch formation [5]. Although jamming in granular materials has previously been discussed in the context of the gravitational stress induced by the weight of the grains, it can result from any compressive stress. For example, a solid object being pulled slowly through a dense granular medium is resisted by local jamming, and can only advance with large scale reorganizations of the grains. The force needed to induce such reorganizations produces a drag force with strong fluctuations which qualitatively distinguish it from the analogous drag in fluids [6].

We have made an experimental study of the dynamic evolution of jamming in granular media through these fluctuations in the granular drag force. The successive collapse and formation of jammed states give a stick-slip character to the force and the slip events are nucleated in the bulk of the grains opposing its motion. While the fluctuations are remarkably periodic for small depths, they undergo a transition to "stepped" motion at large depth which we attribute to the long-range nature of the force chains. Many of the results and details of the apparatus will be published elsewhere [7].

EXPERIMENTAL DETAILS

The experimental apparatus, shown in figure 1, consists of a vertical cylinder of diameter d_c inserted to a depth H in a bed of glass spheres moving with constant speed [6]. The cylinder is attached to a fixed force cell [8], which measures the force F(t) acting on the cylinder as function of time. The bearings on the cylinder's support structure allow it to advance freely only in the direction of motion so that the force cell alone is opposing the drag force from the grains. We incorporate a spring of known spring constant, k, between the cylinder and the force cell -- choosing k so that this spring dominates the elastic response of the cylinder and all other parts of the apparatus. We vary the speed (v) from 0.04 to 1.4 mm/s, the depth of insertion (H) from 20 to 190 mm, and the cylinder diameter (d_c) from 8 to 24 mm, studying grains of diameter (d_g) 0.3, 0.5, 0.7, 0.9, and 1.1 mm [9]. The force is recorded at 150 Hz and the response time of the force cell and the amplifier are < 0.2 ms.

EXPERIMENTAL RESULTS

As shown in figure 2 and consistent with earlier results [6], we find that the average drag force on the cylinder is independent of v and k, and is given by $F_{avg} = \eta \rho g d_c H^2$, where η characterizes the grain properties (surface friction, packing fraction, etc.), ρ is the density of the glass beads, and g is gravitational acceleration. As shown in figure 2, however, F(t) displays large stick-slip fluctuations consisting of linear rises associated with a compression of the spring and sharp drops associated with the reorganization of the jammed grains opposing the motion. The linear rises in F(t) correspond to the development of an increasingly compressed jammed state of the grains opposing the motion. We find that

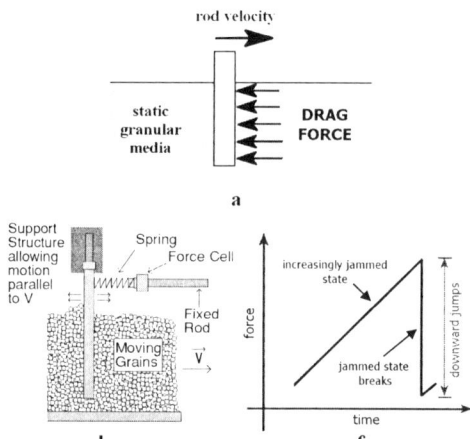

Figure 1. a. A schematic representation of the drag force in a static granular medium. **b.** A cross-sectional drawing of the experimental apparatus used in these measurements. **c.** A sketch of the fluctuations in the drag force as a function of time showing the increase in force corresponding to the compression of the jammed state, and the drop in the force as the jammed state collapses in a large-scale reorganization of the grains.

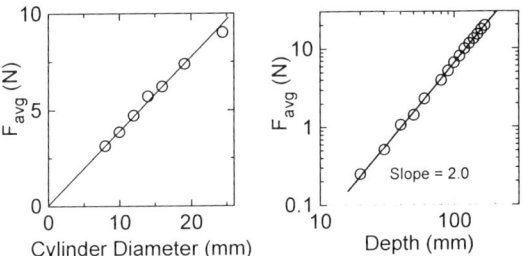

Figure 2. The average drag force on the cylinder as a function of cylinder diameter (d_c) and depth of insertion into the granular medium (H) demonstrating that $F_{avg} = \eta \rho g d_c H^2$.

the slopes of the rises are given by $v^{-1}dF/dt = k$ for all springs with k<100 N/cm, confirming that the spring dominates the elasticity of the apparatus. This result also implies that the jammed grains opposing the cylinder's motion do not move relative to each other or the cylinder except during the large-scale reorganizations. The power spectra, P(f) (the squared amplitudes of the Fourier components of F(t)), can be scaled by plotting kvP(f) vs. f/kv so that they collapse in the low frequency regime (f < 10 Hz) as shown in figure 4 [10]. This indicates that the fluctuations reflect intrinsic properties of the development and collapse of the jammed state rather than details of the measurement process. The power spectra also exhibit a distinct power law, $P(f) \sim f^{-2}$, a phenomenon which has been reported in other stick-slip processes and is intrinsic to random sawtooth signals [11].

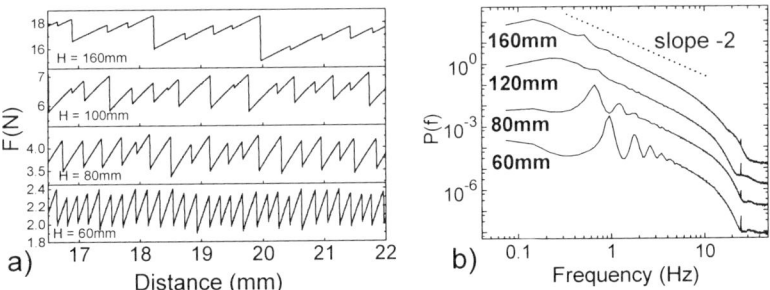

Figure 3. The change in the qualitative nature of the fluctuations with depth of insertion. **a.** The drag force as a function of time at 4 different values of H for d_c =10 mm. Note the transition from purely periodic fluctuations H ≤ 60 mm to stepped fluctuations with increasing depth H ≥ 100 mm in which the fluctuations change qualitatively not by any simple scaling. b. The power spectrum of the fluctuations in F(t) at different depths of insertion for d_c = 16 mm(offset for clarity). Note that in both the periodic and the stepped regimes the spectrum has a long f^{-2} regime.

During each fluctuation the force first rises to a local maximum value, and then drops sharply, corresponding to a collapse of the jammed state as illustrated in figure 1c. Such a collapse allows the cylinder to advance relative to the granular reference frame, with a corresponding decompression of the spring and a drop in the measured force F(t). The interparticle forces in the jammed state are largest at the cylinder's surface, and one might thus expect that the reorganization is nucleated among grains in contact with the cylinder, but we find little change either in F_{avg} or in the fluctuations when we vary the coefficient of friction between the grains and the cylinder by a factor of 2.5 (substituting a teflon-coated cylinder for the usual steel cylinder). We find that the power spectra are also essentially unchanged even by substituting a half-cylinder (i.e. a cylinder bisected along a plane through its axis and oriented so that the plane is normal to the grain flow) for a full cylinder of the same size, indicating that the geometric factors do not play a significant role either [7]. These results indicate that the fluctuations are not determined by the interface between the dragged object and the medium, but rather that the failure of the jammed state is nucleated within the bulk of the medium. In this respect, the fluctuations are rather different from either ordinary frictional stick-slip processes which originate at a planar interface between moving objects, or the motion of a frictional plate on top of a granular medium [12].

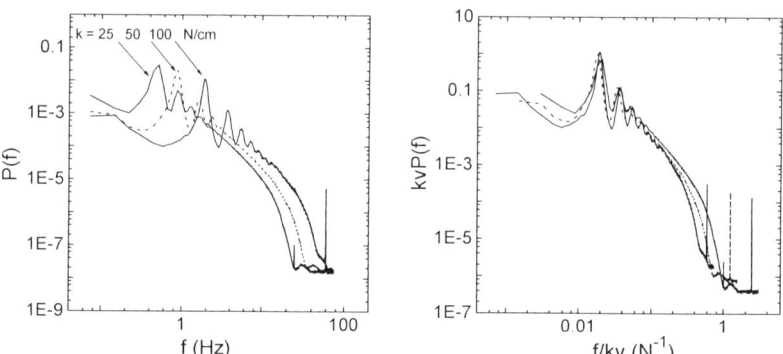

Figure 4. The scaling of the power spectra with spring constant, plotting kvP(f) vs. f/kv. That the scale curves collapse demonstrates that the fluctuations are intrinsic to the granular medium rather than dependent on the details of the measurement process.

A striking feature of the data which can be seen in figure 3 is that the fluctuations change character with depth. For $H < H_c$ (where $H_c \sim 80$ mm for the conditions in the figure) the fluctuations are quite periodic, but as the depth increases, however, we observe a change in $F(t)$ to a "stepped" signal: instead of a long linear increase followed by a roughly equal sudden drop, $F(t)$ rises in small linear increments, followed by small drops until $F(t)$ reaches a characteristic high value, at which point a large drop is observed. As seen in figure 3b, for low depths the power spectra display a distinct peak characteristic of periodic fluctuations, but these peaks are suppressed for large depths in correlation with the changes in the qualitative character of $F(t)$. The transition from a periodic to a "stepped" signal is rather unexpected, since it implies qualitative changes in the failure and reorganization process as H increases and the existence of a critical depth, H_c. An explanation for H_c could be provided by Janssen's law [13] which states that the average pressure (which correlates directly with the local failure process) should become depth independent below some critical depth in containers with finite width. This should not occur in our container, however, which has a diameter of 25 cm, much larger than H_c. Furthermore we see no deviation in the behavior of the average drag force from $F_{avg} \sim H^2$, which depends on the pressure increasing linearly with the depth (see figure 2). Since there are no other length scales in this problem, we hypothesize that the transition arises from the finite size of the container. This explanation, which is discussed in detail elsewhere, is supported by the transition occurring at smaller H in a smaller container, and also by the transition occurring at a larger H for smaller diameter beads [7].

CONCLUSION

In conclusion, we have characterized the fluctuations in the drag force resisting the motion of an object moving slowly through a dense granular medium. We find that these stick-slip type fluctuations are related to failure in the bulk of the jammed grains impeding the motion and that their character is strongly dependent on the depth of the medium in our finite-sized container.

ACKNOWLEDGEMENTS

We gratefully acknowledge the support of the Petroleum Research Foundation administered by the ACS, the Alfred P. Sloan Foundation, and NSF grants PHYS95-31383 and DMR97-01998, and NASA grant NAG3-2384.

*Author to whom correspondence should be addressed. Current address: Dept. of Physics, 104 Davey Laboratory, Pennsylvania State University, University Park, PA 16802
schiffer@phys.psu.edu

REFERENCES

1. M. E. Cates, J. P. Wittmer, J.-P. Bouchaud, and P. Claudin, Phys. Rev. Lett. **81**, 1841 (1998) and cond-mat/9901009 (1999); M. E. Cates and J. P. Wittmer, Physica A **263**, 354 (1999).
2. A. J. Liu and S. R. Nagel, Nature (London) **396**, 21 (1998).
3. H. M. Jaeger, S. R. Nagel, and R. P. Behringer, Rev. Mod. Phys. **68**, 1259

(1996); L. P. Kadanoff, Rev. Mod. Phys. **71**, 435 (1999).
4.   C. Liu et al., Science **269**, 513 (1995); S. N. Coppersmith et al., Phys. Rev. E **53**, 4673 (1996); D.M. Mueth, H. M. Jaeger, and S. R. Nagel, Phy. Rev. E **57**, 3164 (1998); X.Jia, C. Caroli, B. Velicky, Phys Rev. Lett. **82**, 1863 (1999); M. L. Nguyen and S. N. Coppersmith, Phys. Rev. E **59**, 5870 (1999); A. V. Tkachenko and T. Witten, Phys. Rev. E **60** 687 (1999); D. Howell, R. P. Behringer, and C. Veje, Phys. Rev. Lett. **82**, 5241 (1999); L. Vanel et al. Phys. Rev. E **60**, 5040 (1999) and Phys. Rev. Lett **84**, 1439 (2000).
5.   P. Claudin and J.-P. Bouchaud, Phys. Rev. Lett. **78**, 231 (1997) and J. E. Scott, V. M. Kenkre, and A. J. Hurd, Phys. Rev. E **57**, 5850 (1998).
6.   R. Albert, M. A. Pfeifer, A.-L. Barabasi, and P. Schiffer, Phys. Rev. Lett. **82**, 205 (1999).
7   I. Albert et al., Phys. Rev. Lett. **84**, 5122 (2000) and in preparation.
8.   The apparatus was a modified version of that described in reference 6, to which we added the spring on the force cell and improved the bearings.
9.   Unless otherwise noted, data are shown for $d_g = 0.9$ mm, $k = 25$ N/cm and $v = 0.2$ mm/s.
10.   Similar scaling has been reported for fluctuations in the normal force from sheared granular materials in an annular Couette geometry. B. Miller, C. O'Hern, and R. P. Behringer, Phys. Rev. Lett **77**, 3110 (1996).
11.   A. L. Demirel and S. Granick, Phys. Rev. Lett. **77**, 4330 (1996); H. J. S. Feder and J. Feder, Phys. Rev. Lett. **66**, 2669 (1991); E. Kolb, T. Mazozi, E. Clement, and J. Duran, Eur. Phys. J. B. **8**, 483 (1999).
12.   S. Nasuno, A. Kudrolli, and J. P. Gollub, Phys. Rev. Lett. **79**, 949 (1997); S. Nasuno, A. Kudrolli, A. Bak, and J. P. Gollub, Phys. Rev. E **58**, 2161 (1998); J.-C. Geminard, W. Losert, and J. P. Gollub, Phys Rev. E **59**, 5881 (1999).
13.   See, for example, R. L. Brown and J. C. Richards, *Principles of Powder Mechanics* (Pergamon Press, Oxford, 1970).

AUTHOR INDEX

SUBJECT INDEX